Inverse Problems and Zero Forcing for Graphs

Mathematical
Surveys
and
Monographs

Volume 270

Inverse Problems and Zero Forcing for Graphs

Leslie Hogben
Jephian C.-H. Lin
Bryan L. Shader

AMERICAN
MATHEMATICAL
SOCIETY
Providence, Rhode Island

2020 *Mathematics Subject Classification.* Primary 05C50, 15-02, 05-02, 15A29, 15A18, 05C69, 05C22, 15B57.

For additional information and updates on this book, visit
www.ams.org/bookpages/surv-270

Library of Congress Cataloging-in-Publication Data

Cataloging-in-Publication Data has been applied for by the AMS.
See http://www.loc.gov/publish/cip/.
DOI: https://doi.org/10.1090/surv/270

∞ The paper used in this book is acid-free and falls within the guidelines
established to ensure permanence and durability.
Visit the AMS home page at `https://www.ams.org/`

10 9 8 7 6 5 4 3 2 1 27 26 25 24 23 22

Contents

Preface

The primary driver for writing this book is the desire to provide an on-ramp to the new, developing, and mathematically diverse research related to the inverse eigenvalue problem for graphs (IEP-G) and the related area of zero forcing, propagation, and throttling. Inverse problems play a central role in mathematics and naturally arise in applications. In many instances the inverse problem reduces to a question about the existence of a matrix with a prescribed set of eigenvalues and prescribed structure. The IEP-G studies such questions.

Due to the lack of effective tools for the IEP-G, much early research focused on ancillary problems (subquestions) that could lead to progress on the IEP-G. The most important of these is the study of the maximum possible multiplicity of an eigenvalue among matrices described by G, or equivalently maximum possible nullity or minimum possible rank. Since the birth of the minimum rank problem in a 1997 paper by Peter Nylen, the IEP-G and related questions have proven to be intriguing, but difficult problems. The 2006 American Institute of Mathematics Workshop 'Spectra of families of matrices described by graphs, digraphs, and sign patterns' and the small research group that grew out of it, 'Minimum rank of symmetric matrices described by a graph,' catalyzed two new research areas, which we call Strong Properties of Matrices and Zero-Forcing of Graphs, related to the IEP-G. Since then there has been a rapid, expansive development of the areas, which has resulted in many deep results for the IEP-G. Because of the growth of these emerging areas, we believe that there is a need for a book like this which provides the essential concepts, techniques and results in a unified way, and which suggests topics for future research.

A pleasing aspect of these new topics is their interplay with other areas of mathematics and applications. The topic of Strong Properties of Matrices is closely related to the Implicit Function Theorem and the transversal intersection of manifolds. It provides algebraic conditions on a matrix with a certain spectral property and graph that guarantee the existence of a matrix with the same spectral property for each graph in a related family of graphs. Additionally, strong properties provide interesting combinatorial matrix theoretic and graph minor problems.

Zero forcing originated independently in both the IEP-G setting and in the study of control of quantum systems (where it is often referred to as graph infection). Zero forcing is a game on a graph in which vertices are initially filled or unfilled, and at each stage certain vertices change from unfilled to filled according to some rule. The goal of the game is to fill all vertices. Zero forcing gives a graph-theoretic method to bound the multiplicity of an eigenvalue of a symmetric matrix with a given zero-nonzero structure. Zero forcing is closely related to the power domination process that provides a model for placing monitoring units in an electric network, and has connections to graph searching, including the well-studied Cops and Robbers game in graph theory. Additionally, variants of zero forcing have arisen in various applications. Related graph parameters, such as the number of rounds needed to fill all vertices (propagation time) and minimizing a combination of the number of initially filled vertices and number of rounds needed (called throttling) are active areas of study.

The book is partitioned into four parts. Part 1 comprises Chapters 1 and 2 and provides a gentle introduction to the IEP-G, zero forcing, and some ancillary problems of the IEP-G. Chapter 1 also provides motivation and describes applications and connections with other areas of mathematics.

Part 2 focuses on various strong properties. Chapter 3 describes how the Implicit Function Theorem can be used to show that under certain conditions a solution to an inverse problem implies a solution to all 'nearby' inverse problems. Each strong property can be phrased as a matrix theoretic criterion. Chapter 4 shows how the strong properties can be used and establishes key consequences of the various strong properties. Chapter 5 provides a unified and technical theoretical derivation of the strong properties. This is done for both for completeness and in case readers need to develop other strong properties. However, we note that researchers may take the strong properties for granted and immediately use them in research.

Part 3 contains additional discussion of subproblems of the IEP-G and applies strong properties to these problems. In particular, Chapter 6 discusses ordered multiplicity lists for spectra of matrices described by G, Chapter 7 discusses rigid linkages, which provide additional information about eigenvalue multiplicities, and Chapter 8 discusses $q(G)$, the minimum number of distinct eigenvalues among matrices with off-diagonal nonzero pattern described by the edges of G.

Part 4 comprises Chapters 9–11 and focuses on zero forcing and its variants, related problems, and applications. Chapter 9 discusses zero forcing, its variants, connections to various ancillary problems of the IEP-G, and other graph coloring/searching parameters, such as power domination and Cops and Robbers. Chapter 10 discusses 'time' to complete coloring all the vertices, called propagation time or capture time. Chapter 11 discusses minimizing some combination of number of initially filled vertices and time.

Some graph and linear algebra terminology and notation is introduced throughout the book. Since there is more ambiguity in graph terminology (starting with the definition of a graph), a more complete summary is provided in Appendix A. Background material from linear algebra can be found in any standard linear algebra reference, such as *Handbook of Linear Algebra* [**178**], Horn and Johnson's *Matrix Analysis* [**196**], or Zhang's *Matrix Theory* [**271**].

Partial drafts of this book has been used as background material for two research communities, the American Mathematical Society Mathematics Research Community *Finding Needles in Haystacks: Approaches to Inverse Problems using Combinatorics and Linear Algebra* [**12**], and the American Institute of Mathematics Research Community *Inverse eigenvalue problems for graphs and zero forcing* [**9**]. We thank the Haystacks MRC and IEPG-ZF ARC participants, especially Ben Small, for their feedback. We also thank the reviewers for their suggestions. All of these comments have helped to improve the book.

A few places in the book refer to websites that contain some applications or videos that may be of interest to the reader. These urls were active at the time of publication, but may change in the future, and readers will need to search for and find these or similar sites.

Part 1

Introduction to the Inverse Eigenvalue Problem of a Graph and Zero Forcing

This book concerns recent significant developments for the Inverse Eigenvalue Problem of a Graph, known as the IEP-G, and an important related concept known as zero forcing, and connections between these and other areas of mathematics and applications. In this part we introduce the IEP-G in the context of applications in which it arises, and introduce zero forcing as an upper bound for the maximum multiplicity of an eigenvalue of a graph. The IEP-G and motivations for studying it are discussed in Chapter 1. Due to the challenging nature of the IEP-G, historically it has been approached through the study of graph parameters that provide partial solutions, called ancillary problems. One of these ancillary problems, maximum eigenvalue multiplicity or maximum nullity, led to the introduction of zero forcing in the context of the IEP-G, and we introduce zero forcing in maximum eigenvalue multiplicity in Chapter 2. Zero forcing was also introduced independently in other applications; this is discussed in Chapter 9 where zero forcing is discussed more thoroughly. Ancillary problems of the IEP-G are discussed further in Part 3.

CHAPTER 1

Introduction to and Motivation for the IEP-G

In this chapter, we give a gentle overview of the origins of one of our main topics, highlight some of the most important early results and techniques, and provide a brief non-technical overview of some applications and connections to other areas of mathematics.

1.1. Forward Problems

There is a rich history of modeling the dynamics of certain (physical, biological, economic, etc.) phenomena with systems of algebraic or differential equations. Deriving the model, and then finding solutions to the resulting equations is known as a *forward problem.*

EXAMPLE 1.1. The motion of the simple spring-mass system (see Figure 1.1) can be modeled by

$$\ddot{y} = -\frac{k}{m}y,$$

where k is the spring constant, m is the mass of the object, and $y(t)$ is the displacement from rest at time t. The general solution to this system is known to be

$$y(t) = c_1 \cos(\sqrt{\tfrac{k}{m}}t) + c_2 \sin(\sqrt{\tfrac{k}{m}}t), \tag{1.1}$$

and (1.1) represents the general solution to the forward problem for the simple spring-mass system.

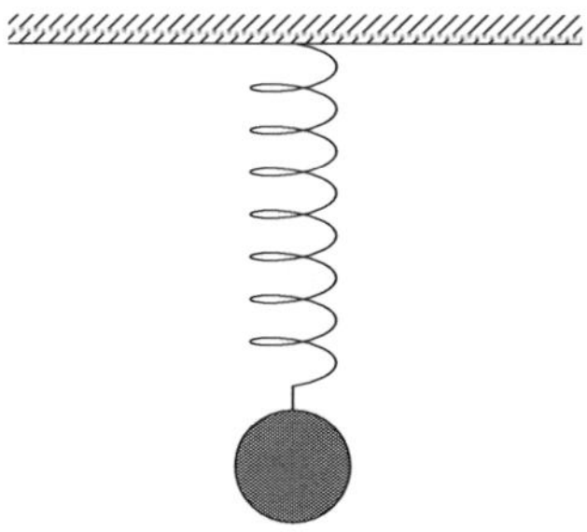

FIGURE 1.1. A spring-mass system with mass m and spring-constant k.

Another interesting forward problem is the application of graph energy to molecular chemistry, which was one of the motivations for the study of eigenvalues of the adjacency matrix; see Section 1.3.

1.2. Inverse Problems

Mathematical models and solutions to the corresponding forward problems allow one to investigate the behavior of the modeled phenomena. Such investigations naturally lead to questions of the form: Is it possible for a certain behavior to occur in the model? Such problems are called *inverse problems.* A simple instance of a inverse problem is given in the next example.

EXAMPLE 1.2. For a fixed spring constant k, is it possible to find a mass m such that the spring-mass system oscillates periodically with period p?

From the forward solution (1.1), for given k and m, the mass oscillates periodically with period $2\pi\sqrt{\frac{m}{k}}$. Hence setting $m = \frac{kp^2}{4\pi^2}$ results in a system that oscillates periodically with period p.

1.3. Matrices, Forward Problems, and Structure

In many instances, a model of a given phenomenon can be expressed as an algebraic or differential matrix equation, and the explicit solutions can be expressed in terms of the spectral parameters (e.g., eigenvalues, eigenvectors, generalized eigenvectors) of a pertinent matrix. In other words, a solution of the forward problem is determined by the spectral parameters of a matrix.

EXAMPLE 1.3. Multi-story buildings are subject to lateral movement due to seismic events, nearby vehicle traffic, wind, etc. A simple model of the lateral movement when the floors of an n-story building are slightly displaced considers the building as consisting of rigid, massive floors that are supported by flexible, massless columns (see Figure 1.2).

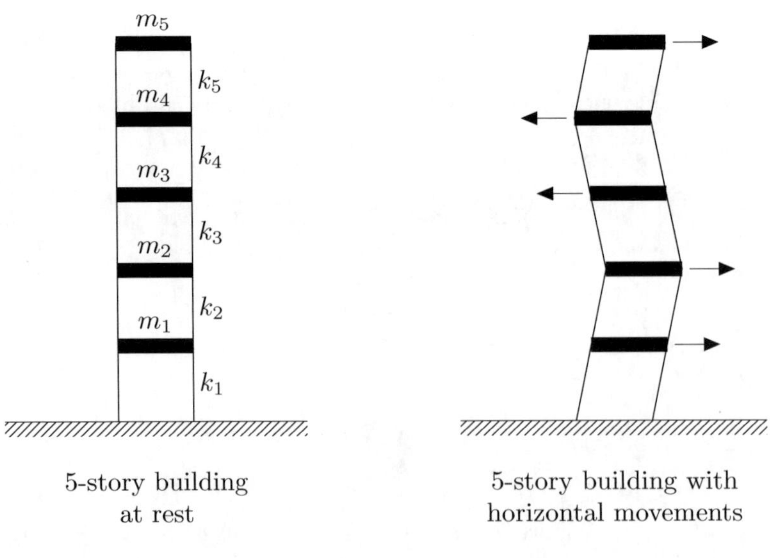

FIGURE 1.2. A 5-story building.

Newton's and Hooke's Laws lead to the following system of equations that model the lateral movement of the building:

$$M\ddot{\mathbf{x}} = -K\mathbf{x}, \tag{1.2}$$

where

$$M = \begin{bmatrix} m_1 & & & \\ & m_2 & & \\ & & \ddots & \\ & & & m_n \end{bmatrix}$$

is the diagonal matrix whose ith diagonal is the mass of the ith floor,

$$K = \begin{bmatrix} k_1 + k_2 & -k_2 & & & \\ -k_2 & k_2 + k_3 & -k_3 & & \\ & \ddots & \ddots & \ddots & \\ & & -k_{n-1} & k_{n-1} + k_n & -k_n \\ & & & -k_n & k_n \end{bmatrix}$$

is a symmetric, tridiagonal matrix formed from the constants k_i that measure the lateral stiffness of the ith column, and

$$\mathbf{x}(t) = \begin{bmatrix} x_1(t) \\ \vdots \\ x_n(t) \end{bmatrix}$$

is the vector of lateral displacements of the n stories at time t.

As M is a diagonal matrix and its diagonal entries are positive, we can define $\sqrt{M}$ to be

$$\sqrt{M} = \begin{bmatrix} \sqrt{m_1} & & & \\ & \sqrt{m_2} & & \\ & & \ddots & \\ & & & \sqrt{m_n} \end{bmatrix}.$$

Pre-multiplying (1.2) by $\sqrt{M}^{-1}$ and using the substitution $\mathbf{y} = \sqrt{M}\mathbf{x}$, we see that (1.2) is equivalent to $\ddot{\mathbf{y}} = -L\mathbf{y}$, where $L = \sqrt{M}^{-1}K\sqrt{M}^{-1}$. Note that L is a symmetric, tridiagonal matrix. Indeed, L is an M-matrix.[1] Hence, each eigenvalue of L is nonnegative.

Thus the model of the lateral movement of the building can be expressed as a simple linear system of differential equations whose coefficient matrix is symmetric and positive semidefinite. It is known that there exists a basis $\mathbf{v}_1, \ldots, \mathbf{v}_n$ of $\mathbb{R}^n$ of eigenvectors of L. Letting λ_j be the eigenvalue of L corresponding to $\mathbf{v}_i$ ($i = 1, 2, \ldots, n$). Then the general solution to (1.2) is

$$c_1 \cos(\sqrt{\lambda_1}t)\mathbf{v_1} + d_1 \sin(\sqrt{\lambda_1}t)\mathbf{v_1} + \cdots + d_n \cos(\sqrt{\lambda_n}t)\mathbf{v_n} + d_n \sin(\sqrt{\lambda_n}t)\mathbf{v_n}$$

for $c_i, d_i \in \mathbb{R}$. This solves the forward problem for the horizontal motion of an n-story building.

We note that Example 1.3 is essentially the same as a model that would arise in the study of n vibrating beads on a string. As we discuss later, this phenomena is of historical importance in the theory of vibrations.

---

[1]An *M-matrix* has the form $L = sI - Z$ where Z is an entrywise nonnegative matrix, and s is a real number that is at least as large as the largest eigenvalue of Z.

FIGURE 1.3. The path P_n on n vertices.

Some typical features are illustrated by Example 1.3. First, the models often result in a corresponding coefficient matrix that is real and symmetric. For this reason, this book will primarily focus on developing a theory in the real, symmetric matrix setting. Second, the structure problem at hand is manifested in the combinatorial structure of the corresponding coefficient matrix. To make this more precise, the *graph of the* $n \times n$ *matrix* A, denoted by $\mathcal{G}(A)$, is defined to have vertex set $\{1, 2, \ldots, n\}$ and edge set $\{\{i, j\} : i \neq j \text{ and } a_{ij} \neq 0\}$. Thus, under the mild assumption each constant k_i in Example 1.3 is nonzero, $\mathcal{G}(L)$ is the path on n vertices which is illustrated in Figure 1.3. Throughout we denote the path with n vertices by P_n. The next two examples illustrate how different graphs of matrices naturally arise.

EXAMPLE 1.4. Figure 1.4 shows two n-story skyscrapers with a skybridge joining them at level ℓ. Here the nodes correspond to the various floors of the buildings. Modeling as in Example 1.3 we arrive at a system of the equations of the form $\ddot{\mathbf{y}} = -N\mathbf{y}$ where N is a symmetric matrix and $\mathcal{G}(N)$ is obtained by inserting an edge between two disjoint copies of the path P_n that joins two vertices, one from each path at the same distance from a pendent vertex.

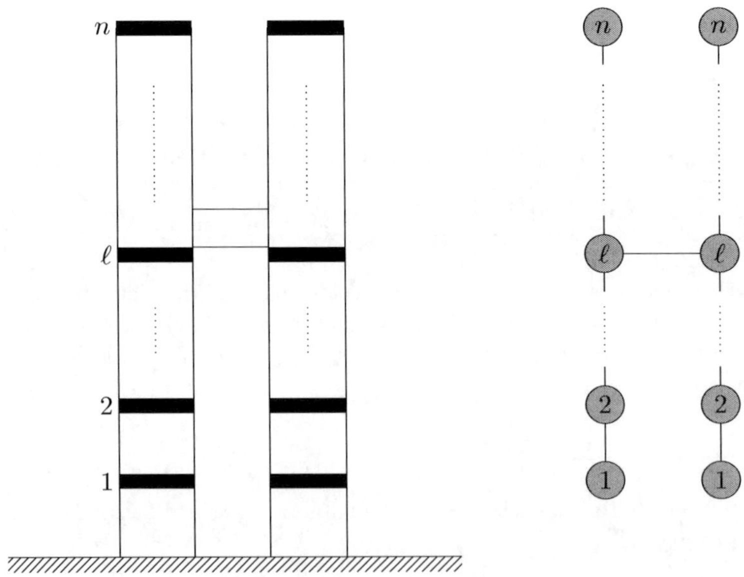

FIGURE 1.4. Twin towers with n floors and a skywalk at level ℓ and the corresponding graph.

EXAMPLE 1.5. The use of tuned mass dampers is one way that architects reduce sway in the upper floors of a skyscraper.[2] One example of a tuned mass damper is the placement of a pendulum between floors as illustrated in Figure 1.5.[3] The basic idea is that when a lateral force hits the building, the walls of the building will move slightly, which in turn causes the pendulum to oscillate. When designed correctly, the pendulum's period is different than the building's and the forces of the pendulum and building counteract each other, thereby lessening the sway. The video at `https://www.youtube.com/watch?v=GzMuF-LMGaM` shows a tuned mass damper in action.

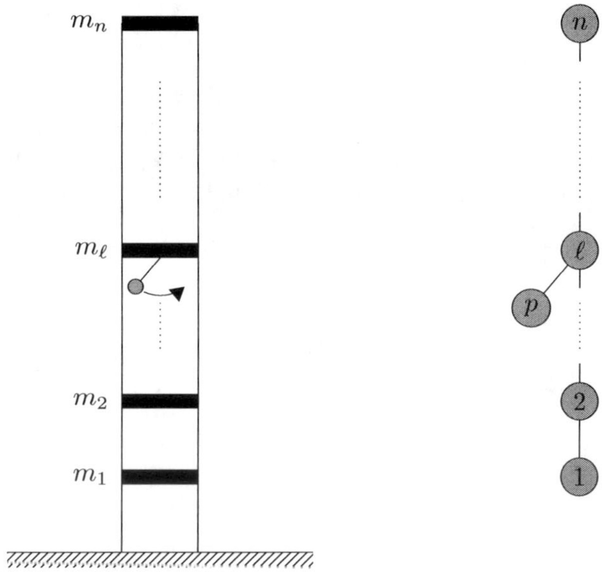

FIGURE 1.5. An n-story skyscraper with a tuned mass damper pendulum at level ℓ and the corresponding graph.

EXAMPLE 1.6. The study of vibrations of an elastic membrane provides another historically important example. The Helmholtz equation (in two dimensions) is given by

$$\frac{\partial^2 z}{\partial t^2} = c^2 \left(\frac{\partial^2 z}{\partial x^2} + \frac{\partial^2 z}{\partial y^2} \right), \tag{1.3}$$

and is defined on a region Ω of the real plane, where $z(t)\colon \Omega \to \mathbb{R}$ represents vertical displacement of the points in Ω at time t and c is a constant. When Ω is thought of as a homogeneous elastic membrane, solutions to the Helmholtz equation (that is, functions satisfying (1.3)) represent the possible motions of vibration of Ω. If there is a nonzero function f satisfying (1.3), then the corresponding resonant frequency

---

[2]Taipei 101, which stands 1667 feet tall, is the tallest building in Taiwan and was the tallest building in the world from 2004 until 2009. It has an 18-foot diameter, 660-metric ton steel sphere suspended by eight cables in its upper stories; see `https://www.atlasobscura.com/places/tuned-mass-damper-of-taipei-101` for more details.

[3]Here, for simplicity, the pendulum fits between just one floor. Often times the pendulum spans several floors.

of Ω is $\omega = c\sqrt{\frac{k}{m}}$, where the membrane's tension constant (respectively, mass per unit area) is k (respectively, m) and f has the form

$$g_1(x,y)\cos(c\sqrt{\tfrac{k}{m}}t) + g_2(x,y)\sin(c\sqrt{\tfrac{k}{m}}t)$$

for some real-valued functions $g_1(x,y)$ and $g_2(x,y)$ defined on Ω.

As noted by many (see, e.g., [**236**]), if one approximates Ω by a discrete grid of particles with equal spacing ω and assumes that the elastic membrane behaves like a spring according to Hooke's Law, then the Helmholtz equation on Ω can be approximated by

$$L\mathbf{Z} = \frac{k\omega^2}{c^2}\mathbf{Z},$$

where L is the Laplacian matrix[4] of the graph of the discrete grid, and $\mathbf{Z}$ is the vector of values of $z(x,y)$ ordered according to the labeling of vertices of the grid. Figure 1.6 illustrates such a discretization for a square. In particular, a nonzero $\mathbf{Z}$ will be an eigenvector of the Laplacian matrix of the grid graph with eigenvalues $k\omega^2/c^2$. The website `http://www.math.ucsd.edu/~fan/hear/index.html` created by Rob Ellis allows one to "hear" the eigenvalues of the Laplacian of several graphs.

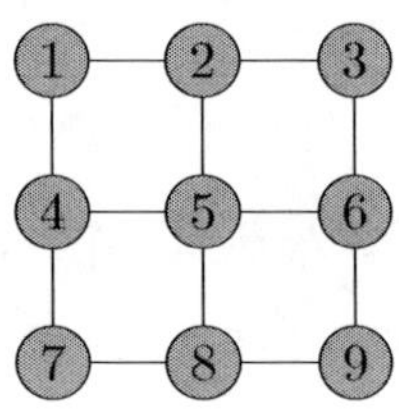

$$\begin{bmatrix} 2 & -1 & 0 & -1 & 0 & 0 & 0 & 0 & 0 \\ -1 & 3 & -1 & 0 & -1 & 0 & 0 & 0 & 0 \\ 0 & -1 & 2 & 0 & 0 & -1 & 0 & 0 & 0 \\ -1 & 0 & 0 & 3 & -1 & 0 & -1 & 0 & 0 \\ 0 & -1 & 0 & -1 & 4 & -1 & 0 & -1 & 0 \\ 0 & 0 & -1 & 0 & -1 & 3 & 0 & 0 & -1 \\ 0 & 0 & 0 & -1 & 0 & 0 & 2 & -1 & 0 \\ 0 & 0 & 0 & 0 & -1 & 0 & -1 & 3 & -1 \\ 0 & 0 & 0 & 0 & 0 & -1 & 0 & -1 & 2 \end{bmatrix}$$

FIGURE 1.6. A 3×3 grid graph and its Laplacian matrix.

The concept of graph energy arose in molecular chemistry where certain numerical quantities, such as the heat of formation of a hydrocarbon, are related to total π-electron energy that can be calculated as the energy of a molecular graph. Gutman in [**164**] defined the *energy* of a graph G in 1978 to be the sum of the absolute values of the eigenvalues of its adjacency matrix,

$$E(G) = \sum_{i=1}^{n} |\lambda_i|$$

where $\operatorname{spec}(A_G) = \{\lambda_1, \ldots, \lambda_n\}$. In his 2001 survey [**165**], Gutman explains how total π- electron energy is calculated from graph energy.

1.4. Inverse Eigenvalue Problems and Matrices with a Given Graph

Now that we have seen several examples of forward problems that can be modeled by matrix equations, we turn our attention to the related inverse problems that provide the historical underpinnings of this book.

[4]The *adjacency matrix* A_G of the graph G with vertex set $\{1,\ldots,n\}$ is the $n \times n$ matrix whose (i,j)-entry is 1 if ij is an edge of G and is 0 otherwise. The *Laplacian matrix* of G is the difference of the diagonal degree matrix and the adjacency matrix.

Inverse problems either take on a "qualities" or a "design" point of view. A prime example of a quality perspective is M. Kac's famous question [**208**]: "Can you hear the shape of a drum?" The sound is produced by the vibrations of a thin elastic membrane corresponding to the area Ω in the plane, and is related to specifically the eigenfunctions and eigenvalues (that is, the solutions) of the corresponding Helmholtz equation (1.3), because the eigenvalues correspond to the frequencies of the region's fundamental modes of vibration. Kac asked if there exist non-congruent[5] regions Ω_1 and Ω_2 of the plane each of whose boundaries is a smooth curve such that their frequencies are identical (under the assumption that the membranes are clamped on the boundary)? Kac noted that the area of the membrane Ω is determined by the frequencies; this was conjectured by Lorentz and proven by Weyl [**264**]. In 1992, Gordon, Webb and Wolpert [**161**] provided a pair of non-congruent concave polygons with the same frequencies; thus in general one can't hear the shape of a drum. We note that one may generalize Kac's question to higher dimensions; soon after Kac posed his question, John Milnor showed the existence of two 16-dimensional, non-congruent tori with the same spectrum [**229**]. Also, under additional assumptions (such as convexity, no holes and analyticity of the boundary), it is known that one can hear the shape of a drum [**270**].

An example of an inverse problem taking on the design perspective is related to the mass damper pendulum described in Example 1.5. To provide some comfort to the tenants in the upper floor of a skyscraper architects will design the building (e.g., select the masses of the floor, the flexibility of the columns, the mass and length of the pendulum) so that for typical events (e.g., prevailing wind gusts, vehicle traffic or seismic activity) high-frequency solutions don't exist, or their corresponding eigenvectors have entries corresponding to the upper floors of small magnitude. In terms of the spectra of the corresponding coefficient matrix N this is asking for all eigenvalues to be small; or to have the eigenvectors corresponding to large eigenvalues to have entries corresponding to the higher floors of small magnitude.

As the above examples indicate, many inverse problems reduce to problems about the eigenvalues and eigenvectors of a real, symmetric matrix whose structure reflects the inherent structure of the underlying problem. This observation is the foundation of the Inverse Eigenvalue Problem for Graphs, which is one of the two main topics of this book. The other is zero forcing, which arose in numerous contexts, including in the study of maximum possible multiplicity of an eigenvalue of a matrix described by a given graph. Zero forcing is introduced in Chapter 2, together with subproblems of the IEP-G such as maximum multiplicity of an eigenvalue. In the remainder of this section, we introduce the notation and concepts for the Inverse Eigenvalue Problem for Graphs (for short, the IEP-G), a few important historical results for the IEP-G, and additional connections with other mathematics.

Unless otherwise stated, graphs in this book are simple (that is, they have neither loops nor multiple edges), undirected, and finite, and matrices are real. Thus, a *graph* G consists of a finite subset $V(G)$ of *vertices* and a finite set $E(G)$ of $\{\{i,j\} : i,j \in V(G) \text{ and } i \neq j\}$. An *edge* of G is an element of $E(G)$, and $E(G)$ is the *edge set* of G. The *order of* G equals the cardinality $|V(G)|$ of its vertex set. We often say ij *is an edge of* G, or write i–j *is in* G to indicate that $\{i,j\} \in E(G)$. Appendix A presents graph terminology pertinent to this book, and includes an

[5] I.e., there is no isometry that maps Ω_1 bijectively to Ω_2.

annotated list of the families of graphs that frequently occur in the book. Each family will also be defined when it first appears in the book. A good supplementary graph theory (respectively, matrix theory) reference is [**112**] (respectively, [**196**]).

As previously noted, an $n \times n$ symmetric matrix $A = [a_{ij}]$ determines a graph $\mathcal{G}(A)$ called the *graph of* A; namely, $\mathcal{G}(A)$ has vertex set $\{1, \ldots, n\}$ and ij is an edge of $\mathcal{G}(A)$ if and only if $i \neq j$ and $a_{ij} \neq 0$. Note that $\mathcal{G}(A)$ is independent of the values of the diagonal entries of A.

DEFINITION 1.7. Let G be a graph of order n with vertex set $\{v_1, \ldots, v_n\}$ and let $S_n(\mathbb{R})$ denote the set of $n \times n$ real symmetric matrices. Then the set of matrices A such that $\mathcal{G}(A) = G$ is denoted by $\mathcal{S}(G)$; that is,

$$\begin{aligned} \mathcal{S}(G) = \{A = [a_{ij}] \in S_n(\mathbb{R}) \quad : \quad & a_{ij} \neq 0 \text{ if } v_iv_j \text{ is an edge of } G \text{ and} \\ & a_{ij} = 0 \text{ if } i \neq j \text{ and } v_iv_j \text{ is not an edge of } G\}. \end{aligned}$$

EXAMPLE 1.8. Here we give $\mathcal{S}(G)$ for several examples of graphs G.

If P_n is the path 1–2– $\cdots$ –n, then $\mathcal{S}(P_n)$ consists of all (real symmetric) matrices of the form

$$\begin{bmatrix} a_1 & b_1 & & & \\ b_1 & a_2 & b_2 & & \\ & \ddots & \ddots & \ddots & \\ & & b_{n-2} & a_{n-1} & b_{n-1} \\ & & & b_{n-1} & a_n \end{bmatrix},$$

where the entries not displayed are zero, and each of $b_1, \ldots, b_{n-1}$ is nonzero. In particular,

$$\begin{bmatrix} 1 & 2 & 0 & 0 \\ 2 & 0 & 1 & 0 \\ 0 & 1 & -1 & 1 \\ 0 & 0 & 1 & 2 \end{bmatrix} \in \mathcal{S}(P_4), \text{ but } \begin{bmatrix} 1 & 2 & 0 & 0 \\ 2 & 1 & 0 & 0 \\ 0 & 0 & -1 & 1 \\ 0 & 0 & 1 & 2 \end{bmatrix} \notin \mathcal{S}(P_4).$$

When P_n is the path 1–2– $\cdots$ – n, matrices in $\mathcal{S}(P_n)$ are also known as irreducible, symmetric tridiagonal matrices or symmetric Jacobi matrices. These matrices are central in several areas of mathematics, e.g., orthogonal polynomials and numerical analysis.

A *complete graph* is a graph in which each pair of distinct vertices is joined by an edge; when the order is n it is denoted by K_n. If G is a graph of order n, then the *complement of* G is denoted by $\overline{G}$ and is the graph with the same vertex set having ij as an edge if and only if $i \neq j$ and ij is not an edge of G. The *empty graph* of order n, $\overline{K_n}$, is the complement of K_n. It follows that $\mathcal{S}(K_n)$ consists of all $n \times n$ symmetric matrices with each off-diagonal entry nonzero and $\mathcal{S}(\overline{K_n})$ consists of all $n \times n$ diagonal matrices.

The *star* $K_{1,n-1}$ of order $n \geq 1$ is a graph having $n - 1$ edges each of which is incident to the same vertex, called the *center* of the star (when $n = 2$ either vertex is a center vertex). Taking the vertices of $K_{1,n-1}$ to be $1, \ldots, n$ with 1 being the center of the star, we have that $\mathcal{S}(K_{1,n-1})$ consists of all matrices of the form

$$\left[\begin{array}{c|cccc} a_1 & b_1 & b_2 & \cdots & b_{n-1} \\ \hline b_1 & a_2 & & & \\ b_2 & & a_3 & & \\ \vdots & & & \ddots & \\ b_{n-1} & & & & a_n \end{array}\right],$$

where each b_i is nonzero.

As a final example, let G be the graph representing dual five story buildings joined by a skywalk at level 3 described in Example 1.4. Then with an appropriate labeling of the vertices, $\mathcal{S}(G)$ consists of all matrices of the form

$$\left[\begin{array}{ccccc|ccccc} a_1 & b_1 & 0 & 0 & 0 & 0 & 0 & 0 & 0 & 0 \\ b_1 & a_2 & b_2 & 0 & 0 & 0 & 0 & 0 & 0 & 0 \\ 0 & b_2 & a_3 & b_3 & 0 & 0 & 0 & b_{10} & 0 & 0 \\ 0 & 0 & b_3 & a_4 & b_4 & 0 & 0 & 0 & 0 & 0 \\ 0 & 0 & 0 & b_4 & a_5 & 0 & 0 & 0 & 0 & 0 \\ \hline 0 & 0 & 0 & 0 & 0 & a_6 & b_6 & 0 & 0 & 0 \\ 0 & 0 & 0 & 0 & 0 & b_6 & a_7 & b_7 & 0 & 0 \\ 0 & 0 & b_{10} & 0 & 0 & 0 & b_7 & a_8 & b_8 & 0 \\ 0 & 0 & 0 & 0 & 0 & 0 & 0 & b_8 & a_9 & b_9 \\ 0 & 0 & 0 & 0 & 0 & 0 & 0 & 0 & b_9 & a_{10} \end{array}\right],$$

where each b_i is nonzero.

The *spectrum* of a square matrix A is the multiset of eigenvalues of A, where the multiplicity of an eigenvalue in the spectrum is its algebraic multiplicity. In other words the spectrum of A is the multiset of zeros of the *characteristic polynomial* $p_A(x) = \det(xI - A)$ of A. We denote the spectrum of A by $\operatorname{spec}(A)$ and denote the multiplicity of eigenvalue λ by $\operatorname{mult}_A(\lambda)$.$\operatorname{mult}_A(\lambda)$ As customary, when listing the $\operatorname{spec}(A)$ we use $\lambda^{(m)}$ to indicate that λ is an eigenvalue of A of multiplicity m. E.g., $\operatorname{spec}(A) = \{2^{(3)}, 1, 4^{(3)}\}$ means that the eigenvalues of A are 2 of multiplicity 3, 1, and 4 of multiplicity 3.

DEFINITION 1.9. The **Inverse Eigenvalue Problem for Graphs** is the following:

> Given a graph G of order n, determine the multisets Λ of n real numbers such that there is a matrix in $\mathcal{S}(G)$ with spectrum Λ.

For convenience, we shorten the phrase "Inverse Eigenvalue Problem for Graphs" to just IEP-G.

As we study spectra, it is useful to have notation for submatrices. Given a subset α of row indices, and a subset β of column indices, $A[\alpha, \beta]$ (respectively, $A(\alpha, \beta)$) is the submatrix of A obtained from A by keeping (respectively, removing) the rows with index and columns with index in β). When A is square and $\alpha = \beta$, we shorten the notation to $A[\alpha]$ (respectively, $A(\beta)$). When $\alpha = \{i\}$ and $\beta = \{j\}$, we abbreviate to $A[i, j]$ (respectively, $A(i, j)$).

1.5. Initial Results for the IEP-G

A specific instance of the IEP-G takes the design point of view, namely, given a graph G of order n and a multiset Λ of n real numbers does there exist a symmetric matrix whose graph is G having Λ as its spectrum? Additionally, as we shall see, some results related to the IEP-G take on a qualities point of view. For example, if G is a tree and $A \in \mathcal{S}(G)$, then the largest and smallest eigenvalues of A are simple (see Theorem 1.14).

In the remainder of this section, we give some examples and basic results that show that the choice of G places restrictions upon the possible spectrum that can be achieved. The more substantial proofs of the results are deferred to later in the book.

As $\mathcal{S}(\overline{K_n})$ is the set of $n \times n$ diagonal matrices, every set of n real numbers is the spectrum of a matrix whose graph is the empty graph on n vertices. Additionally, since the only $n \times n$ symmetric matrix with spectrum $\{\lambda, \lambda, \dots, \lambda\}$ is the $n \times n$ λI_n, and $\mathcal{G}(\lambda I_n) = \overline{K_n}$ we have the following result.

PROPOSITION 1.10. *The empty graph of order n is the only graph of order n for which every multiset of n real numbers is the spectrum of some matrix with that graph.*

We note that Proposition 1.10 is heavily dependent upon the matrices being real. Indeed, a result of Friedland [**145**] asserts that A is any $n \times n$ complex matrix, and Λ is any multiset of n complex numbers, then there exists a complex diagonal matrix D such $D + A$ has spectrum Λ.

As observed above, for $n \geq 2$ there is no matrix in $\mathcal{S}(K_n)$ having all equal eigenvalues. The following asserts that this is the only type of multiset for which the IEP-G does not have a solution for K_n.

PROPOSITION 1.11. [**35**] *Let $n \geq 2$ and let $\lambda_1 \leq \lambda_2 \leq \cdots \leq \lambda_n$. Then there exists an $A \in \mathcal{S}(K_n)$ with eigenvalues $\lambda_1, \lambda_2, \dots, \lambda_n$ if and only if $\lambda_1 < \lambda_n$.*

The IEP-G for the path P_n is closely related to the vibrations of an n-story building, or the vibrations of a string. The theory of vibrations, along with the theory of Partial Differential Equations (PDEs) originated in the late 18th century with the work of d'Alembert [**94, 95, 96**] and Euler [**125**]. They were able to find all solutions to the one-dimensional equation, and that the natural frequencies of a vibrating spring of length ℓ are $\frac{nv}{2\ell}$ where v is the velocity of the wave, and n is a positive integer. In particular, the frequencies are distinct.

The following analog for the discrete case asserts that the eigenvalues of each matrix in $\mathcal{S}(P_n)$ are distinct.

PROPOSITION 1.12. *Let $A \in \mathcal{S}(P_n)$. Then the eigenvalues of A are distinct.*

This is well known for any irreducible tridiagonal matrix, and one can see that the multiplicity of any real eigenvalue λ is at most one by deleting the first row and last column of $A - \lambda I$ to obtain an $(n-1) \times (n-1)$ matrix of rank $n-1$ (since $(A - \lambda I)(1, n)$ is upper triangular with nonzero diagonal entries).

A *tree* is a graph that is connected and contains no cycles. For example, both paths and stars are trees; but K_n is not a tree for $n \geq 3$. For a tree T, the largest and smallest eigenvalues are necessarily simple. One way to see this is via a simple result about scalings of matrices in $\mathcal{S}(T)$ and the theory of nonnegative matrices.

LEMMA 1.13. [**109**] *Let T be a tree, and $A \in \mathcal{S}(T)$. Then there exists a diagonal matrix D each of whose diagonal entries is ± 1 such that each off-diagonal entry of DAD^{-1} is nonnegative.*

Let $A \in \mathcal{S}(T)$ and choose D such that all the off-diagonal entries of $B = DAD^{-1}$ are nonnegative. Choose $c \in \mathbb{R}$ sufficiently large such that $cI + B \geq 0$ and $cI + B$ is positive semidefinite. Then the Perron-Frobenius theorem for irreducible matrices implies that the greatest eigenvalue of $cI + B$ (and hence of A) is simple. For the least eigenvalue, start with $-A$.

THEOREM 1.14. *Let T be a tree and $A \in \mathcal{S}(T)$. Then the greatest and least eigenvalue of A each have multiplicity* 1.

It is now reasonable to ask: Is every set of n distinct eigenvalues the spectrum of some matrix in $\mathcal{S}(P_n)$? The answer is yes; and was first observed by Gantmacher and Krein [**148**] and later by Hochstadt [**176**].

THEOREM 1.15. [**176**] *The multiset Λ of n real numbers is the spectrum of a matrix in $\mathcal{S}(P_n)$ if and only if Λ has no repeated element.*

One direction of Theorem 1.15, that Λ has no repeated element implies Λ is the spectrum of a matrix in $\mathcal{S}(P_n)$, is known to be true more generally. Duarte [**119**] gives a constructive proof for trees. We shall see a proof based on the Implicit Function Theorem that can be extended to arbitrary graphs, and hence the next result holds. This was first proven, with a more complicated technique, in [**233**]. See the discussion preceding Theorem 3.1.

THEOREM 1.16. *Let G be a graph of order n, and let Λ be a set of n distinct real numbers. Then there is a matrix $A \in \mathcal{S}(G)$ with spectrum Λ.*

The IEP-G has been solved for graphs of order at most five as discussed in Section 4.3 and for the graph families listed in the next theorem.[6] Note that the graph in Example 1.4 is a double path and the graph in Example 1.5 is a generalized star.

THEOREM 1.17. [**23, 35, 134, 138, 176, 201, 202, 206**] *The IEP-G has been solved for the following families of graphs:*

- *Complete graphs.*
- *Cycles.*
- *Paths.*
- *Double paths.*
- *Generalized stars.*
- *Double generalized stars.*
- *Linear trees.*

1.6. Connections of the IEP-G to Nodal Domains

We conclude this chapter with some results about how the structure of a matrix forces certain qualitative properties on its eigenvalues and eigenvectors and connections to nodal domains and Chaldni figures. We note that if $A \in \mathcal{S}(P_n)$, then A is similar by a diagonal matrix D to a matrix $B \in \mathcal{S}(P_n)$ with each off-diagonal entry nonpositive.

[6]See Appendix A or Chapter 6 for definitions of double paths, generalized stars, and double generalized stars.

THEOREM 1.18. [**139**] *Let P_n be the path 1–2–$\cdots$ –n and let $B \in \mathcal{S}(P_n)$ with all off-diagonal entries nonnegative and having eigenvalues $\lambda_1 < \cdots < \lambda_n$. Let $\mathbf{x} = [x_i]$ be an eigenvector of B corresponding to λ_j. Then*

$$j = n - |\{i \in \{1, \ldots, n-1\} : x_i x_{i+1} < 0 \text{ or } x_i = 0\}|.$$

We note that Fiedler [**139**] established a result of this type for all trees. Theorem 1.18 is a discrete analog of classical nodal domain theorems, which we now briefly describe. Let Ω be an open bounded subset of $\mathbb{R}^n$ (or more generally an n-dimensional Riemannian manifold with boundary $\partial\Omega$) and let $\triangle$ be a differential operator on Ω. The eigenvalue equation on Ω is

$$-\triangle u = \lambda u \text{ on } \Omega, \text{ and } \triangle = 0 \text{ on } \partial\Omega. \tag{1.4}$$

Nonzero u satisfying (1.4) are called eigenfunctions of $\triangle$ corresponding to the eigenvalue λ. For many $\triangle$ the eigenvalues are known to be countable list of real numbers that is bounded below and they are taken to be $\lambda_1 \leq \lambda_2 \leq \cdots$. Given an eigenfunction u, the nodes of u are the points $x \in \mathbb{R}^n$ such that $u(x) = 0$, and a nodal domain of u is a maximally connected subset of $\mathbb{R}^n$ on which u does not have a sign change.

When $n = 2$, Ω represents the area of a thin flexible membrane, and (1.4) is its wave-equation. Then a plot of the nodes of u is known as a Chaldni figure. Two Chaldni figures for a square are illustrated in Figure 1.7. Chaldni figures arise physically by placing sand on the membrane, and then exciting the membrane to have it vibrate according to λ and u. The grains of sand will naturally move to places on the drum head that have no vertical movement, that is, to the nodes of u. This can be seen at `https://www.youtube.com/watch?v=1RFysSAxWxI`.

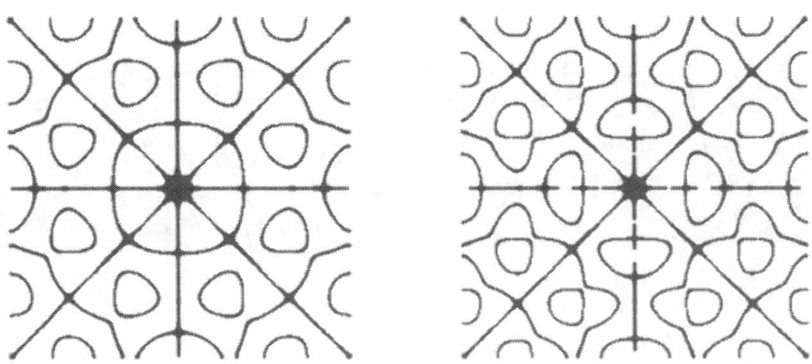

FIGURE 1.7. Various Chaldni figures for a square plate.

The following are some classical results about nodal domains for the Laplace operator. We direct the reader to [**254**] for Sturm's Oscillation theorem, [**91**] for Courant's Nodal Domain Theorem, and [**245**] for Pleijel's extension.

THEOREM 1.19. **Sturm's Oscillation Theorem**
For the vibrating string on the interval (a, b) with endpoints stationary, the nodes of the kth eigenfunction divide (a, b) into k nodal domains.

THEOREM 1.20. **Courant's Nodal Domain Theorem**
Let u be an eigenfunction for the wave-equation on an open, bounded and connected set Ω of $\mathbb{R}^n$ having a sufficiently nice boundary. If u corresponds to the kth smallest eigenvalue, then u has at most k nodal domains.

THEOREM 1.21. **Pleijel's Nodal Domain Theorem**
Let Ω be an open, bounded connected set in $\mathbb{R}^2$, and let $\triangle$ be the Laplace operator on Ω. Then for each positive integer k there are only finitely many eigenvalues of (1.4) *having a corresponding eigenfunction with exactly k nodal regions.*

CHAPTER 2

Zero Forcing and Maximum Eigenvalue Multiplicity

Prior to the mid 2000s, general results for the IEP-G were rare. There was some very limited success in focusing on the IEP-G for special families of graphs (see Theorem 1.17), and more progress by studying related less complicated graph parameters that provide information to aid the solution of the IEP-G, here called *ancillary problems*. Perhaps the most important of these ancillary problems, maximum eigenvalue multiplicity or maximum nullity, is a focus of this chapter. Finding the maximum nullity of a set of matrices is equivalent to finding the minimum rank of the same set; rank minimization and the study of low rank matrices has many applications, including to compressed sensing and recommender algorithms. The study of maximum nullity led to the introduction of zero forcing (see Chapter 9 for information on how zero forcing was introduced independently in other areas).

In this chapter we describe the maximum nullity problem, highlighting some historically important results, and introduce zero forcing, which is the second main theme of the book. Zero forcing and additional graph parameters derived from it are discussed in more detail in Part 4. In this chapter we also introduce a variant of the maximum nullity problem where the matrices considered are restricted to positive semidefinite matrices, and describe related Colin de Verdière type parameters, which provide an introduction to the strong properties discussed in Part 2. Variants of the maximum nullity problem where the matrices considered are symmetric with zero diagonal or skew symmetric are introduced in Section 9.4 in conjunction with the related zero forcing parameter. Further discussion of ancillary problems, including the minimum number of distinct eigenvalues, is presented in Part 3. While some graph terms are defined when introduced, a more complete summary of graph terminology and notation used is provided in Appendix A. Linear algebra terminology is more uniform and can be found in any standard reference, such as *Handbook of Linear Algebra* [**178**], Horn and Johnson's *Matrix Analysis* [**196**], or Zhang's *Matrix Theory* [**271**].

2.1. Introduction to Maximum Nullity and Minimum Rank

Evidence suggests that the problem of finding a matrix $A \in \mathcal{S}(G) = \{A \in S_n(\mathbb{R}) : \mathcal{G}(A) = G\}$ with spectrum Λ is more difficult when Λ has an element of high multiplicity (e.g., Theorems 1.14 and 1.16). To study this, the next concept was introduced.

DEFINITION 2.1. The *maximum multiplicity* of the graph G is denoted by $\mathrm{M}(G)$ and is defined to be the largest integer k such that there is a matrix $A \in \mathcal{S}(G)$ having an eigenvalue of multiplicity k. Since A is a real symmetric matrix, the multiplicity of λ as an eigenvalue of A equals the nullity of $\lambda I - A$. Since $\mathcal{G}(A) = \mathcal{G}(\lambda I - A)$,

$\mathrm{M}(G)$ is equal to the largest integer k such that there exists a matrix $A \in \mathcal{S}(G)$ of nullity k. Thus

$$\mathrm{M}(G) = \max\{\operatorname{null} A : A \in \mathcal{S}(G)\}$$

and we usually refer to $\mathrm{M}(G)$ as the *maximum nullity* of G.

DEFINITION 2.2. The *minimum rank* of G is defined to be the smallest integer k such that there is a matrix in $\mathcal{S}(G)$ of rank k:

$$\operatorname{mr}(G) = \min\{\operatorname{rank} A : A \in \mathcal{S}(G)\}.$$

Since $\operatorname{rank} A + \operatorname{null} A = n$ for every $n \times n$ matrix,

$$\operatorname{mr}(G) + \mathrm{M}(G) = n$$

for each graph G of order n.

Historically, the maximum multiplicity or maximum nullity problem was originally studied primarily from the perspective of minimum rank.

Note that every strictly diagonally dominant[1] matrix $A \in \mathcal{S}(G)$ has rank $|V(G)|$. Hence the maximum rank of a symmetric matrix with graph G is not very interesting; it is always $|V(G)|$. Also note that changing any diagonal entry of a matrix changes its rank by at most one. Thus, starting with $A \in \mathcal{S}(G)$ with $\operatorname{rank} A = \operatorname{mr}(G)$ and changing one diagonal entry at time until one reaches a diagonally dominant matrix, one encounters matrices in $\mathcal{S}(G)$ of each rank in $\{\operatorname{mr}(G), \dots, |V(G)|\}$.

REMARK 2.3. For any graph G, $\mathrm{M}(G) \geq 1$ and $\operatorname{mr}(G) \leq n-1$ because $\mathcal{S}(G)$ contains singular matrices: For example, choose any matrix $B = [b_{ij}] \in \mathcal{S}(G)$, and then add $-\sum_{j=1}^{n} b_{ij}$ to b_{ii}, so that every row sum is zero. (If B is chosen as the negative of the adjacency matrix of G, this results in the *Laplacian matrix* of G.)

Next we compute the minimum rank of some families of graphs.

EXAMPLE 2.4. Assume P_n is the path 1–2– $\cdots$ –n. Let $A \in \mathcal{S}(P_n)$. Then the matrix obtained from A by deleting row n and column 1 is a lower triangular matrix each of whose diagonal entries is nonzero. Hence $\operatorname{rank} A \geq n-1$, so $\operatorname{mr}(P_n) \geq n-1$. Remark 2.3 implies $\operatorname{mr}(P_n) \leq n-1$.

EXAMPLE 2.5. Let $n \geq 3$. Since the smallest and largest eigenvalues of a matrix whose graph is a tree are simple (see Theorem 1.14), $\mathrm{M}(K_{1,n-1}) \leq n-2$ for $n \geq 3$. The adjacency matrix of $K_{1,n-1}$ has nullity $n-2$.

One can view the minimum rank of a graph geometrically: Let G be a graph of order with vertex set $\{1, \dots, n\}$. A *representation of G of dimension d* is a (symmetric) nondegenerate[2] bilinear form B, and a collection of vectors $\mathbf{v_1}, \dots, \mathbf{v_n}$ in $\mathbb{R}^d$ such that for $i \neq j$ we have $B(\mathbf{v_i}, \mathbf{v_j}) \neq 0$ if and only if ij is an edge of G. It is not difficult to see that a representation of G of dimension d corresponds to the n columns of a $d \times n$ matrix V and an invertible $d \times d$ symmetric matrix C such that $V^T C V \in \mathcal{S}(G)$. In fact, by expressing C as $Q^T D Q$ where D is an invertible diagonal matrix and Q is an orthogonal[3] matrix, a representation of G of dimension d corresponds to a $d \times n$ matrix $W = QV$ and an invertible diagonal matrix D such that $W^T D W \in \mathcal{S}(G)$. The next result is well known.

---

[1]A matrix is *strictly diagonally dominant* if for every row, the magnitude of the diagonal entry in that row exceeds the sum of the magnitudes of the other entries.

[2]A bilinear form B is *nondegenerate* if $B(\mathbf{x}, \mathbf{y}) = 0$ for all $\mathbf{y}$ only if $\mathbf{x} = \mathbf{0}$.

[3]A matrix $Q \in \mathbb{R}^{n \times n}$ is an *orthogonal matrix* if $Q^T Q = I_n$.

PROPOSITION 2.6. *Let G be a graph. Then the smallest d such that G has a representation in dimension d is $d = \mathrm{mr}(G)$.*

PROOF. Suppose that G has a representation in dimension d. Then there exists a $d \times n$ matrix V and an invertible $d \times d$ symmetric matrix C such that $V^TCV \in \mathcal{S}(G)$. As $\mathrm{rank}(V^TCV) \leq d$, we see that $\mathrm{mr}(G) \leq d$.

To complete the proof, we show that G has a representation of dimension $\mathrm{mr}(G)$. Let $A \in \mathcal{S}(G)$ with $r = \mathrm{rank}\, A = \mathrm{mr}(G)$. By the Spectral Theorem for Symmetric Matrices, there exist an orthogonal matrix Q and an invertible diagonal $r \times r$ matrix D such that

$$A = Q^T \begin{bmatrix} D & O \\ O & O \end{bmatrix} Q.$$

Taking V to be the matrix consisting of the first r rows of Q gives $A = V^TDV$ which is a representation of G of dimension r. □

Intuitively, Proposition 2.6 says that graphs with small minimum rank (or equivalently large maximal nullity) are in some sense low-dimensional objects.

EXAMPLE 2.7. Let G be a complete multipartite graph $K_{2,\ldots,2}$ with t parts each of size 2. Let $\theta_1, \ldots, \theta_t$ be distinct real numbers in $[0, \pi/2)$. Then

$$V = \begin{bmatrix} \cos\theta_1 & -\sin\theta_1 & \cdots & \cos\theta_t & -\sin\theta_t \\ \sin\theta_1 & \cos\theta_1 & \cdots & \sin\theta_t & \cos\theta_t \end{bmatrix}$$

and $C = I_2$ is a representation of G. Hence $\mathrm{mr}(K_{2,\ldots,2}) \leq 2$.

2.2. Introduction to Zero Forcing

Zero forcing is an important combinatorial concept that provides an upper bound on $\mathrm{M}(G)$ and is the second major focus of this book. The motivation behind zero forcing is the observation that in some instances when $A \in \mathcal{S}(G)$ has a null vector $\mathbf{x} = [x_i]$, the structure of the graph G and the location of some zero entries in $\mathbf{x}$ force other entries to be zero. More precisely, if

- $B \subseteq V(G)$
- $A \in \mathcal{S}(G)$,
- $A\mathbf{x} = \mathbf{0}$,
- $x_i = 0$ for all $i \in B$,
- $i \in B$, $j \notin B$, and j is the only vertex not in B such that $ij \in E(G)$

then $x_j = 0$. This can be seen by noting that the assumptions imply the ith coordinate equation of $A\mathbf{x} = \mathbf{0}$ yields $a_{ij}x_j = 0$. As $a_{ij} \neq 0$, we conclude $x_j = 0$. For example, if $A \in \mathcal{S}(P_n)$ and P_n is $1\text{–}2\text{–}\cdots\text{–}n$, $A\mathbf{x} = \mathbf{0}$, and $x_1 = 0$, then iteratively applying the above idea we see that $x_1 = 0$ forces $x_2 = 0$, which in turns forces $x_3 = 0$, etc. Thus no $A \in \mathcal{S}(P_n)$ has a nonzero null vector with first coordinate zero. It is not hard to see that this implies that $\mathrm{M}(P_n) \leq 1$ (see the proof of Lemma 2.13 for more explanation).

One can represent the process of forcing zeros in a null vector just described as a game on a given graph G. Initially each vertex is filled or unfilled.[4] One then

[4] Most early papers colored a vertex black when the corresponding entry in the null vector is known to be zero and white when the entry is unknown, and later black was replaced by blue. Recent work replaces the terms 'blue' and 'white' by 'filled' and 'unfilled,' which has the advantage of displaying well in all media. The earlier use of colors explains terms such as 'color change rule,' which continue in common usage.

proceeds to fill some unfilled vertices by the following *color change rule*: A filled vertex u can change an unfilled vertex w to filled if w is the unique unfilled neighbor of u. This is sometimes referred to as the *standard color change rule* or Z-*color change rule* to distinguish it from variants discussed in Chapter 9. When u is used to change w to filled, we say u *forces* w and write $u \to w$. We use gray to indicate a vertex is filled in illustrations in this book.

The process of applying the color change rule to a grid graph is illustrated in Figure 2.1. An animated version is available at
`https://aimath.org/~hogben/4x7gridanimate.gif`.

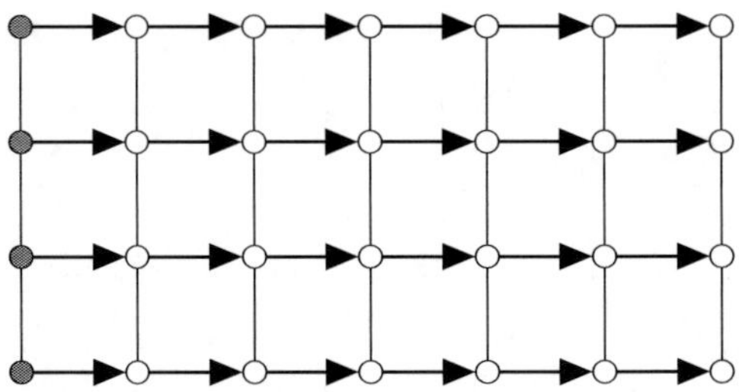

FIGURE 2.1. The zero forcing process on a 4×7 grid.

DEFINITION 2.8. Let G be a graph. A subset $B \subseteq V(G)$ defines an initial set of filled vertices (with all vertices not in B unfilled); this is called an *initial coloring of G*. Given an initial coloring B of G, the *final coloring* of B is the set of filled vertices obtained by applying the color change rule until no more changes are possible.[5] A *zero forcing set* for G is a subset of vertices B such that the final coloring of B is $V(G)$. The *zero forcing number* is

$$\mathrm{Z}(G) = \min\{|B| : B \text{ is a zero forcing set of } G\}.$$

REMARK 2.9. In order for some vertex u to perform the first force, u itself must be filled and at least $\delta(G) - 1$ neighbors must be filled (where $\delta(G)$ denotes the minimum degree of G). Thus $\delta(G) \le \mathrm{Z}(G)$. Certainly $\mathrm{Z}(G) \le |G|$, and $\mathrm{Z}(G) \le |G| - 1$ if G contains at least one edge because all vertices except one that has a neighbor is a zero forcing set.

EXAMPLE 2.10. Let G be the path P_n labelled 1–2–$\cdots$–n with $n \ge 3$. Since $\delta(P_n) = 1$ and $B = \{1\}$ is a zero forcing set (with the forces $1 \to 2$, $2 \to 3$, $\dots$, $n-1 \to n$), $\mathrm{Z}(P_n) = 1$. But $\{2\}$ is not a zero forcing set, as there are no forces.

A *cycle* C_n on n vertices is the graph obtained from a path v_1–v_2–$\cdots$–v_n by adding the edge $v_1 v_n$.

EXAMPLE 2.11. Since $\delta(C_n) = 2$ and two adjacent vertices of a cycle are a zero forcing set of C_n, $\mathrm{Z}(C_n) = 2$.

EXAMPLE 2.12. For the complete graph, $\delta(K_n) = n - 1$ with $n \ge 2$, and any set of $n-1$ vertices is a zero forcing set, so $\mathrm{Z}(K_n) = n - 1$.

[5] Other terms that have been used for the final coloring include the original term *derived set* and the more recent term *closure*.

Theorem 2.14 is the fundamental result relating zero forcing to the maximum nullity problem. It was introduced at the 2006 American Institute of Mathematics' Workshop "Spectra of families of matrices described by graphs, digraphs, and sign patterns" and relies on the next lemma.

LEMMA 2.13. [**6**] *Suppose* $\alpha \subseteq \{1,\dots,n\}$ *with* $|\alpha| < k$ *and* U *is a subspace of* $\mathbb{R}^n$ *of dimension* k*. Then* U *contains a nonzero vector* $\mathbf{v}$ *with* $\mathbf{v}[\alpha] = \mathbf{0}$.

PROOF. It is a standard fact from linear algebra that if U and V are subspaces of $\mathbb{R}^n$ with $\dim V > n - \dim U$, then V interests U non-trivially. The subspace V of all vectors $\mathbf{v} = [v_i]$ having $v_i = 0$ for $i \in \alpha$ has $\dim V > n-k$. Hence, U contains a nonzero vector $\mathbf{v}$ with $\mathbf{v}[\alpha] = \mathbf{0}$. □

THEOREM 2.14. [**6**] *Let* G *be a graph. Then* $\mathrm{M}(G) \le \mathrm{Z}(G)$.

PROOF. It was argued at the beginning of this section that if B is a zero forcing set of G, $A \in \mathcal{S}(G)$, $\mathbf{x} = [x_i] \in \ker A$, and $x_i = 0$ for all $i \in B$, then $\mathbf{x} = 0$. Then Lemma 2.13 implies that no set of vertices of G of cardinality less than $\operatorname{null} A$ is a zero forcing set, and the inequality follows. □

Theorem 2.14 has been applied to compute values of $\mathrm{M}(G)$ and $\mathrm{Z}(G)$ for many graphs G for which $\mathrm{Z}(G) = \mathrm{M}(G)$. A common proof method is to exhibit a matrix $A \in \mathcal{S}(G)$ and a zero forcing set B such that $|B| = \operatorname{null} A$. We illustrate this in the proof of the next (well-known) result. Values of $\mathrm{Z}(G)$ and $\mathrm{M}(G)$ for many more graph families are also presented in Theorem 9.5.

PROPOSITION 2.15. *The following statements hold.*

(1) *For* $n \ge 2$, $\mathrm{mr}(K_n) = 1$ *and* $\mathrm{M}(K_n) = \mathrm{Z}(K_n) = n - 1$.
(2) *For positive* p *and* q *such that* $p+q \ge 3$, $\mathrm{mr}(K_{p,q}) = 2$ *and* $\mathrm{M}(K_{p,q}) = \mathrm{Z}(K_{p,q}) = p + q - 2$.
(3) *For* $n \ge 3$, $\mathrm{mr}(C_n) = n - 2$ *and* $\mathrm{M}(C_n) = \mathrm{Z}(C_n) = 2$.
(4) *For* $n \ge 4$, $\mathrm{mr}(W_n) = n - 3$ *and* $\mathrm{M}(W_n) = \mathrm{Z}(W_n) = 3$, *where* $W_n = C_{n-1} \vee K_1$ *is the wheel on* n *vertices.*

PROOF. (1): The all-ones matrix J_n is in $\mathcal{S}(K_n)$ and has rank 1, and $\mathrm{Z}(K_n) = n - 1$ (Example 2.12). Thus $\mathrm{M}(K_n) = \mathrm{Z}(K_n) = n - 1$

(2): The adjacency matrix of $K_{p,q}$ has rank 2, and a set consisting of all but one vertex in each part is a zero forcing set. Thus $\mathrm{M}(K_{p,q}) = \mathrm{Z}(K_{p,q}) = p + q - 2$.

(3): As noted in Example 2.11, any two consecutive vertices are a zero forcing set of the cycle. The eigenvalues of the adjacency matrix A_{C_n} are $\omega^j + \omega^{n-j}$ for $j = 1, 2\dots, n$, where ω is a primitive nth root of unity. Thus $\lambda = 2\cos(\frac{2\pi}{n})$ is an eigenvalue of A_{C_n} with multiplicity 2, so $\mathrm{M}(C_n) \ge 2$. Thus $\mathrm{M}(C_n) = \mathrm{Z}(C_n) = 2$.

(4) Let A_{n-1} denote the adjacency matrix of C_{n-1}, $\lambda = 2\cos\left(\frac{2\pi}{n-1}\right)$, and $B_{n-1} = A_{n-1} - \lambda I_{n-1}$, so $\operatorname{rank} B_{n-1} = n-3$. Since $\lambda < 2$, every entry of $B_{n-1}\mathbb{1}_{n-1}$ is nonzero, where $\mathbb{1} \in \mathbb{R}^{n-1}$ is an all-ones vector. Define $B = \begin{bmatrix} B_{n-1} & B_{n-1}\mathbb{1} \\ \mathbb{1}^T B_{n-1} & \mathbb{1}^T B_{n-1}\mathbb{1} \end{bmatrix}$, so $B \in \mathcal{S}(W_n)$ and note that $\operatorname{rank} B = \operatorname{rank} B_{n-1} = n - 3$ and $\operatorname{null} B = 3$. Observe that a set consisting of the vertex of degree $n-1$ and two adjacent vertices of degree three is a zero forcing set. □

The proof of (4) illustrates a common matrix construction technique. Other methods used to establish the common value $\mathrm{Z}(G) = \mathrm{M}(G)$ were illustrated in the proof of Proposition 2.15 and in Examples 2.4 and 2.10 for the path. The AIM minimum rank catalog [**7**] also contains recent information about the minimum rank, maximum nullity, and zero forcing number for various families of graphs.[6] DeLoss et al. determined the $\mathrm{mr}(G)$ for all graphs of up to order 7 in [**111**], and showed that $\mathrm{M}(G) = \mathrm{Z}(G)$ if the order of G is less than eight.

Zero forcing was introduced in multiple different contexts between 2000 and 2015. In this section, we focused on its connection with maximum nullity, which is the source of the name, and connections to the IEP-G, since maximum nullity equals maximum eigenvalue multiplicity. Other origins are discussed in Chapter 9, where there is a much more extensive discussion of the techniques and results for zero forcing, as well as discussion of many natural variants and related parameters. Each zero forcing variant is distinguished by its color change rule, which defines when a vertex can change another vertex from unfilled to filled. Some of these color change rules are motivated by the maximum nullity of a family of matrices described by a graph, such as positive semidefinite matrices (see Section 2.5, where positive semidefinite zero forcing is introduced). However, translation by a scalar matrix may remove the matrix from the class being studied (such as positive semidefinite matrices described by a graph), so the maximum nullity is not the maximum multiplicity of an eigenvalue. In such a situation, there remains a linear algebraic motivation but the most straightforward connection to an inverse eigenvalue problem is lost.

2.3. Historical Background to the Minimum Rank Problem

The study of the minimum rank of symmetric matrices with a given graph formally arose with Nylen's 1996 paper [**238**]. In this section we first summarize some earlier results that foreshadowed the minimum rank problem and then discuss Nylen's work.

Parter's paper in 1960 [**241**] appears to be the first paper to concentrate on the spectral properties of matrices with a given graph. Parter attributes the concept of the graph of a matrix to Rosenblatt [**249**] and Symanzik [**258**] and Harary. Parter studied sign-symmetric matrices[7] and showed that for a sign-symmetric matrix A with graph $\mathcal{G}(A)$, "certain properties of the Jordan canonical form of A can be expressed in terms of the topology of its $\mathcal{G}(A)$." Parter was motivated by the fact that (in today's terminology) each sign-symmetric matrix whose underlying graph is a path is diagonally similar to a symmetric matrix and does not have a multiple eigenvalue. Thus, he was led to studying a sign-symmetric matrix whose underlying graph is a tree and that has a multiple eigenvalue. He proved that each such matrix is diagonally similar to a symmetric matrix whose graph is the same tree. More importantly, he established the next result.

Given a tree T and a vertex v of T, a *branch of T at v* is a connected component of $T - v$. If $A \in \mathcal{S}(T)$, each branch of T at v determines a principal submatrix of A, namely, $A[W]$ where W is the vertex set of the branch.

THEOREM 2.16. [**241**] *Let T be a tree and let $A \in \mathcal{S}(T)$. Then* $\mathrm{null}\, A \geq 2$ *if and only if there a vertex v of T and at least three branches $T[W_i]$ of T at v such that the principal submatrix $A[W_i]$ is singular.*

[6] This catalog is not always up to date.

[7] The square matrix $A = [a_{ij}]$ is *sign-symmetric* provided for each i and j, a_{ij} and a_{ji} are both positive, both negative, or both zero.

Fiedler's 1975 paper [**139**] studied eigenvectors of matrices whose graph is a tree.

THEOREM 2.17. [**139**] *Let T be a tree and let $A \in \mathcal{S}(T)$ with* $\operatorname{null} A \geq 2$. *Then there exists an index i such every null vector of A has ith coordinate equal to* 0.

An index i with the property that every eigenvector of A corresponding to an eigenvalue λ has ith coordinate zero is called a *Fiedler index* (or vertex) of A corresponding to λ.

Additionally, Fiedler showed in [**139**] that often one can determine the location of a eigenvalue of $A \in \mathcal{S}(T)$ among the ordering of its eigenvalues from a eigenvector of A corresponding to λ. For the specialization of this to paths see Theorem 1.18.

THEOREM 2.18. [**139**] *Let T be a tree, let $A \in \mathcal{S}(T)$, and let $\mathbf{y} = [y_j]$ be an eigenvector of A corresponding to the eigenvalue λ. Let p be the number of edges ij of T such that $a_{ij}y_iy_j < 0$, q be the number of edges ij of T such that $a_{ij}y_iy_j > 0$, and r be the number of coordinates of y that are zero. If there does not exist an index i such that $y_i = 0$ and $y_j = 0$ for all j adjacent to i in T, then there are $p+r$ eigenvalues of A greater than λ and $q+r$ eigenvalues of A less than λ (including multiplicity).*

Wiener's 1984 paper [**266**] strengthened Theorem 2.16 as described in the next result. Note that if A is a matrix, then $\operatorname{null} A - 1 \leq \operatorname{null} A(i) \leq \operatorname{null} A + 1$ where $A(i)$ is the submatrix obtained from A by deleting row and column i (because $\operatorname{rank} A - 2 \leq \operatorname{rank} A(i) \leq \operatorname{rank} A$ as shown in Proposition 2.20(4)).

THEOREM 2.19 (Parter-Wiener Theorem). [**241, 266**] *Let T be a tree, $A \in \mathcal{S}(T)$ and i be a vertex of T such that both A and $A(i)$ are singular. Then*

(1) *There is a vertex j such that* $\operatorname{null} A(j) = \operatorname{null} A + 1$.
(2) *if* $\operatorname{null} A \geq 2$, *then the j in (a) can be chosen so that the principal submatrices of at least three branches of T at j are singular.*
(3) *if* $\operatorname{null} A = 1$, *then the j in (a) can be chosen so that the principal submatrices of two branches of T at j are singular.*

By replacing A by $A - \lambda I$, and nullity by the multiplicity of λ, Theorem 2.19 applies to the multiplicities of any eigenvalue λ of A. Theorem 2.19 is now referred to as the *Parter-Wiener Theorem*, and the vertex i for which $\operatorname{null}(A - \lambda I) = \operatorname{null}((A - \lambda I)(i)) + 1$ is called a *Parter vertex* or *Parter-Wiener vertex* for A corresponding the eigenvalue λ. Over the subsequent years, the Parter-Wiener theorem has been further developed and there are many very beautiful results (see Chapter 2 of [**205**] and the references therein). We do not discuss these because they take us too far away from this book's focus.

Nylen's introduction of the minimum rank problem. As noted previously, Nylen's 1996 paper [**238**] appears to be the first paper to formally introduce the study of the minimum rank problem for graphs. Nylen made the basic observations in Remark 2.3, Example 2.4, and the next result (the proof is presented to illustrate techniques).

PROPOSITION 2.20. [**238**] *Let G be a graph of order $n \geq 1$. Then*

(1) *If $A \in \mathcal{S}(G)$ and $v \in V(G)$, then the principal submatrix $A(v)$ corresponds to the induced subgraph $G - v$;*

(2) *If G is disconnected with connected components $G_1, \ldots, G_k$, then*

$$\operatorname{mr}(G) = \sum_{j=1}^{k} \operatorname{mr}(G_j);$$

(3) *If a graph G of order n is disconnected with k connected components, then* $\operatorname{mr}(G) \leq n - k$.
(4) $\operatorname{mr}(G) - 2 \leq \operatorname{mr}(G - v) \leq \operatorname{mr}(G)$ *for each vertex v of G;*
(5) $\operatorname{mr}(G) - 1 \leq \operatorname{mr}(G - e) \leq \operatorname{mr}(G) + 1$ *for each edge e of G; and*

PROOF. (1) is immediate. For (2), the rank of a direct sum of matrices is the sum of the ranks of the summands, and (3) then follows since $\operatorname{mr}(H) \leq |V(H)| - 1$ for any graph H.

The first inequality in (4) follows from the fact that appending a new row or column to a matrix can increase the rank by at most one. The second inequality in (4) comes from the fact that the rank of a submatrix is a lower bound on the rank.

The inequalities in (5) follow from the subadditivity of the rank function and the observation that each $A \in \mathcal{S}(G)$ (respectively, $B \in \mathcal{S}(G - e)$) there is a symmetric rank 1 matrix C such that $A + C \in \mathcal{S}(G - e)$ (respectively, $B + C \in \mathcal{S}(G)$). □

A more substantial result is the following.

THEOREM 2.21. [**238**] *Let G be a graph and $A \in \mathcal{S}(G)$ with* $\operatorname{rank} A = \operatorname{mr}(G)$, *and let i be a vertex of G. Then* $\operatorname{rank} A(i) = \operatorname{mr}(G) - 2$ *or* $\operatorname{rank} A(i) = \operatorname{mr}(G)$*; that is,*

$$\operatorname{rank} A(i) \neq \operatorname{mr}(G) - 1.$$

PROOF. Note that $\operatorname{rank} A \geq \operatorname{rank} A(i) \geq \operatorname{rank} A - 2$. Suppose that $\operatorname{rank} A(i) > \operatorname{mr}(G) - 2$. The ith column of A is in the span of the remaining columns of A; otherwise we are led to the contradiction that $\operatorname{rank} A = 2 + \operatorname{rank} A(i) > \operatorname{mr}(G)$. For notational simplicity we assume $i = 1$, so

$$A = \left[\begin{array}{c|c} a_{11} & \mathbf{v}^\top A(1) \\ \hline A(1)\mathbf{v} & A(1) \end{array}\right].$$

Since A minimizes the rank of matrices in $\mathcal{S}(G)$, $a_{11} = \mathbf{v}^\top A(1)\mathbf{v}$ and $\operatorname{rank} A = \operatorname{rank} A(1)$. Hence, $\operatorname{mr}(G) = \operatorname{rank} A(i)$. □

The rest of Nylen's paper focuses on $\mathcal{S}(T)$ for trees T. In particular he proved the next result, which provides the following algorithm for determining the minimum rank of a tree. A *high-degree vertex* of a graph is a vertex of degree at least 3. Let F be a forest. In honor of Nylen, we define a *Nylen path* of F to be a path in F that has at most one high-degree vertex of F, and whose initial and terminal vertices each have degree one in F. Note that a path that is a connected component of F is a Nylen path. If v is a high-degree vertex, then each high-degree vertex that lies furthest from v is the high-degree vertex of a Nylen path of F. If P is a Nylen path, then $F - V(P)$ denotes the forest obtained from F by removing the vertices of P and all edges incident to vertices of P.

PROPOSITION 2.22. *Let P be a Nylen path of the forest F. Then*

(1) *If P has a high-degree vertex i, then* $\operatorname{mr}(F) = \operatorname{mr}(F - i) + 2$.
(2) *If P does not have a high-degree vertex then* $\operatorname{mr}(F) = |V(P)| - 1 + \operatorname{mr}(F - V(P))$.

PROOF. To prove (1), assume that P has i as a high-degree vertex, let $A \in \mathcal{S}(F)$ with $\operatorname{rank} A = \operatorname{mr}(F)$, let P^1 and P^2 be the two paths that result from deleting i from P, and let $U_j = V(P^j), j = 1, 2$. Suppose to the contrary that $\operatorname{mr}(F) \neq \operatorname{mr}(F - i) + 2$. Then by Theorem 2.21, $\operatorname{mr}(F) = \operatorname{rank} A(i)$. It follows that column i of A is in the span of the remaining columns of A and $\operatorname{rank} A(i) = \operatorname{mr}(F - i)$. This implies that column space of $A[U_i]$ contains a standard basis vector with a 1 in a position corresponding to a pendant vertex of P_i. This then implies that $A[U_i]$ is nonsingular. But $\operatorname{rank} A(i) = \operatorname{mr}(F - i)$ implies $A[U_i]$ is a singular matrix, a contradiction. Thus $\operatorname{mr}(F) = \operatorname{mr}(F - i) + 2$.

Part (2) follows from Proposition 2.20 (2) and Example 2.4. □

ALGORITHM 2.23 (Nylen's Algorithm for the minimum rank of a forest).
Input: *A forest F.*
Output: $\operatorname{mr}(F)$.

Count := 0.
While $V(T) \neq \emptyset$ *is not empty do:*
(1) *Find a Nylen path P of T.*
(2) **If** *P contains a high-degree vertex i:*
Count := Count + 2
$F := F - i$.
Else:
Count := Count + $|V(P)| - 1$.
$F := F - V(P)$.
Return: $\operatorname{mr}(F)$:= Count.

It was later noted the removal of a Nylen path decreases the path cover number of a forest by 1, and hence the above algorithm can be modified to compute a minimum path cover of a tree (see also Algorithm 2.26), giving a nice combinatorial interpretation of the maximum nullity for trees.

COROLLARY 2.24. *For a forest F,* $\mathrm{P}(F) = \mathrm{M}(F)$.

The next algorithm computes $\mathrm{M}(T)$ by finding a minimum path cover. The method of computation of $\mathrm{P}(T)$ in Algorithm 2.26 uses pendent generalized stars and works from the outside in (the same approach as Nylen's Algorithm). For algorithms with different approaches, see [**127**] and the references therein. A path P in T is a *pendent path* of vertex v if P is a component of $T - v$ and (in T) P is connected to v by one of its end-points.

A *generalized star* or *spider*[8] is a tree that has at most one high-degree vertex; a path is a generalized star. A *pendent generalized star* of T is a connected induced subgraph S of T such that: (1) there is exactly one high-degree vertex v of T in S (v is called the *center* of S); (2) $\deg_T(v) = k + 1$ and exactly k of the components of $T - v$ are pendent paths of v; (3) S is induced by the vertices of these k pendent paths and v. The next result enables the algorithm to work.

LEMMA 2.25. [**127**] *Any tree T contains a pendent generalized star or is a generalized star.*

[8]The term 'generalized star' is used in the inverse eigenvalue problem/minimum rank literature, whereas the term 'spider' is used in graph theory. We follow the literature, so we use the term 'generalized star' in Parts 1, 2, and 3, whereas we use the term 'spider' in Part 4.

ALGORITHM 2.26. *Computation of* $\mathrm{M}(T)$, $\mathrm{Z}(T)$ *and* $\mathrm{P}(T)$ *for a tree* T.
Input: *A tree* T.
Output: $\mathrm{M}(T)$, $\mathrm{Z}(T)$, $\mathrm{P}(T)$, *and a path cover* $\mathcal{P}$ *of* T.

(1) $\mathcal{P} := \emptyset$
(2) $T' := T$.
(3) **While** T' *has a pendent generalized star* S:
  (a) *Let* v *be the center of* S.
  (b) *Choose* 2 *pendent paths of* v *and combine with* v *to create a new path* R_1.
  (c) *Denote any other pendent paths of* v *by* $R_2, \dots, R_k$
  (d) $\mathcal{P} := \mathcal{P} \cup \{R_1, \dots, R_k\}$.
  (e) $T' := T' - V(S)$.
(4) **If** $T' \neq \emptyset$:
  Choose a minimum path cover for the generalized star T' *and add these paths to* $\mathcal{P}$.
(5) **Return** $\mathcal{P}$ *(minimum path cover) and* $\mathrm{M}(T) = \mathrm{Z}(T) = \mathrm{P}(T) = |\mathcal{P}|$.

Any minimum path cover of a tree yields minimum zero forcing set(s) by choosing one endpoint of each path in the cover [**6**]. Note that the algorithms above give an effective way to calculate the minimum rank for a tree T. In contrast, the minimum rank problem for general graphs does not seem to have an easy solution and the inverse eigenvalue problem for trees is only solved for a small number of families of trees. Indeed, we will see in Chapter 8 that for most trees T even the minimum number of distinct eigenvalues of a matrix in $\mathcal{S}(T)$ is not yet known.

2.4. Further Properties of Minimum Rank

In this section we provide additional information about minimum rank, including discussing how minimum rank behaves with respect to various graph operations, and report characterizations of extreme values of minimum rank. Values of minimum rank for additional families can be found in Theorem 9.5.

Computation. When computing the maximum nullity (or equivalently the minimum rank) of a graph, the maximum of the nullities is take over an infinite family of matrices, and until recently there was no method to obtain an exact answer in a finite amount of time even for graphs of relatively small orders. In 2019 Brimkov and Scherr introduced and implemented an exact algorithm in [**64**] that uses polynomial equation solvers to compute the minimum value of k such that the determinants of all $k \times k$ submatrices of a matrix of variables are zero. When this method is applied to a matrix of variables described by a graph G with additional equations used to ensure the variables corresponding to edges of G are nonzero, this determines that $\mathrm{mr}(G) = k - 1$. Some of the ideas for exact computation of minimum rank applied in [**64**] are present in [**41**] form a theoretical point of view without a practical implementation. For more than ten years there has been code available [**71, 183**] for implementing various techniques such as cut-vertex reduction (described later in this section) and bounds on minimum rank, such as the zero forcing number. This code was introduced in Deloss et al. [**111**] where it was used to determine the minimum rank of 1210 of 1252 (nonisomorphic) graphs of order 7 (because for these 1210 graphs the best upper and lower bounds agreed). In that paper, the minimum ranks of the of remaining 42 "exceptional" graphs of order 7 were determined by construction of examples achieving the known lower bound or applying a fact specific to the given graph. The use of this code to compute the minimum rank of various graphs, together with code for other variants of zero forcing and related parameters, is illustrated in [**183**, Chapter 2]. Brimkov and Scherr apply their algorithm to automatically compute the minimum rank of the 22 "exceptional" graphs of order 7 with 12 or fewer edges [**64**].

Graph Operations. The next theorem summarizes the effect of various graph operations on minimum rank. Suppose $G_i, i = 1, \dots, h$ are graphs of order at least two, there is a vertex v such that for all $i \neq j$, $G_i \cap G_j = \{v\}$, and $G = \bigcup_{i=1}^h G_i$; if $h \geq 2$, then v is a *cut-vertex* of G, and any graph with a cut-vertex can be expressed as such a decomposition with $h \geq 2$. When all $G_i - v$ are required to be connected, $G = \bigcup_{i=1}^h G_i$ is called the *cut-vertex decomposition* at v.

THEOREM 2.27. *Let G be a graph.*

(1) [**238**] **Vertex deletion** *If v is a vertex of G, then* $\operatorname{mr}(G) - 2 \leq \operatorname{mr}(G - v) \leq \operatorname{mr}(G)$.
(2) [**238**] **Edge deletion** *If e is an edge of G, then* $\operatorname{mr}(G) - 1 \leq \operatorname{mr}(G - e) \leq \operatorname{mr}(G) + 1$.
(3) [**127**] **Unions** *Let G_i be graphs for $i = 1, \dots, t$. Then*
$$\operatorname{mr}\left(\bigcup_{i=1}^{t} G_i\right) \leq \sum_{i=1}^{t} \operatorname{mr}(G_i).$$
(4) [**127**] **Clique covers** *Recall that the (edge) clique cover number of G is denoted by* $\operatorname{cc}(G)$. *Then*
$$\operatorname{mr}(G) \leq \operatorname{cc}(G).$$
(5) [**25, 197**] **Cut-vertex reduction** *Let G be a graph having a cut-vertex v and let $G = \bigcup_{i=1}^h G_i$ be the cut-vertex decomposition at v. Then*
$$\operatorname{mr}(G) = \sum_{i=1}^{h} \operatorname{mr}(G_i - v) + \min\left\{\sum_{i=1}^{h} \operatorname{mr}(G_i) - \operatorname{mr}(G_i - v), 2\right\}.$$
In particular, if $\operatorname{mr}(G_i) - \operatorname{mr}(G_i - v) = 0$ *for all but at most one of the G_i, then* $\operatorname{mr}(G) = \sum_{i=1}^{b} \operatorname{mr}(G_i)$.

PROOF. For (1) and (2) see Proposition 2.20.

For (3), let $G = \cup_{i=1}^t G_i$. Construct a matrix A of rank at most $\sum_{i=1}^t \operatorname{mr}(G_i)$ having $\mathcal{G}(A) = G$ as follows: Choose a matrix A_i such that $\operatorname{rank} A_i = \operatorname{mr}(G_i)$ for for each $i = 1, \dots, t$ and embed A_i in a matrix $\tilde{A}_i$ of size $|V(G)|$. Then choose $c_i \in \mathbb{R}$ such that no cancellation of nonzero entries occurs and let $A = \sum_{i=1}^t c_i \tilde{A}_i$. Then $\operatorname{rank} A \leq \sum_{i=1}^t \operatorname{mr}(G_i)$ follows from the subadditivity of rank.

Statement (4) follows immediately from (3) and $\operatorname{mr}(K_s) = 1$ for $s \geq 2$.

A nice proof of statement (5) can be found in [**127**]. □

Each of the inequalities in Theorem 2.27 is tight, and it is a nice exercise to find examples illustrating this.

Perhaps surprisingly, the minimum rank for a join of graphs is not yet completely understood. However, the minimum rank of a join under additional assumptions has been studied in [**22**]. We now discuss some of the results from that work. The *join minimum rank* of the graph G is $\text{jmr}(G) = \text{mr}(G \vee K_1)$. Note that $\text{mr}(G) \leq \text{jmr}(G) \leq \text{mr}(G) + 2$. Let c_1 denote the number of isolated vertices of G.

PROPOSITION 2.28. [**22**] *Let G be a graph. Then*

$$\text{jmr}(G) = \begin{cases} \text{mr}(G) & \text{if } c_1 = 0; \\ \text{mr}(G) + 1 & \text{if } c_1 = 1; \\ \text{mr}(G) + 2 & \text{if } c_1 \geq 2. \end{cases}$$

A symmetric matrix A has *balanced inertia* if $i_-(A) \leq i_+(A) \leq i_-(A) + 1$, where $i_-(A)$ (respectively, $i_+(A)$) is the number of negative (positive) eigenvalues of A. A graph G is *inertia-balanced* if there exists $A \in \mathcal{S}(G)$ that satisfies

- $\text{rank}\, A = \text{mr}(G)$, and
- A has balanced inertia.

A *cograph* or a *decomposable graph* is a graph that can be constructed from isolated vertices by a sequence of joins or disjoint unions. The cograph $G = \vee_{i=1}^r G_i$ is *anomalous* if

- $\text{jmr}(G) \leq 2$ for $i = 1, \ldots, r$, and
- $K_{3,3,3} = \overline{K}_3 \vee \overline{K}_3 \vee \overline{K}_3$ is a subgraph of G

($K_{3,3,3}$ is shown in Figure 2.2).

THEOREM 2.29. [**22**] *Suppose $G = \vee_{i=1}^r G_i$ is a cograph with $r > 1$ and each G_i is inertia-balanced. Then G is inertia-balanced and*

$$\text{mr}(G) = \begin{cases} \max_i\{\text{jmr}(G_i)\} & \text{if } G \text{ is not anomalous;} \\ 3 & \text{if } G \text{ is anomalous.} \end{cases}$$

EXAMPLE 2.30. (Complete multipartite graph) [**127**] Let $n_1 \geq n_2 \geq \cdots \geq n_r > 0$. Then we let $G = K_{n_1,n_2,\ldots,n_r} := \overline{K}_{n_1} \vee \overline{K}_{n_2} \vee \cdots \vee \overline{K}_{n_r}$. Using Theorem 2.29 and additional results in [**22**] it follows that

$$\text{mr}(G) = \begin{cases} 0 & \text{if } k = 1; \\ 1 & \text{if } n_1 = n_2 = 1, n_3 = 0; \\ 2 & \text{if } n_1 > 1, n_3 < 3; \\ 3 & \text{if } n_3 \geq 3; \end{cases}$$

There are techniques involving Kronecker products of matrices for constructing matrices with a desired nullity that are particularly well-suited to graphs that are formed from graph products. If A is an $s \times s$ real matrix and B is a $t \times t$ real matrix, then the *Kronecker product* of A and B, denoted by $A \otimes B$, is the $s \times s$ block matrix whose ijth block is the $t \times t$ matrix $a_{ij}B$. The following result is standard (see, for example, [**159**, §9.7]).

OBSERVATION 2.31. *Let G be a graph of order n and let G' be a graph of order n'. If $A \in \mathcal{S}(G)$ and $A' \in \mathcal{S}(G')$, then $A \otimes I_{n'} + I_n \otimes A' \in \mathcal{S}(G \,\square\, G')$ where $G \,\square\, G'$ is the Cartesian product of G and G' (see Appendix A). If $\mathbf{x}$ is an eigenvector of A for eigenvalue λ and $\mathbf{x}'$ is an eigenvector of A' for eigenvalue λ', then $\mathbf{x} \otimes \mathbf{x}'$ is an eigenvector $A \otimes I_{n'} + I_n \otimes A'$ for eigenvalue $\lambda + \lambda'$.*

This observation is applied in Section 9.1 (see, for example, Proposition 9.95). Results for the tensor product are presented in Section 9.4 because they involve matrices with all diagonal entries required to be zero.

Graphs with Extreme Minimum Rank. Graphs with extremely high or low minimum rank have been characterized.

THEOREM 2.32. [**138**] *Let G be a graph of order $n \geq 1$. Then* $\operatorname{mr}(G) = n - 1$ *if and only if P is a path on n vertices.*

A *polygonal path* is a graph that can be constructed by starting with a cycle and adding one cycle at a time by identifying an edge of a new cycle with an edge of the most recently added cycle that has a vertex of degree 2.

THEOREM 2.33. [**185**] *Let G be a 2-connected graph of order $n \geq 3$. Then* $\operatorname{mr}(G) = n - 2$ *if and only if G is a polygonal path.*

Observe that Theorem 2.33 and cut-vertex reduction can be used to determine whether a given graph G has $\operatorname{mr}(G) = |V(G)| - 2$. Graphs having $\operatorname{mr}(G) = |V(G)| - 2$ have also been characterized in [**203**] (without the assumption that G is 2-connected).

Recall that a graph G has $\operatorname{mr}(G) = 0$ if and only if $G = rK_1$. The next result is well known.

PROPOSITION 2.34. *A graph G of order n has* $\operatorname{mr}(G) = 1$ *if and only if* $G = K_r \sqcup (n-r)K_1$ *where $r \geq 2$.*

PROOF. Recall that $\operatorname{rank} J_n = 1$ for the all-ones matrix $J_n \in \mathcal{S}(K_n)$. If A is a symmetric matrix of rank 1, then $A = \mathbf{x}\mathbf{x}^T$ for some real vector x and it follows that $\mathcal{G}(A)$ consists of a complete graph of order at least two along with isolated vertices. □

THEOREM 2.35. [**34**] *A connected graph G has* $\operatorname{mr}(G) \leq 2$ *if and only if G does not contain as an induced subgraph any of P_4, $K_{3,3,3}$, Dart, or $\ltimes$ (all shown in Figure 2.2).*

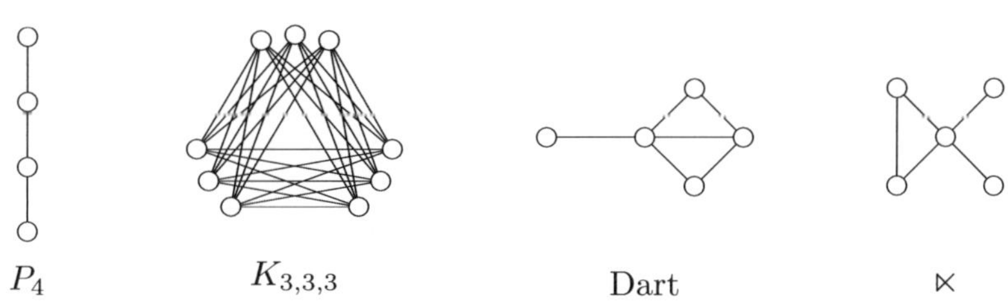

FIGURE 2.2. Forbidden induced subgraphs for $\operatorname{mr}(G) \leq 2$.

THEOREM 2.36. [**34**] *A graph G has* $\operatorname{mr}(G) \leq 2$ *if and only if the complement of G is of the form*

$$(K_{s_1} \cup K_{s_2} \cup K_{p_1,q_1} \cup \cdots \cup K_{p_k,q_k}) \vee K_r$$

for appropriate nonnegative integers $k, s_1, s_2, p_1, q_1, \ldots, p_k, q_k, r$ with $p_i + q_i > 0$ for all $i = 1, \ldots, k$.

Additional Topics. There are numerous other results on minimum rank and maximum nullity; see [**128**] for a more comprehensive list. Here we have given only a limited selection to keep the focus on the Inverse Eigenvalue Problem and the strong properties. From this perspective, the one glaring omission so far is a discussion of maximum nullity parameters having the strong Arnol'd property, and that is addressed in Section 2.6.

2.5. Minimum Positive Semidefinite Rank and Zero Forcing Number

In this section we consider the question of minimum rank, maximum nullity, and related zero forcing number for positive semidefinite matrices.

DEFINITION 2.37. The *minimum positive semidefinite rank* (respectively, *maximum positive semidefinite nullity* of G is the smallest rank (respectively, largest nullity) of a positive semidefinite matrix in $\mathcal{S}(G)$. That is,

$$\operatorname{mr}_+(G) = \min\{\operatorname{rank} A : A \in \mathcal{S}_+(G)\} \text{ and } \operatorname{M}_+(G) = \max\{\operatorname{null} A : A \in \mathcal{S}_+(G)\}$$

where $\mathcal{S}_+(G) = \{A \in \mathbb{R}^{n\times n} : A \text{ is positive semidefinite and } \mathcal{G}(A) = G\}$.

It is immediate that $\operatorname{mr}_+(G) + \operatorname{M}_+(G) = |V(G)|$. Because a PSD matrix must have nonnegative eigenvalues, arbitrary shifting is not allowed. Thus the maximum nullity of G need not be the maximum multiplicity of an eigenvalue of a PSD matrix described by G, which is the same as maximum multiplicity of an eigenvalue of a symmetric matrix described by G, i.e., $\operatorname{M}(G)$. Maximum positive semidefinite nullity is the maximum multiplicity of the least eigenvalue of a matrix $A \in \mathcal{S}_+(G)$ (and also the maximum multiplicity of the greatest eigenvalue of A by considering $\rho I - A$ where ρ is the spectral radius of A); as such it can be considered an ancillary problem of the IEP-G.

OBSERVATION 2.38. *From the definitions,* $\operatorname{mr}_+(G) \geq \operatorname{mr}(G)$ *and* $\operatorname{M}_+(G) \leq \operatorname{M}(G)$ *for every graph* G.

REMARK 2.39. The proofs of parts (3) and (4) of Theorem 2.27 remain valid for PSD minimum rank (choosing positive constants in the linear combination of matrices), so $\operatorname{mr}_+\left(\cup_{i=1}^t G_i\right) \leq \sum_{i=1}^t \operatorname{mr}_+(G_i)$ where $G = \cup_{i=1}^c G_i$ and $\operatorname{mr}_+(G) \leq \operatorname{cc}(G)$.

For PSD matrices, we specialize the concept of a representation (see Section 2.1). A function $\psi : V(G) \to \mathbb{R}^d$ is a *orthogonal representation* of G if for all $v, w \in V(G)$,

(1) $v \not\sim w \implies \psi(v)\cdot\psi(w) = 0$.
(2) $v \sim w \implies \psi(v)\cdot\psi(w) \neq 0$.

For a graph G of order n, a d-dimensional orthogonal representation of G can be viewed as a $d\times n$ matrix X such that $X^TX \in \mathcal{S}(G)$, where $\psi(v_i)$ is the ith column of X. What is here called an orthogonal representation is sometimes called a *faithful orthogonal representation.* Similar to minimum rank and representations (see Theorem 2.6), $\operatorname{mr}_+(G)$ is the smallest dimension of an orthogonal representation of G.

PROPOSITION 2.40. [**226**] *For any graph* G, $\operatorname{mr}_+(G)$ *equals the smallest* d *for which* G *has a* d*-dimensional orthogonal representation.*

Positive semidefinite zero forcing was designed as the zero forcing upper bound for maximum positive semidefinite nullity $\mathrm{M}_+(G)$.

DEFINITION 2.41. Let G be a graph. The Z_+*-color change rule* or *PSD color change rule* is: Let B be the set of filled vertices and let $W_1, \dots, W_k$ be the sets of vertices of the $k \geq 1$ components of $G - B$. If $u \in B$, $w \in W_i$, and w is the only white neighbor of u in $G[W_i \cup B]$, then fill w. Note that it is possible that there is only one component of $G - B$, and in that case Z_+-forcing is the same as standard zero forcing. The *positive semidefinite zero forcing number* or *PSD forcing number*, denoted by $\mathrm{Z}_+(G)$, is the minimum cardinality of a PSD forcing set. Forcing using the PSD color change rule is also called *PSD forcing*, etc.

As expected, the PSD zero forcing number is an upper bound for the maximum positive semidefinite nullity. The proof is more sophisticated than for standard zero forcing and is reproduced here (in [**18**] a slightly stronger result is established, namely that the PSD zero forcing number is an upper bound for the maximum Hermitian PSD nullity).

As we saw earlier, zero forcing captures the observation that the graph of A and zeros in certain positions of a null vector $\mathbf{x}$ of A force other entries of $\mathbf{x}$ to be zero. When A is positive definite, the following result gives more entries that are forced to be zero.

LEMMA 2.42. *Let $A \in \mathcal{S}_+(G)$, $\mathbf{x}$ be a nonzero null vector of A and α be a subset of vertices for which $x_i = 0$ for all $i \in \alpha$. Let j be a vertex of G not in α such that there is a vertex $k \in \alpha$ such that j is the unique neighbor of k in the connected component of $G - \alpha$ that contains j. The $x_j = 0$.*

PROOF. We may assume that A and $\mathbf{x}$ have the form

$$\left[\begin{array}{c|c|c} A_1 & O & C_1 \\ \hline O & A_2 & C_2 \\ \hline C_1^\top & C_2^\top & D \end{array}\right], \text{ and } \mathbf{x} = \left[\begin{array}{c} \mathbf{x}_1 \\ \hline \mathbf{x}_2 \\ \hline \mathbf{0} \end{array}\right]$$

where $D = A[\alpha]$ and A_1 corresponds to the connected component of $G - \alpha$ containing j. By the column inclusion property of Hermitian positive semidefinite matrices [**196**], $C_1 = A_1 W$ for some matrix W. Hence $C_1^\top = W^\top A_1$. Since $A\mathbf{x} = \mathbf{0}$, $A_1\mathbf{x} = \mathbf{0}$, which implies that $C_1^\top \mathbf{x}_1 = \mathbf{0}$. The row of $C_1^\top$ corresponding to vertex k has exactly one nonzero entry, namely the entry in column j. Hence $x_j = 0$. □

THEOREM 2.43. [**18**] *For every graph G, $\mathrm{M}_+(G) \leq \mathrm{Z}_+(G)$.*

PROOF. Let $A \in \mathcal{S}_+(G)$ with $\operatorname{null} A = \mathrm{M}_+(G)$. By Lemma 2.13, for each subset α of $\operatorname{null} A - 1$ vertices there is a nonzero null vector $\mathbf{x}$ with $x_j = 0$ for $j \in \alpha$. By Lemma 2.42, each such α is not a PSD zero forcing set. Thus $\mathrm{M}_+(G) \leq \mathrm{Z}_+(G)$ by Lemma 2.13.

□

The bound in Theorem 2.43 is tight (examples of graphs G for which $\mathrm{M}_+(G) = \mathrm{Z}_+(G)$ include trees, cycles, complete graphs, complete bipartite graphs, and outerplanar graphs [**28**]). The Möbius Ladder of order eight, also known as V_8, provides a contrasting example, since $\mathrm{M}_+(V_8) < \mathrm{Z}_+(V_8)$ [**18, 232**]. Some of the known values of $\mathrm{Z}_+(G)$ and $\mathrm{M}_+(G)$ are listed in Theorem 9.44 below.

Maximum nullity of a tree is its path cover number, which can be very large (think of a star). The situation for maximum positive semidefinite nullity is very different.

THEOREM 2.44. [**192**] *Let G be a graph of order n. Then $\mathrm{M}_+(G) = 1$ if and only if G is a tree.*

THEOREM 2.45. [**51, 194**] *Let G be a graph having a cut-vertex v and let $G = \bigcup_{i=1}^{h} G_i$ be the cut-vertex decomposition at v. Then*

$$\mathrm{mr}_+(G) = \sum_{i=1}^{h} \mathrm{mr}_+(G_i) \text{ and } \mathrm{M}_+(G) = \left(\sum_{i=1}^{h} \mathrm{M}_+(G_i)\right) - h + 1.$$

COROLLARY 2.46. *If G is a graph and $\deg_G(v) = 1$, then*

$$\mathrm{M}_+(G - v) = \mathrm{M}_+(G).$$

The next result, which relates vertex connectivity and maximum positive semidefinite nullity, was originally stated in terms of orthogonal representations; its proof is extremely deep. Recall that $\kappa(G)$ denotes the vertex connectivity (see Appendix A)

THEOREM 2.47. [**226**] *For every graph G, $\kappa(G) \leq \mathrm{M}_+(G)$.*

Maximum semidefinite nullity and minimum semidefinite rank are discussed further in Section 9.3, where their values for many graphs are presented in Theorem 9.44.

2.6. Colin de Verdière Type Parameters

Corollary 2.24, which asserts that for a forest the maximum nullity is the path cover number, perhaps misleads one to think that the computation of the minimum rank is easy. However, the minimum rank, or equivalently the maximum nullity, for graphs that have cycles is not so well-behaved. For example, $\mathrm{M}(K_n) = n - 1$ whereas $\mathrm{P}(K_n) = \lceil \frac{n}{2} \rceil$.

One might also think that inserting an edge into a graph gives more freedom[9] and hence believe that $\mathrm{M}(G) \geq \mathrm{M}(H)$ whenever H is subgraph of G. This is far from true, as the example $H = \overline{K_n}$ and $G = P_n$ demonstrate (since $\mathrm{M}(H) = n$ and $\mathrm{M}(G) = 1$). The next example is more subtle.

EXAMPLE 2.48. Consider the star $K_{1,4}$ with center vertex 1, and let H be the graph obtained from $K_{1,4}$ by inserting the edge 4–5 (see Figure 2.3). Then we can see that $\mathrm{M}(H) = 2$: Since H can be covered by three cliques (two K_2 and one K_3), $\mathrm{mr}(H) \leq 3$ and $\mathrm{M}(H) \geq 2$. Since $B = \{3, 4\}$ is a zero forcing set, $\mathrm{M}(H) \leq 2$. As $\mathrm{M}(K_{1,4}) = 3$, we see that inserting an edge increases the maximum multiplicity.

[9] After all, there is one additional variable available to tinker with in the set of symmetric matrices determined by the graph!

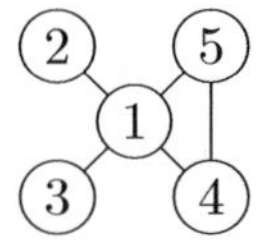

FIGURE 2.3. The graph H.

The Colin de Verdière Number μ. Colin de Verdière's work offers a way of remedying this undesirable behavior of the maximum multiplicity with respect to subgraphs. In [**88**] (English translation [**89**]), he introduced the notion of the strong Arnol'd property and used it to define a new parameter $\mu(G)$ for graphs that connects the spectral properties and the topological properties of a graph. The parameter $\mu(G)$ is now called the Colin de Verdière number. The entrywise (or Hadamard product) of two matrices A and B of the same dimensions is denoted by $A \circ B$.

DEFINITION 2.49. A symmetric matrix A has the *strong Arnol'd property* (or the *SAP*) if $X = O$ is the only symmetric matrix A such that

- $A \circ X = O$, $I \circ X = O$, and
- $AX = O$.

DEFINITION 2.50. The *Colin de Verdière number* of a graph G, also called the *Colin de Verdière parameter* of G, and denoted by $\mu(G)$, is defined to be maximum nullity among matrices $A \in \mathcal{S}(G)$ such that

(1) A has the SAP,
(2) every off-diagonal entry of A is negative or zero, and
(3) A has exactly one negative eigenvalue (counting the multiplicity).

By definition, $\mathrm{M}(G) \geq \mu(G)$ for any graph G.

EXAMPLE 2.51. The matrix $-J$ (the negative of the all-ones matrix) has the SAP since $-J \circ X = O$ implies $X = O$, and $-J$ has exactly one negative eigenvalue. Thus $\mu(K_n) = n - 1$ since $\mu(K_n) \leq \mathrm{M}(K_n) = n - 1$.

EXAMPLE 2.52. Let A be the negative of the adjacency matrix of $K_{1,3}$. The nullity of A is 2. We may write

$$A = \begin{bmatrix} 0 & -1 & -1 & -1 \\ -1 & 0 & 0 & 0 \\ -1 & 0 & 0 & 0 \\ -1 & 0 & 0 & 0 \end{bmatrix} \text{ and } X = \begin{bmatrix} 0 & 0 & 0 & 0 \\ 0 & 0 & x_{23} & x_{24} \\ 0 & x_{23} & 0 & x_{34} \\ 0 & x_{24} & x_{34} & 0 \end{bmatrix}$$

so that $A \circ X = I \circ X = O$. Then compute

$$AX = \begin{bmatrix} 0 & -x_{23} - x_{24} & -x_{23} - x_{34} & -x_{24} - x_{34} \\ 0 & 0 & 0 & 0 \\ 0 & 0 & 0 & 0 \\ 0 & 0 & 0 & 0 \end{bmatrix}$$

and solve $AX = O$. The only solution is $x_{23} = x_{24} = x_{34} = 0$, so A has the SAP. Furthermore, $\operatorname{spec}(-A) = \{-\sqrt{3}, 0, 0, \sqrt{3}\}$. Thus $\mu(K_{1,3}) = 2$ since $\mu(K_{1,3}) \leq \mathrm{M}(K_{1,3}) = 2$.

The situation is different for $K_{1,4}$.

EXAMPLE 2.53. Let $A = [a_{ij}] \in \mathcal{S}(K_{1,4})$ such that $\operatorname{null} A = \mathrm{M}(K_{1,4}) = 3$. Then

$$A = \begin{bmatrix} a_{11} & a_{12} & a_{13} & a_{14} & a_{15} \\ a_{12} & 0 & 0 & 0 & 0 \\ a_{13} & 0 & 0 & 0 & 0 \\ a_{14} & 0 & 0 & 0 & 0 \\ a_{15} & 0 & 0 & 0 & 0 \end{bmatrix} \text{ and } X = \begin{bmatrix} 0 & 0 & 0 & 0 & 0 \\ 0 & 0 & x_{23} & x_{24} & x_{25} \\ 0 & x_{23} & 0 & x_{34} & x_{35} \\ 0 & x_{24} & x_{34} & 0 & x_{45} \\ 0 & x_{25} & x_{35} & x_{45} & 0 \end{bmatrix}$$

(note that $\operatorname{null} A = 3$ forces $a_{ii} = 0$ for $i = 2, 3, 4, 5$). We solve $AX = O$ to obtain

$$\begin{aligned} x_{25} &= -\frac{a_{13}}{a_{15}} x_{23} - \frac{a_{14}}{a_{15}} x_{24}, \; x_{34} = -\frac{a_{12}}{a_{13}} x_{24} - \frac{a_{12}}{a_{14}} x_{23}, \\ x_{35} &= \frac{a_{12}a_{14}}{a_{13}a_{15}} x_{24}, \; x_{45} = \frac{a_{12}a_{13}}{a_{14}a_{15}} x_{23}. \end{aligned}$$

So A does not have the SAP and $\mu(K_{1,4}) < \mathrm{M}(K_{1,4})$.

A graph parameter ζ with the property $\zeta(H) \le \zeta(G)$ whenever H is a minor of G, denoted by $H \preceq G$, is called *minor monotone* (see Appendix A for the definition of minor). For a minor monotone parameter ζ, the graph minor theorem [**112**] asserts that for any given k, there is a finite set $\mathbf{Forb}_k$ of graphs such that $\zeta(G) \le k$ if and only if G does not contain any graph in $\mathbf{Forb}_k$ as a minor. Colin de Verdière showed that μ is minor monotone.

THEOREM 2.54. [**89**] *If $H \preceq G$, then $\mu(H) \le \mu(G)$.*

Note that if G is obtained from H by adding an edge, then $H \prec G$, so $\mu(H) \le \mu(G)$ and we have obtained the desired behavior for edge addition. Thus we can conclude that $\mu(K_{1,4}) < 3 = \mathrm{M}(K_{1,4})$ from Example 2.48 without doing the computations in Example 2.53.

If K_h is a minor of G, then

$$\mathrm{M}(G) \ge \mu(G) \ge \mu(K_h) = h - 1$$

by the minor monotonicity. Thus, the Colin de Verdière number gives an immediate application to the minimum rank problem. Recall that the *Hadwiger number* $\eta(G)$ of a graph is the largest h such that K_h is a minor of G.

PROPOSITION 2.55. [**19**] *For any graph G, $\mathrm{M}(G) \ge \mu(G) \ge \eta(G) - 1$.*

EXAMPLE 2.56. Another immediate consequence of minor monotonicity of the Colin de Verdière number is to provide an alternate proof of Theorem 2.32, that a graph is a path if and only if $\mathrm{M}(G) = 1$. If G is a path, then $\mathrm{M}(G) = 1$ was easily established (e.g., Example 2.4 or from $\mathrm{Z}(P_n) = 1$). If G is not a path, then either G is disconnected, or G contains a K_3 or a $K_{1,3}$ as its minor. When G is disconnected, then there is a matrix in $\mathcal{S}(G)$ with nullity at least two since each connected component can contribute at least nullity one (cf. Proposition 2.20(3)). When G contains K_3 or $K_{1,3}$ as a minor, $\mathrm{M}(G) \ge \mu(G) \ge 2$ by Example 2.52 and Proposition 2.55.

Much of the importance of the Colin de Verdière number comes from its connection to the topological properties of graphs.

THEOREM 2.57. [**89**] *Let G be a graph of order at least two. Then,*

(1) *$\mu(G) \le 1$ if and only if G is a disjoint union of one or more paths,*
(2) *$\mu(G) \le 2$ if and only if G is an outerplanar graph, and*

(3) $\mu(G) \leq 3$ *if and only if* G *is a planar graph.*

Many more results have been established for the Colin de Verdière number; see, e.g., [**195**] for a survey. Theorem 2.57 can be rephrased by forbidden minors. That is, $\mu(G) \leq 1$ if and only if G does not contain K_3 or $K_{1,3}$ as a minor; $\mu(G) \leq 2$ if and only if G does not contains K_4 or $K_{2,3}$ as a minor; and $\mu(G) \leq 3$ if and only if G does not contain K_5 or $K_{3,3}$ as a minor.

Many interesting problems remain open regarding the Colin de Verdière number. Here we list one; another one appears later as Conjecture 2.70.

CONJECTURE 2.58. [**89**] *For any graph* G, $\mu(G) + 1 \geq \chi(G)$.

There are other graph parameters defined through the SAP, and they are referred to as *Colin de Verdière type parameters.* One strength of a Colin de Verdière type parameter is that when the set of forbidden minors is known, then the value of the parameter becomes computable by deterministic algorithms, though finding the forbidden minors can be a hard task.

Another impact of the Colin de Verdière type parameters is the notion of using a perturbation and the Implicit Function Theorem to construct matrices. The SAP can be viewed as a necessary condition for perturbing a matrix without changing its nullity. Inspired by the SAP, there are other strong properties through which a perturbation that preserves the spectrum, or other spectral properties, are made possible; these are discussed in Chapter 3.

The Colin de Verdière Type Parameter for Maximum Nullity, ξ. Most pertinent to the minimum rank/maximum multiplicity problem is the graph parameter by Barioli, Fallat and Hogben in [**27**].

DEFINITION 2.59.

$$\xi(G) = \max\{\operatorname{null} A : A \in \mathcal{S}(G) \text{ and } A \text{ has the SAP}\}.$$

Like μ, ξ is minor monotone. A simpler version, namely that if H is a subgraph of G, then $\xi(H) \leq \xi(G)$, represented a major turning point for research on the IEP-G.

THEOREM 2.60. [**27**] *Let* H *be a minor of* G, *and let* B *be a matrix in* $\mathcal{S}(H)$ *having the strong Arnol'd Property. Then there exists a matrix* $A \in \mathcal{S}(G)$ *such that* $\operatorname{null} A = \operatorname{null} B$ *and* A *has the SAP. Thus* $\xi(H) \leq \xi(G)$.

A related proof is given for Lemma 3.11 in Chapter 3, in the context of other strong properties, and the full proof of Theorem 2.60 (specific to ξ) can be found in [**27**].

COROLLARY 2.61. [**27**] *Let* $H \preceq G$. *If* B *is a matrix in* $\mathcal{S}(H)$ *having an eigenvalue* λ *and* $B - \lambda I$ *has the SAP, then there exists a matrix* $A \in \mathcal{S}(G)$ *such that* $\operatorname{mult}_\lambda A = \operatorname{mult}_\lambda B$ *and* A *has the SAP.*

The Colin de Verdière Type Parameter for Positive Semidefinite Maximum Nullity, ν. Colin de Verdière also introduced another parameter based on positive semidefinite matrices.

DEFINITION 2.62. For a graph G, the parameter $\nu(G)$ is defined as the maximum nullity over positive semidefinite matrices $A \in \mathcal{S}_+(G)$ with the SAP.

Again, ν is minor monotone, so by the graph minor theorem, the graphs with $\nu(G) \leq k$ for a fixed k can be characterized by a finite set of forbidden minors.

THEOREM 2.63. [**90**] *If $H \preceq G$, then $\nu(H) \leq \nu(G)$.*

Although introduced independently by Colin de Verdière before the positive semidefinite maximum nullity $\mathrm{M}_+(G)$ was defined, $\nu(G)$ provides a lower bound for maximum positive semidefinite nullity:

PROPOSITION 2.64. [**19**] *For any graph G, $\mathrm{M}_+(G) \geq \nu(G) \geq \eta(G) - 1$.*

The result of Lovász, Saks, and Schrijver that for every graph G, $\kappa(G) \leq \mathrm{M}_+(G)$ [**226**] (Theorem 2.47), extends to ν.

THEOREM 2.65. [**193**] *For every graph G, $\kappa(G) \leq \nu(G)$.*

2.7. The δ-Conjecture and the Graph Complement Conjecture

In this section we discuss two long-standing conjectures related to minimum rank, maximum nullity, and Colin de Verdière type parameters.

The δ-Conjecture. Lovász, Saks, and Schrijver reported in [**226**] that Maehara conjectured in 1987 that $\delta(G) \leq \mathrm{M}_+(G)$ (described in terms of orthogonal representations). A slightly weaker form, $\delta(G) \leq \mathrm{M}(G)$ was stated at the 2006 AIM workshop [**11**]. Subsequently the stronger version $\delta(G) \leq \nu(G)$ was conjectured. Collectively these variants have become known as the δ-Conjecture. Since 2011, H. Tracy Hall has presented his proof of the strong δ-Conjecture $\delta(G) \leq \nu(G)$ several times (e.g., [**167**]), but it has not yet appeared in written form.

The Graph Complement Conjecture and Nordhaus-Gaddum problems. The *complement* $\overline{G}$ of G is the graph with the same vertex set as G and $e \in E(\overline{G})$ if and only if $e \notin E(G)$. The next conjecture, which was first posed at the 2006 AIM workshop [**11**], is called the *Graph Complement Conjecture* or *GCC*.

CONJECTURE 2.66 (GCC). *For any graph G of order n,*

$$\mathrm{mr}(G) + \mathrm{mr}(\overline{G}) \leq n + 2,$$

or equivalently,

$$\mathrm{M}(G) + \mathrm{M}(\overline{G}) \geq n - 2.$$

The bound in Conjecture 2.66 is tight because $\mathrm{mr}(\overline{P_n}) = 3$ for $n \geq 4$ ([**6**], see also Theorem 9.5(23)). A *Nordhaus-Gaddum (sum) problem* for a graph parameter ζ is to establish tight upper and lower bounds on $\zeta(G) + \zeta(\overline{G})$ where $\overline{G}$ is the complement of G. Thus the Graph Complement Conjecture is a Nordhaus-Gaddum problem. The related Nordhaus-Gaddum upper bound on M for a graph G of order n is $\mathrm{M}(G) + \mathrm{M}(\overline{G}) \leq 2n - 1$ and this is tight with $G = K_n$.

There are many natural variants of Conjecture 2.66 using related parameters, such as the next three conjectures. The Graph Complement Conjecture for the zero forcing number has been proved; see Theorem 9.30.

CONJECTURE 2.67 (GCC$_+$). *For any graph G of order n,*

$$\mathrm{mr}_+(G) + \mathrm{mr}_+(\overline{G}) \leq n + 2,$$

or equivalently,

$$\mathrm{M}_+(G) + \mathrm{M}_+(\overline{G}) \geq n - 2.$$

CONJECTURE 2.68 (GCC_ξ). **[20]** *For any graph G of order n,*

$$\xi(G) + \xi(\overline{G}) \geq n - 2.$$

CONJECTURE 2.69 (GCC_ν). **[20]** *For any graph G of order n,*

$$\nu(G) + \nu(\overline{G}) \geq n - 2.$$

CONJECTURE 2.70 (GCC_μ). **[214]** *For any graph G of order n,*

$$\mu(G) + \mu(\overline{G}) \geq n - 2.$$

The Nordhaus-Gaddum upper bound for maximum positive semidefinite nullity is again $\mathrm{M}_+(G) + \mathrm{M}_+(\overline{G}) \leq 2n - 1$ and this is tight with $G = K_n$. The upper bounds for Colin de Verdière type parameters are at most $\sqrt{2}n$; see **[32]** for more information.

Many of the partial results for the GCC and its variants listed in the next theorem have been proved using Colin de Verdière type parameters. Clearly GCC_ν implies GCC_ξ implies GCC, GCC_μ implies GCC_ξ, and GCC_ν implies GCC_+; each result in the next theorem is stated for the strongest relevant form of GCC and not for other forms that follow from that. See Appendix A for definitions of graph theory terms such as tree-width, k-tree, etc.

THEOREM 2.71.

(1) **[255]** *GCC_ν is true for graphs of tree-width at most* 3.
(2) **[255]** *If G is a k-connected partial k-tree, then G satisfies GCC_ν. In particular, if G is a k-tree, then G satisfies GCC_ν.*
(3) **[20]** *If G is a graph of order $n \leq 10$, then G satisfies GCC_ξ.*
(4) **[20]** *If G is a graph order $n \leq 8$, then G satisfies GCC_ν.*
(5) **[179]** *If* $\mathrm{mr}(G) \leq 4$ *or* $\mathrm{mr}(\overline{G}) \leq 4$, *then G satisfies GCC. If* $\mathrm{mr}_+(G) \leq 4$ *or* $\mathrm{mr}_+(\overline{G}) \leq 4$, *then G satisfies GCC_+.*
(6) **[128]** *If $\nu(G) \geq |V(G)| - 4$ or $\nu(\overline{G}) \geq |V(G)| - 4$, then G satisfies GCC_ν. If $\xi(G) \geq |V(G)| - 4$ or $\xi(\overline{G}) \geq |V(G)| - 4$, then G satisfies GCC_ξ.*
(7) **[20, 214]** *If $\mu(G) \geq |V(G)| - 6$ or $\mu(\overline{G}) \geq |V(G)| - 6$, then G satisfies GCC_μ.*
(8) **[20]** *If* $\mathrm{mr}(G) \geq |V(G)| - 3$ *or* $\mathrm{mr}(\overline{G}) \geq |V(G)| - 3$, *then G satisfies GCC. If* $\mathrm{mr}_+(G) \geq |V(G)| - 3$ *or* $\mathrm{mr}_+(\overline{G}) \geq |V(G)| - 3$, *then G satisfies GCC_+. If $\nu(G) < 2$ or $\nu(\overline{G}) < 2$, then G satisfies GCC_ν.*
(9) **[20, 214]** *GCC_μ is true for all planar graphs.*
(10) **[20]** *If G is a cograph, then G satisfies GCC_ν.*
(11) **[230]** *If G is chordal, then G satisfies GCC_ν.*
(12) **[20]** *If G and H are graphs that satisfy GCC_ν, then the join $G \vee H$ and disjoint union $G \cup H$ satisfy GCC_ν.*

Part 2

Strong Properties, Theory, and Consequences

An essential part of the IEP-G is to show the existence of a matrix with a prescribed spectrum and with a prescribed pattern of zero and nonzero entries. Often, a spectrum is easy to be realized for one pattern, but we wish to realize it for another closely related pattern. A natural way to do so is to apply an orthogonal similarity to a matrix. The similarity preserves the spectrum but may create many additional nonzero entries. For example, every matrix with at least two distinct eigenvalues is orthogonally similar to a matrix whose entries are all nonzero, which is described by a complete graph. Consequently, the complete graph with at least two vertices allows any spectrum with at least two distinct eigenvalues. However, this method is not so applicable when we want certain entries to stay as zero. Manifold theory provides a more flexible solution.

For a given graph, the set of symmetric matrices described by the graph is a manifold. For a given spectrum, the set of symmetric matrices with this spectrum is also known to be a manifold. The spectrum can be realized by a matrix of the graph if and only if the two manifolds intersect. The Strong Spectral Property, one of the strong properties to be introduced, is a condition that guarantees the robustness of the intersection. As a consequence, when the Strong Spectral Property holds, one may perturb the pattern manifold slightly without losing the intersection with the spectrum manifold. In other words, the matrix can be perturbed to another prescribed pattern while preserving the same spectrum.

This part will introduce the strong properties, their applications, and their theoretical underpinnings. Chapter 3 introduces the strong properties and provides a survey of how the strong properties are used in various contexts. In particular, the Supergraph Lemma (Lemma 3.15) provides a way to prove the existence of cospectral matrices for each supergraph. Chapter 4 introduces more advanced usage of the strong properties, including the Augmentation Lemma, the Matrix Liberation Lemma, and the Minor Monotonicity Theorem. Chapter 5 reviews all the results and gives a gentle introduction to the technical details required, in the hope of providing a concrete understanding for the reader.

CHAPTER 3

Implicit Function Theorem and Strong Properties

3.1. Implicit Function Theorem

A useful way of viewing the Implicit Function Theorem is in terms of the robustness of the solvability of a system of equations. Consider a system of equations

$$\begin{cases} f_1(\mathbf{x},\mathbf{y}) &= 0, \\ f_2(\mathbf{x},\mathbf{y}) &= 0, \\ &\vdots \\ f_n(\mathbf{x},\mathbf{y}) &= 0, \end{cases}$$

where $\mathbf{x} \in \mathbb{R}^m$, $\mathbf{y} \in \mathbb{R}^n$, and $f_i \colon \mathbb{R}^{m+n} \to \mathbb{R}^n$ is continuously differentiable $(i = 1, 2, \ldots, n)$. More simply, we can write the system of equations above as $f(\mathbf{x},\mathbf{y}) = \mathbf{0}$ where $f = (f_1, \ldots, f_n)$ and $\mathbf{0}$ is the origin in $\mathbb{R}^n$. Suppose that $\mathbf{x} = \mathbf{a}$ and $\mathbf{y} = \mathbf{b}$ satisfy $f(\mathbf{x},\mathbf{y}) = \mathbf{0}$ and the $n \times n$ Jacobian matrix $[\frac{\partial f_i}{\partial \mathbf{y}_j}]\big|_{\substack{\mathbf{x}=\mathbf{a}\\ \mathbf{y}=\mathbf{b}}}$ is invertible. Loosely speaking the Implicit Function Theorem says that given a small perturbation $\hat{\mathbf{a}}$ of $\mathbf{a}$, the system of equations $f(\hat{\mathbf{a}},\mathbf{y}) = \mathbf{0}$ still has a (unique) solution for $\mathbf{y}$, and that this solution is a continuous function of $\hat{\mathbf{a}}$. More precisely, the Implicit Function Theorem asserts that there is an open neighborhood U of $\mathbf{a}$ and an open neighborhood V of $\mathbf{b}$ such that for each $\mathbf{x} \in U$ there is a unique $\mathbf{y} \in V$ such that $f(\mathbf{x},\mathbf{y}) = \mathbf{0}$. Furthermore, there is a continuous function $g \colon U \to V$ such that $f(\mathbf{x}, g(\mathbf{x})) = \mathbf{0}$ for all $\mathbf{x} \in U$. See, e.g., [**114, 146, 215**] for more details about the Implicit Function Theorem.

A simple example is the equation $f(x,y) = 0$ in $\mathbb{R}^2 = \mathbb{R}^{1+1}$ defined by $f(x,y) = x^2 + y^2 - 1$, as shown in Figure 3.1. The solution set to this system is the set of points on the unit circle of radius 1 centered at the origin. At $(x,y) = (0,1)$, the partial derivative $\frac{\partial f}{\partial y}$ is $2y\big|_{\substack{x=0\\ y=1}} = 2$ and geometrically we know y is a function of x for (x,y) near the point $(0,1)$. In contrast, at $(x,y) = (1,0)$, the derivative $\frac{\partial f}{\partial y}$ is $2y\big|_{\substack{x=1\\ y=0}} = 0$ and y cannot be written as a function of x in a neighborhood of $(1,0)$ since the circle has a vertical tangent line at $(1,0)$.

Many problems related to the IEP-G reduce to whether or not a system of equations has a solution. For example, the question "does there exist a matrix

$$M = \begin{bmatrix} y_1 & x_1 & x_2 \\ x_1 & y_2 & 0 \\ x_2 & 0 & y_3 \end{bmatrix} \in \mathcal{S}(K_{1,2})$$

with spectrum $\{1,2,3\}$" is equivalent (by equating coefficients of the characteristic polynomial of M and $(t-1)(t-2)(t-3)$) to "does the system $f(\mathbf{x},\mathbf{y}) = \mathbf{0}$ with

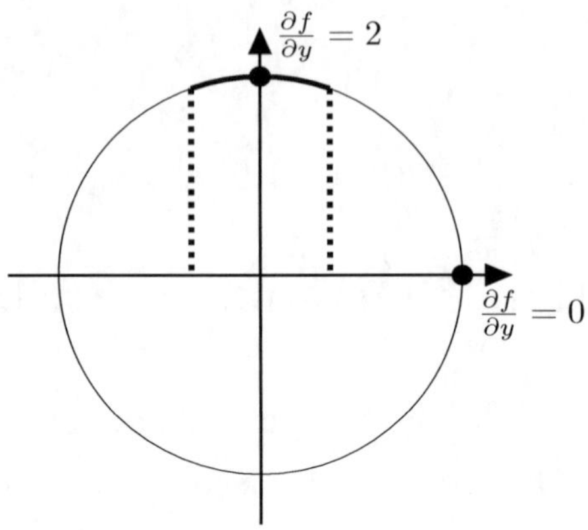

FIGURE 3.1. On a circle, y can be written as a function of x whenever the tangent line is not vertical

$f = (f_1, f_2, f_3)$ and

$$\begin{aligned} f_1(x_1, x_2, y_1, y_2, y_3) &= y_1 + y_2 + y_3 - 6 \\ f_2(x_1, x_2, y_1, y_2, y_3) &= y_1y_2 + y_1y_3 + y_2y_3 - x_1^2 - x_2^2 - 11 \\ f_3(x_1, x_2, y_1, y_2, y_3) &= y_1y_2y_3 - x_1^2y_3 - x_2^2y_2 - 6 \end{aligned}$$

have a solution such that both x_1 and x_2 are nonzero". In some instances, one can use the Implicit Function Theorem to guarantee the existence of a solution. For example, in the above $f(0, 0, 1, 2, 3) = (0, 0, 0)$ and the Jacobian $\frac{\partial f}{\partial \mathbf{y}}$ at $\mathbf{x} = (0, 0)$ and $\mathbf{y} = (1, 2, 3)$ is

$$\begin{bmatrix} 1 & 1 & 1 \\ y_2 + y_3 & y_1 + y_3 & y_1 + y_2 \\ y_2y_3 & y_1y_3 - x_2^2 & y_1y_2 - x_1^2 \end{bmatrix} \Bigg|_{\substack{\mathbf{x}=(0,0) \\ \mathbf{y}=(1,2,3)}} = \begin{bmatrix} 1 & 1 & 1 \\ 5 & 4 & 3 \\ 6 & 3 & 2 \end{bmatrix},$$

which is invertible. Hence by the Implicit Function Theorem, for x_1 and x_2 sufficiently small and positive, one can find (y_1, y_2, y_3) near $(1, 2, 3)$ with

$$f(x_1, x_2, y_1, y_2, y_3) = \mathbf{0}.$$

Thus,

$$\begin{bmatrix} y_1 & x_1 & x_2 \\ x_1 & y_2 & 0 \\ x_2 & 0 & y_3 \end{bmatrix}$$

is a matrix in $\mathcal{S}(K_{1,2})$ with spectrum $\{1, 2, 3\}$.

More generally, we can prove the existence of a matrix of a given graph with its spectrum being a prescribed set of distinct real numbers. To see this, let G be a graph on n vertices, $\Lambda = \{\lambda_1, \ldots, \lambda_n\}$ a set of distinct real numbers, and $\mathbf{y}_\Lambda = (\lambda_1, \ldots, \lambda_n) \in \mathbb{R}^n$. Consider a matrix $M(x, \mathbf{y})$ with $x \in \mathbb{R}$ and $\mathbf{y} = (y_1, \ldots, y_n) \in \mathbb{R}^n$ whose diagonal entries are determined by $\mathbf{y}$ and whose off-diagonal i, j-entry is x whenever $\{i, j\}$ is an edge of G and 0 otherwise. Compute the characteristic polynomial

$$\det(tI - M(x, \mathbf{y})) = t^n + s_1t^{n-1} + \cdots + s_{n-1}t + s_n$$

and define $f(x, \mathbf{y}) = (s_1, \ldots, s_n)$. Thus, f is a continuously differentiable function such that $f(0, \mathbf{y}_\Lambda)$ provides the coefficients (other than the leading one) for the polynomial $p(t) = \prod_{i=1}^n (t - \lambda_i)$. Observe that

$$\frac{\partial \det(tI - M(x, \mathbf{y}))}{\partial y_j}\bigg|_{\substack{x=0\\ \mathbf{y}=\mathbf{y}_\Lambda}} = -p(t)/(t - \lambda_j)$$

and, consequently, $\frac{\partial f}{\partial y_j}$ is the vector formed by the coefficients of $-p(t)/(t - \lambda_j)$. Since the λ_j are distinct, the polynomials $p(t)/(t - \lambda_j)$ for $j = 1, \ldots, n$ form a linearly independent set (they are the basis for the Lagrange interpolation), which implies that the Jacobian matrix

$$\left[\frac{\partial f_i}{\partial \mathbf{y}}\right]_{\substack{x=0\\ \mathbf{y}=\mathbf{y}_\Lambda}}$$

is invertible. By the Implicit Function Theorem, $\mathbf{y}$ can be written as a function of x and there is $\epsilon > 0$ such that $M(\epsilon, \mathbf{y}(\epsilon))$ has the same characteristic polynomial with $M(0, \mathbf{y}_\Lambda)$, so the spectrum of $M(\epsilon, \mathbf{y}(\epsilon))$ is Λ. Meanwhile, since $\epsilon \neq 0$, the matrix $M(\epsilon, \mathbf{y}(\epsilon))$ is in $\mathcal{S}(G)$. Thus, we have proven the existence of a matrix in G with the given discrete spectrum. This argument was given in [**233**], and, indeed, the Implicit Function Theorem is used to prove a stronger statement.

THEOREM 3.1. [**233**] *Let G be a connected graph on n vertices and $v \in V(G)$. Let*

$$\lambda_1 < \mu_1 < \lambda_2 < \cdots < \mu_{n-1} < \lambda_n$$

be distinct real numbers. Then there is a matrix $A \in \mathcal{S}(G)$ such that $\operatorname{spec}(A) = \{\lambda_1, \ldots, \lambda_n\}$ *and* $\operatorname{spec}(A(v)) = \{\mu_1, \ldots, \mu_{n-1}\}$.

The Implicit Function Theorem can be rephrased in the language of manifolds. Two manifolds $\mathcal{M}$ and $\mathcal{N}$ intersect *transversally* at a point $\mathbf{x}$ provided $\mathbf{x} \in \mathcal{M} \cap \mathcal{N}$ and the intersection of the normal spaces of $\mathcal{M}$ at $\mathbf{x}$ and of $\mathcal{N}$ at $\mathbf{x}$ is just the zero vector. In Figure 3.2, the black straight line and the curve on the left intersect transversally, while those on the right does not intersect transversally. The gray lines and curves are small perturbations of the original manifolds. We can see that the intersection still occurs if it is transversal, while the intersection might disappear if it is not transversal.

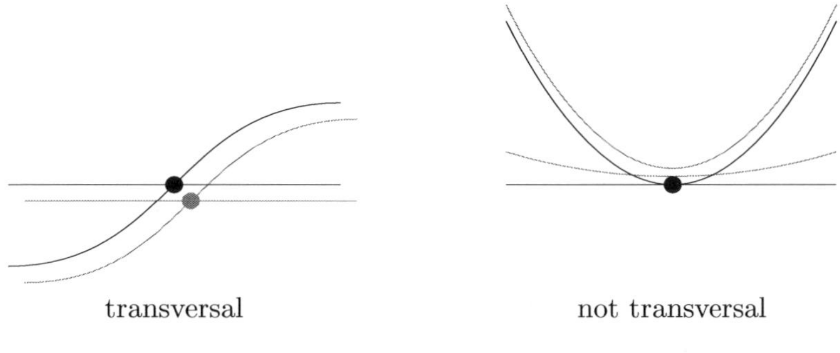

FIGURE 3.2. Transversal and non-transversal intersections of manifolds

The following forms of the Implicit Function Theorem can be found, for example, in [**195**], and we will provide the proofs for completeness in Section 5.3.

THEOREM 3.2. [**195**] *Let $\mathcal{M}_1(s)$ and $\mathcal{M}_2(t)$ be smooth families of manifolds in $\mathbb{R}^d$ for $s \in (-1,1)$ and $t \in (-1,1)$. Suppose $\mathcal{M}_1(0)$ and $\mathcal{M}_2(0)$ intersect transversally at a point $\mathbf{y}_0$. Then there exist a neighborhood $W \subseteq \mathbb{R}^2$ of the origin and a continuous function $f\colon W \to \mathbb{R}^d$ such that $f(0,0) = \mathbf{y}_0$ and $\mathcal{M}_1(s)$ and $\mathcal{M}_2(t)$ intersect transversally at $f(s,t)$ for all $(s,t) \in W$.*

Using Figure 3.1 again as an example, let $\mathcal{M}_1(s)$ be the circle, which is constant with respect to s, and $\mathcal{M}_2(t)$ be the vertical line $x = t$. The two manifolds intersect transversally for all t with $-1 < t < 1$. When $t = 1$, they intersect at $(1,0)$ but not transversally since the normal spaces of both of them equal to $\operatorname{span}\{\begin{bmatrix}1 & 0\end{bmatrix}^\top\}$. Similarly, $\mathcal{M}_1(s)$ and $\mathcal{M}_2(-1)$ intersect at $(-1,0)$ but not transversally. As a consequence, for any t with $-1 < t < 1$, the value of t can be perturbed slightly without losing the intersection between $\mathcal{M}_1(s)$ and $\mathcal{M}_2(t)$. However, for $t = 1$ or $t = -1$, some perturbation of t makes the two manifolds disjoint.

THEOREM 3.3. [**195**] *Assume $\mathcal{M}_1$ and $\mathcal{M}_2$ are smooth manifolds that intersect transversally at $\mathbf{y}$ and let $\mathbf{v}$ be a common tangent to each of $\mathcal{M}_1$ and $\mathcal{M}_2$ with $\|\mathbf{v}\| = 1$. Then for every $\epsilon > 0$ there exists a point $\mathbf{y}' \neq \mathbf{y}$ such that $\mathcal{M}_1$ and $\mathcal{M}_2$ intersect transversally at $\mathbf{y}'$, and*

$$\left\| \frac{1}{\|\mathbf{y} - \mathbf{y}'\|}(\mathbf{y} - \mathbf{y}') - \mathbf{v} \right\| < \epsilon.$$

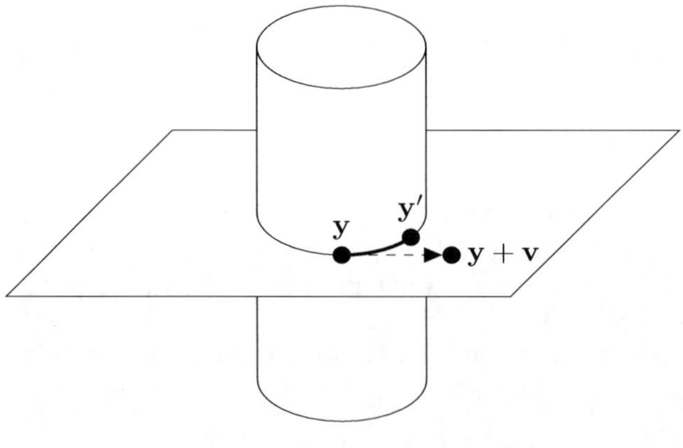

FIGURE 3.3. Illustration of Theorem 3.3

In Figure 3.3, one manifold is a plane and the other manifold is the surface of a cylinder. Focusing at the point $\mathbf{y}$, the dashed arrow $\mathbf{v}$ is a common tangent vector of the two manifolds. By Theorem 3.3, $\mathbf{y}$ can be perturbed into $\mathbf{y}'$ along the intersection of the two manifolds in direction of the initial tangent vector $\mathbf{v}$.

Both Theorems 3.2 and 3.3 are powerful tools for finding matrices with some given spectral properties. For example, we may let $\mathcal{M}_1$ be the set of all matrices of a given spectrum, while $\mathcal{M}_2 = \mathcal{S}(G)$ for some G. Thus, when the intersection of the two manifolds is transversal, we may perturb $\mathcal{M}_2$ slightly to get some more nonzero entries while preserving the intersection with $\mathcal{M}_1$. In fact, each strong property is equivalent to the transversality of certain pairs of manifolds. We will give an intuitive overview in this chapter and provide more applications in Chapter 4. The technical details about the manifolds will be deferred to Chapter 5.

3.2. Strong Properties

The goal of IEP-G is to understand the spectral properties of matrices in $\mathcal{S}(G)$. For a given matrix A, if the spectrum of A is

$$\operatorname{spec}(A) = \{\lambda_1^{(m_1)}, \lambda_2^{(m_2)}, \ldots, \lambda_q^{(m_q)}\} \text{ with } \lambda_1 < \cdots < \lambda_q,$$

where $q = q(A)$ is the number of distinct eigenvalues of A and $m_1, \ldots, m_q$ are the multiplicities, then the *ordered multiplicity list of A* is the tuple $\mathbf{m}(A) = (m_1, \ldots, m_q)$.

Sometimes, the set of matrices with a fixed property (e.g., the set of matrices with a given spectrum) form a manifold in the space of symmetric matrices. Suppose this manifold of matrices with a given property intersect transversally with $\mathcal{S}(G)$. Then we may utilize Theorem 3.2 to perturb $\mathcal{S}(G)$ slightly and get a new matrix for a different graph while preserving the desired spectral property (e.g., the spectrum).

Below we introduce some sets of matrices that are known to be manifolds and recall their tangent spaces and normal spaces. Proving these sets are actually manifolds and finding the tangent and normal spaces can be nontrivial. In this chapter, we will take these properties as granted, and the reader can find more details in [**33**] and alternative treatments in Chapter 5. Here we use E_{ij} for the $0,1$-matrix of order n with a single 1 at the i,j-entry and let $S_{ij}^- = E_{ij} - E_{ji}$. Recall that for two square matrices A and B of the same size, the commutator of A and B is $[A, B] = AB - BA$.

THEOREM 3.4 (Isospectral manifold). *Let $\Lambda = \{\lambda_1, \ldots, \lambda_n\}$ be a set of real numbers and $\mathcal{E}_\Lambda$ the set of all matrices in $S_n(\mathbb{R})$ with spectrum Λ. Then $\mathcal{E}_\Lambda$ is a manifold. Moreover, for a matrix $A \in \mathcal{E}_\Lambda$ the tangent space at A is*

$$\operatorname{span}\left(\{(S_{ij}^-)^\top A + AS_{ij}^-\}_{\substack{i,j\\ i<j}}\right)$$

and the normal space at A is

$$\{X : [A, X] = O\}.$$

THEOREM 3.5 (Iso-oml manifold). *Let $\mathbf{m} = (m_1, \ldots, m_q)$ be a sequence of positive integers with $n = \sum_{k=1}^q m_k$ and $\mathcal{U}_\mathbf{m}$ the set of all matrices in $S_n(\mathbb{R})$ with ordered multiplicity list $\mathbf{m}$. Then $\mathcal{U}_\mathbf{m}$ is a manifold. Moreover, for a matrix $A \in \mathcal{U}_\mathbf{m}$ the tangent space at A is*

$$\operatorname{span}\left(\{(S_{ij}^-)^\top A + AS_{ij}^-\}_{\substack{i,j\\ i<j}} \cup \{A^k\}_{k=0}^{q-1}\right)$$

and the normal space at A is

$$\{X : [A, X] = O,\ \operatorname{tr}(A^k X) = 0 \text{ for } k = 0, \ldots, q-1\}.$$

THEOREM 3.6 (Isorank manifold). *Let n be a positive integer, r a nonnegative integer with $r \le n$, and $\mathcal{R}_r$ the set of all matrices in $S_n(\mathbb{R})$ with rank r. Then $\mathcal{R}_r$ is a manifold. Moreover, for a matrix $A \in \mathcal{R}_r$ the tangent space at A is*

$$\operatorname{span}(\{AE_{ij} + E_{ij}A\}_{i,j})$$

and the normal space at A is

$$\{X : AX = O\}.$$

We call $\mathcal{E}_\Lambda$, $\mathcal{U}_{\mathbf{m}}$, and $\mathcal{R}_r$ the *isospectral manifold*, the *iso-oml manifold*, and the *isorank manifold*, respectively. For convenience, when A is a symmetric matrix, we let $\mathcal{E}_A = \mathcal{E}_{\text{spec}(A)}$, $\mathcal{U}_A = \mathcal{U}_{\mathbf{m}(A)}$, and $\mathcal{R}_A = \mathcal{R}_{\text{rank}\,A}$.

We call $\mathcal{S}(G)$ the *pattern manifold* of G. The structure of a pattern manifold is simpler since it becomes a subspace of $S_n(\mathbb{R})$ if we include the limit points. Let $\mathcal{S}^{\text{cl}}(G)$ be the topological closure of $\mathcal{S}(G)$. Then $\mathcal{S}^{\text{cl}}(G)$ is a subspace of $S_n(\mathbb{R})$ of dimension $|V(G)| + |E(G)|$.

THEOREM 3.7 (Pattern manifold). *Let G be a graph. Then $\mathcal{S}(G)$ is a manifold. Moreover, for a matrix $A \in \mathcal{S}(G)$ the tangent space at A is $\mathcal{S}^{\text{cl}}(G)$ and the normal space at A is*

$$\{X : A \circ X = I \circ X = O\}.$$

Thus, for any graph G, a matrix $A \in \mathcal{S}(G)$ has the SAP if and only if the isorank manifold $\mathcal{R}_A$ and the pattern manifold intersect transversally at A. Indeed, the strong properties are the translations of the transversality for different spectral properties.

The original strong property, the SAP, has been discussed in Section 2.6. There are at least two more known strong properties, which were introduced in [**33**].

DEFINITION 3.8. The real symmetric matrix A has the *Strong Spectral Property* (or the *SSP* for short) if the only real symmetric matrix X such that

$$\begin{aligned} A \circ X &= O, \\ I \circ X &= O, \text{ and} \\ [A, X] &= O \end{aligned}$$

is $X = O$.

DEFINITION 3.9. The real symmetric matrix A has the *Strong Multiplicity Property* (or the *SMP* for short) if the only real symmetric matrix such that

$$\begin{aligned} A \circ X &= O, \\ I \circ X &= O, \\ [A, X] &= O \text{ and} \\ \operatorname{tr}(A^k X) &= 0 \text{ for } k = 0, \dots, q-1 \end{aligned}$$

is $X = O$, where q is the number of distinct eigenvalues of A.

REMARK 3.10. When A and X are symmetric, the condition $[A, X] = AX - XA = O$ is equivalent to AX is symmetric, which is also equivalent to the inner product of the i-th row of A and the j-th column of X equals the inner product of the j-th row of A and the i-th column of X for all possible i and j.

Similar to the intuition around the SAP, the SSP and the SMP are equivalent to the transversality of some manifolds. Let G be a graph and $A \in \mathcal{S}(G)$. Then A has the SSP if and only if the isospectral manifold $\mathcal{E}_A$ and the pattern manifold $\mathcal{S}(G)$ intersect transversally at A, while A has the SMP if and only if $\mathcal{U}_A$ and $\mathcal{S}(G)$ intersect transversally at A.

By definition, the SSP implies the SMP, and the SMP implies the SAP. Lemma 3.11 shows that the SSP, SMP, and SAP allow a matrix to be perturbed without changing its spectrum, ordered multiplicity list, and rank, respectively. From this point of view, the SSP is a stronger condition and it leads to a stronger result (preserving the spectrum).

LEMMA 3.11 (Supergraph Lemma, spanning version [**33, 89**]). *Let G be a graph and H a supergraph of G with $V(G) = V(H)$. If $A \in \mathcal{S}(G)$ is a matrix with the*

(1) *SSP,*
(2) *SMP, or*
(3) *SAP,*

respectively, then there is a matrix $A' \in \mathcal{S}(H)$ such that

(1) $\operatorname{spec}(A') = \operatorname{spec}(A)$ *and A' has the SSP,*
(2) $\mathbf{m}(A') = \mathbf{m}(A)$ *and A' has the SMP, or*
(3) $\operatorname{rank} A' = \operatorname{rank} A$ *and A' has the SAP,*

respectively.

PROOF. We focus on the case of the SSP since the other cases are similar.

Let $H-G$ be the graph with the vertex set $V(G)$ and the edge set $E(H)\setminus E(G)$, and let E be adjacency matrix of $H-G$. Define

$$\mathcal{M}(t) = \mathcal{S}(G) + tE = \{B + tE : B \in \mathcal{S}(G)\}.$$

Thus, $\mathcal{M}(t)$ with $-1 < t < 1$ is a smooth family of manifolds. Since A has the SSP, $\mathcal{E}_A$ and $\mathcal{S}(G) = \mathcal{M}(0)$ intersect transversally at A. By the Implicit Function Theorem (Theorem 3.2), $\mathcal{E}_A$ and $\mathcal{M}(\epsilon)$ intersect transversally for all $\epsilon > 0$ sufficient small, and the intersection A' is continuous in ϵ; see Figure 3.4. We may pick ϵ small enough so that the nonzero entries in A remains nonzero in A'. Thus, $\operatorname{spec}(A') = \operatorname{spec}(A)$ and $A' \in \mathcal{S}(H)$. Moreover, since $\mathcal{M}(\epsilon) \subseteq \mathcal{S}(H)$, $\mathcal{E}_A$ and $\mathcal{S}(H)$ intersect transversally at A', which means A' has the SSP. □

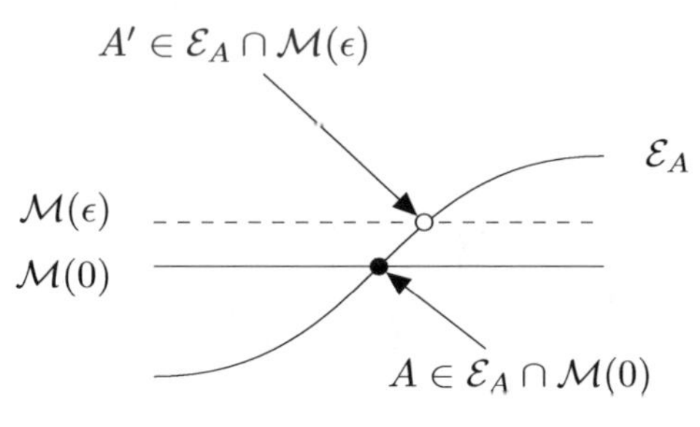

FIGURE 3.4. Illustration of the proof of Lemma 3.11

In the previous section, we proved that for any graph G on n vertices and any n distinct real numbers $\Lambda = \{\lambda_1, \ldots, \lambda_n\}$, there is a matrix $A \in \mathcal{S}(G)$ such that $\operatorname{spec}(A) = \Lambda$. We now show how the language of manifolds significantly simplifies the argument, though the essential idea is the same.

EXAMPLE 3.12. Let G be a graph on n vertices and $\Lambda = \{\lambda_1, \ldots, \lambda_n\}$ a set of distinct real numbers. Consider the diagonal matrix A whose diagonal entries are elements in Λ. Thus, $\operatorname{spec}(A) = \Lambda$ and A has the SSP since $[A, X] = O$ implies X is a diagonal matrix, and, in addition, $I \circ X = O$ implies $X = O$. Since A is a matrix of $\overline{K_n}$ and G is a supergraph of $\overline{K_n}$ on the same vertex set, Lemma 3.11 says that there is a matrix $A' \in \mathcal{S}(G)$ such that $\operatorname{spec}(A') = \operatorname{spec}(A) = \Lambda$ and A' has the SSP.

The applicability of Lemma 3.11 is somewhat limited since it only deals with graphs on the same vertex set. With the following lemmas, there is an easy way to construct larger matrices with the strong properties.

LEMMA 3.13. [**33**] *Let A and B be symmetric matrices with the SSP (or the SMP, respectively). Then $A \oplus B$ has the SSP (or the SMP, respectively) if and only if A and B have no common eigenvalue.*

PROOF. Let A and B be two symmetric matrices with the SSP. Let

$$X = \begin{bmatrix} X_1 & Y \\ Y^\top & X_2 \end{bmatrix}$$

be a matrix whose partition is conformal with $A \oplus B$ such that

$$(A \oplus B) \circ X = O, \ I \circ X = O, \ [A \oplus B, X] = O.$$

Thus, $A \circ X_1 = I \circ X_1 = [A, X_1] = O$, so $X_1 = O$ by the SSP of A. Similarly, $X_2 = O$ by the SSP of B. Moreover, we have $AY = YB$.

If A and B has no common eigenvalue, then $Y = O$ (see, e.g., [**196**, Theorem 2.4.4.1]), so $A \oplus B$ has the SSP. Conversely, if A and B has a common eigenvalue λ such that $\mathbf{a}$ and $\mathbf{b}$ are the corresponding eigenvectors, then $Y = \mathbf{a}\mathbf{b}^\top$ witness the loss of the SSP of $A \oplus B$.

The argument for the SMP is similar. □

LEMMA 3.14. [**195**] *Let A and B be symmetric matrices with the SAP. Then $A \oplus B$ has the SAP if and only if* 0 *is not a common eigenvalue for A and B.*

PROOF. Let A and B be two symmetric matrices with the SAP. Let

$$X = \begin{bmatrix} X_1 & Y \\ Y^\top & X_2 \end{bmatrix}$$

be a matrix whose partition is conformal with $A \oplus B$ such that

$$(A \oplus B) \circ X = O, \ I \circ X = O, \ (A \oplus B)X = O.$$

Thus, $A \circ X_1 = I \circ X_1 = AX_1 = O$, so $X_1 = O$ by the SAP of A. Similarly, $X_2 = O$ by the SAP of B. Thus, $(A \oplus B)X = O$ implies $AY = O$ and $BY^\top = O$.

If at least one of A and B is invertible, then $Y = O$, so $A \oplus B$ has the SSP. Conversely, if both A and B are singular such that $\mathbf{a}$ and $\mathbf{b}$ are the corresponding null vectors, then $Y = \mathbf{a}\mathbf{b}^\top$ witness the loss of the SAP of $A \oplus B$. □

LEMMA 3.15 (Supergraph Lemma [**33**]). *Let G be a graph of order n and H a supergraph of G of order n'. If $A \in \mathcal{S}(G)$ is a matrix with the SSP, then for any set Λ of $n' - n$ distinct real numbers other than those in* $\operatorname{spec}(A)$, *there is a matrix $A' \in \mathcal{S}(H)$ such that* $\operatorname{spec}(A') = \operatorname{spec}(A) \cup \Lambda$.

PROOF. Let Λ be a set of $n' - n$ distinct real numbers other than those in $\operatorname{spec}(A)$ and D the diagonal matrix whose diagonal entries are the elements in Λ. Thus, $A \oplus D$ has the SSP by Lemma 3.13. Then the desired result follows by applying Lemma 3.11 to $A \oplus D \in \mathcal{S}(G \cup \overline{K_{n'-n}})$ and H. □

A matrix with a given strong property secures the corresponding spectral property for all small perturbations of the matrix. The strong properties give much freedom and provide a systematic way to guarantee the existence of new matrices.

We will provide applications of the strong properties to the IEP-G in Chapter 4. Before that, we give a flavor of using the Implicit Function Theorem to general matrices, which are not necessarily symmetric nor square.

3.3. Strong Properties for Sign Patterns

Another application of the Implicit Function Theorem is to sign patterns. A *sign pattern* is a matrix whose entries are in $\{+,-,0\}$. The *qualitative class* $\mathcal{Q}(P)$ of a sign pattern P is the family of all matrices of the same dimensions as P and whose entries follow the signs described by P. The study of sign patterns was motivated by the linearization at an equilibrium of a dynamical system, where the sign of entries in the linearization is known while the exact value is not. For more about the sign patterns, we refer the reader to the book [**67**].

Introduced in [**118**], a *spectrally arbitrary pattern* of order n is a sign pattern P such that matrices $\mathcal{Q}(P)$ achieve all monic real-coefficient polynomials of degree n as their characteristic polynomials. Verifying a sign pattern as a spectrally arbitrary pattern is generally hard. One way to do so is through the Implicit Function Theorem. Example 3.16 illustrates the process.

EXAMPLE 3.16. Let T_3 be a sign pattern and A_3 a matrix in $\mathcal{Q}(T_3)$ with variables a_0, a_1, a_3 such that

$$T_3 = \begin{bmatrix} - & + & 0 \\ - & 0 & + \\ 0 & - & + \end{bmatrix} \text{ and } A_3 = \begin{bmatrix} -a_0 & 1 & 0 \\ -a_1 & 0 & 1 \\ 0 & -a_2 & 1 \end{bmatrix}.$$

Compute

$$\det(tI - A_3) = t^3 + (a_0 - 1)t^2 + (a_1 + a_2 - a_0)t + (a_0a_2 - a_1)$$

and consider the function

$$f(a_0, a_1, a_2, r_0, r_1, r_2) = \begin{bmatrix} a_0 - 1 - r_2 \\ a_1 + a_2 - a_0 - r_1 \\ a_0a_2 - a_1 - r_0 \end{bmatrix}.$$

With this definition, $f(a_0, a_1, a_2, r_0, r_1, r_2) = 0$ if and only if the characteristic polynomial of A_3 is $t^3 + r_2t^2 + r_1t + r_0$. For example, when

$$(a_0, a_1, a_2) = (1, \frac{1}{2}, \frac{1}{2}) \text{ and}$$
$$(r_0, r_1, r_2) = (0, 0, 0),$$

the matrix A_3 with these values of a_0, a_1, a_2 is nilpotent of index 3, and its minimal polynomial t^3.

By direct computation,

$$\frac{\partial f}{\partial(a_0, a_1, a_2)}\bigg|_{\substack{(a_0,a_1,a_2)=(1,\frac{1}{2},\frac{1}{2}) \\ (r_0,r_1,r_2)=(0,0,0)}} = \begin{bmatrix} 1 & 0 & 0 \\ -1 & 1 & 1 \\ a_2 & -1 & a_0 \end{bmatrix}\bigg|_{\substack{(a_0,a_1,a_2)=(1,\frac{1}{2},\frac{1}{2}) \\ (r_0,r_1,r_2)=(0,0,0)}}$$
$$= \begin{bmatrix} 1 & 0 & 0 \\ -1 & 1 & 1 \\ 1/2 & -1 & 1 \end{bmatrix}$$

is invertible. As a consequence of the Implicit Function Theorem, there is $\epsilon > 0$ such that for any (r_0, r_1, r_2) with $0 < r_i < \epsilon$ for $i = 1, 2, 3$, one may solve $f = 0$ for corresponding a_0, a_1, and a_2.

Now every characteristic polynomial $t^3 + r_2t^2 + r_1t + r_0$ with each $0 < r_i < \epsilon$ is achievable by matrices in $\mathcal{Q}(T_3)$. The last observation we have to make is that the characteristic polynomial of A_3 is $t^3 + r_2t^2 + r_1t + r_0$ if and only if the characteristic polynomial of cA_3 is $t^3 + r_2ct^2 + r_1c^2t + r_0c^3$, so we may scale the matrix to realize every monic real-coefficient polynomial of degree n in $\mathcal{Q}(T_3)$.

The technique in Example 3.16 was introduced in [**118**], where it is referred as the *Nilpotent-Jacobian Method* in [**149**]. In general, the sign pattern T_n is defined as

$$T_n = \begin{bmatrix} - & + & 0 & \cdots & 0 \\ - & 0 & + & \ddots & \vdots \\ 0 & \ddots & \ddots & \ddots & 0 \\ \vdots & \ddots & - & 0 & + \\ 0 & \cdots & 0 & - & + \end{bmatrix}$$

for $n \geq 2$. In [**118**], the Nilpotent-Jacobian Method was used to show that T_n is a spectrally arbitrary pattern when $2 \leq n \leq 7$. In 2012, Garnett and Shader [**150**] proved that T_n is a spectrally arbitrary pattern for any $n \geq 2$. Their technique was then formalized as the Nilpotent-Centralizer Method in [**149**]. Here a *superpattern* of P is a sign pattern obtained from P by replacing some (possibly none) 0 with $+$ or $-$.

THEOREM 3.17 (The Nilpotent-Centralizer Method [**149**]). *Let P be an $n \times n$ sign pattern and $A \in \mathcal{Q}(P)$ a nilpotent matrix of index n. If $X = O$ is the only matrix such that $A \circ X = O$ and $[A, X^\top] = O$, then any superpattern of P is a spectrally arbitrary pattern.*

One may notice the similarity of the SSP and the assumption of the Nilpotent-Centralizer Method, which is because both of them are dealing with the spectrum using the Implicit Function Theorem. However, the former condition is on symmetric matrices while the latter condition is for general square matrices.

Another interesting topic related to sign patterns is orthogonality, and a fundamental question is: for which sign pattern P, is there an orthogonal matrix in $\mathcal{Q}(P)$?

A recent paper [**93**] introduces the Strong Inner Product Property and uses it to prove the existence of orthogonal matrices with prescribed sign patterns.

DEFINITION 3.18. An $m\times n$ matrix A with $m \leq n$ has the *Strong Inner Product Property* (or the *SIPP* for short) if rank $A = m$ and the only $m\times m$ symmetric matrix X for which $A \circ (XA) = O$ is $X = O$.

THEOREM 3.19. [**93**] *Let P be an $m \times n$ sign pattern and $Q \in \mathcal{Q}(P)$ such that $QQ^\top = I$. If Q has the SIPP, then the qualitative class of any superpattern of P contains a matrix Q' with $Q'Q'^\top = I$.*

Through the strong properties and the Implicit Function Theorem, we have seen many results on proving the existence of matrices of a supergraph or a superpattern. In fact, there are more flexible usages of the Implicit Function Theorem.

For example, how can we apply the Implicit Function Theorem if the strong property does not hold? Also, the original result by Colin de Verdière contains the minor monotonicity but not just the result on supergraphs. Does the notion of minor monotonicity for SSP and SMP exist? In Chapter 4, we will see even more techniques related to the strong properties.

CHAPTER 4

Consequences of the Strong Properties

4.1. Number of Distinct Eigenvalues

In this section, we focus on $q(G)$, the minimum number of distinct eigenvalues of a graph, and illustrate some immediate applications of the strong properties (other techniques for studying $q(G)$ are discussed in Chapter 8). By Theorem 2.32, $\mathrm{M}(G) = 1$ if and only if G is a path. Here we provide a proof of this characterization using the strong properties.

THEOREM 4.1. **[33]** *Let G be a graph on n vertices. Then the following are equivalent.*
(1) $\mathrm{M}(G) = 1$.
(2) $q(G) = n$.
(3) *G is a path.*

PROOF. The equivalence between (1) and (2) follows from the definition of $\mathrm{M}(G)$ and $q(G)$. Suppose G is a graph with $\mathrm{M}(G) = 1$. As the nullity of a Laplacian matrix equals the number of connected components, G is connected. Example 2.52 and Lemma 3.15 use the SAP to assert that each graph containing $K_{1,3}$ as a subgraph has $\mathrm{M}(G) \geq 2$. Thus, G is either a cycle or a path. However, $\mathrm{M}(C_n) = 2$ for $n \geq 3$. So G is path.

Finally, suppose G is a path. Then Theorem 2.14 and $\mathrm{Z}(G) = 1$ imply $\mathrm{M}(G) = 1$. $\square$

We next characterize the graphs G with $q(G) = |V(G)| - 1$. We will start with some forbidden structures, which are listed in Figure 4.1.

EXAMPLE 4.2. Let A be the adjacency matrix of $K_{1,3}$ and

$$X = \begin{bmatrix} 0 & 0 & 0 & 0 \\ 0 & 0 & a & b \\ 0 & a & 0 & c \\ 0 & b & c & 0 \end{bmatrix}.$$

Suppose $AX - XA = O$. Then $a + b = 0$, $a + c = 0$, and $b + c = 0$, which means $a = b = c = 0$. Therefore, A has the SSP. (We checked that $-A$ has the SAP in Example 2.52.)

PROPOSITION 4.3. **[33]** *Let G be a graph on n vertices. If G contains $K_3 \sqcup K_3$, $K_3 \sqcup K_{1,3}$, $K_{1,3} \sqcup K_{1,3}$ as a subgraph, then $q(G) \leq n - 2$.*

PROOF. Let A be the adjacency matrix of $K_{1,3}$. By Example 4.2, A has the SSP and spectrum $\{-\sqrt{3}, 0, 0, \sqrt{3}\}$. Let B be the all-ones matrix of order 3. Then B is a matrix in $\mathcal{S}(K_3)$ with the SSP and has spectrum $\{0, 0, 3\}$. Therefore, by Lemma 3.13 the direct sum $D = (A+pI)\oplus(B+qI)$ has the SSP and two eigenvalues

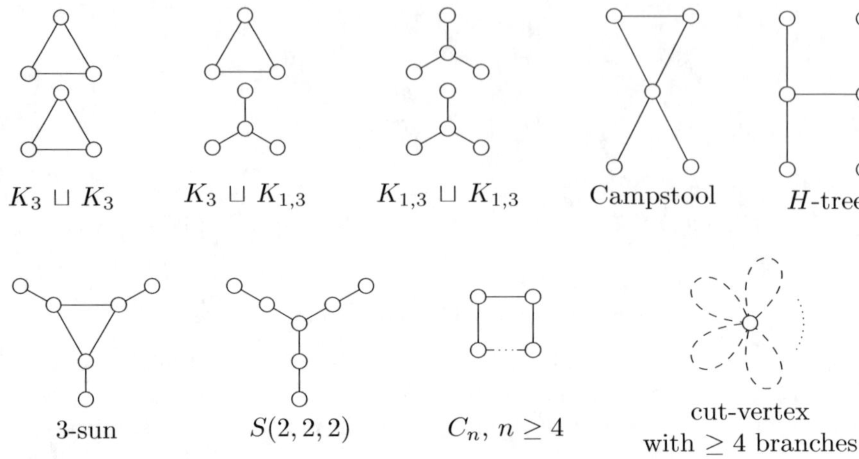

FIGURE 4.1. Forbidden structures for graphs G with $q(G) \geq |V(G)| - 1$.

of multiplicity 2 whenever $A + pI$ and $B + qI$ have no common eigenvalue. By Lemma 3.15, each graph G that contains $K_3 \sqcup K_{1,3}$ allows a matrix with two eigenvalues of multiplicity 2, so $q(G) \leq n-2$. The cases for $K_3 \sqcup K_3$ and $K_{1,3} \sqcup K_{1,3}$ are similar. □

In Figure 4.1, some graphs such as the Campstool, the H-tree, the 3-sun, and the graph $S(2,2,2)$ are defined. We will see in Proposition 4.4 that these graphs are forbidden structures for $q(G) = |V(G)| - 1$.

PROPOSITION 4.4. [**33**] *Let G be a graph on n vertices. If G contains the Campstool, H-tree, 3-sun, or $S(2,2,2)$ as a subgraph, then $q(G) \leq n - 2$.*

PROOF. Let G_1, G_2, G_3, and G_4 be the graphs Campstool, H-tree, 3-sun, and $S(2,2,2)$, respectively. Let

$$A_1 = \begin{bmatrix} 1 & 1 & 1 & 0 & 0 \\ 1 & 1 & -1 & 0 & 0 \\ 1 & -1 & 0 & 1 & 1 \\ 0 & 0 & 1 & 0 & 0 \\ 0 & 0 & 1 & 0 & 0 \end{bmatrix}, \quad A_2 = \begin{bmatrix} 0 & 0 & 1 & 0 & 0 & 0 \\ 0 & 0 & 1 & 0 & 0 & 0 \\ 1 & 1 & -1 & 1 & 0 & 0 \\ 0 & 0 & 1 & 2 & 1 & 1 \\ 0 & 0 & 0 & 1 & 1 & 0 \\ 0 & 0 & 0 & 1 & 0 & 1 \end{bmatrix},$$

$$A_3 = \begin{bmatrix} 2 & 1 & 1 & 1 & 0 & 0 \\ 1 & 2 & 1 & 0 & 1 & 0 \\ 1 & 1 & 2 & 0 & 0 & 1 \\ 1 & 0 & 0 & 1 & 0 & 0 \\ 0 & 1 & 0 & 0 & 1 & 0 \\ 0 & 0 & 1 & 0 & 0 & 1 \end{bmatrix}, \text{ and } A_4 = \begin{bmatrix} 0 & 1 & 0 & 1 & 0 & 1 & 0 \\ 1 & 1 & 1 & 0 & 0 & 0 & 0 \\ 0 & 1 & 1 & 0 & 0 & 0 & 0 \\ 1 & 0 & 0 & 1 & 1 & 0 & 0 \\ 0 & 0 & 0 & 1 & 1 & 0 & 0 \\ 1 & 0 & 0 & 0 & 0 & 1 & 1 \\ 0 & 0 & 0 & 0 & 0 & 1 & 1 \end{bmatrix}.$$

Then $A_i \in \mathcal{S}(G_i)$ and $q(A_i) \leq |V(G_i)| - 2$ for $i = 1, 2, 3, 4$. Also, each of A_i has the SSP, which can be checked by Sage; see, e.g., [**223**]. Therefore, each graph containing G_1, G_2, G_3 or G_4 as a subgraph has $q(G) \leq |V(G)| - 2$. □

Let C_n be the cycle on n vertices. The *flipped-cycle matrix* of order n is the (non-symmetric) square matrix $C = [c_{ij}]$ such that $c_{i,i+1} = 1$ for $i = 1, \dots, n-1$ and $c_{n1} = -1$.

THEOREM 4.5. **[33]** *Let C be the flipped-cycle matrix of order n, $n \geq 3$. Then $C+C^\top$ is a matrix in $\mathcal{S}(C_n)$ with the SMP,* $\operatorname{spec}(C+C^\top) = \left\{2\cos\left(\frac{2\pi(2j+1)}{2n}\right)\right\}_{j=1}^{n}$, *and $q(C+C^\top) = \lceil \frac{n}{2} \rceil$.*

COROLLARY 4.6. **[33]** *Let G be a graph on n vertices. If G contains a cycle of length at least 4 as a subgraph, then $q(G) \leq n-2$.*

LEMMA 4.7. *Let G be a graph on n vertices. If G has a cut-vertex such that its removal results in*

- *at least 4 connected components, or*
- *3 connected components with at least one not isomorphic to a path,*

then $\mathrm{M}(G) \geq 3$ *and* $q(G) \leq n-2$.

PROOF. Let v be the cut-vertex. Since $G-v$ has at least 4 connected components, or has 3 connected components with at least one not isomorphic to a path, $\mathrm{M}(G-v) \geq 4$. By Theorem 2.27, $\mathrm{M}(G) \geq \mathrm{M}(G-v) - 1 \geq 3$ and $q(G) \leq n-2$. □

Before proving the main theorem, we include a theorem regarding a lower bound for $q(G)$. A *unique shortest path* of a graph G is a path between two vertices such that it is the only path between the two vertices of the minimum length. For example, a cycle C_8 has a unique shortest path of length 3 but has no unique shortest path of length 4. A proof of Theorem 4.8 can be found in **[3]** and in Theorem 8.4.

THEOREM 4.8. **[3]** *Let G be a graph. If G has a unique shortest path on k vertices, then $q(G) \geq k$.*

THEOREM 4.9. **[33]** *A graph G has $q(G) \geq |V(G)| - 1$ if and only if G is only of the following:*

- *a path,*
- *the disjoint union of a path and an isolated vertex,*
- *a path with one leaf attached to an interior vertex, or*
- *a path with an extra edge joining two vertices at distance 2.*

PROOF. Suppose G is a graph with $q(G) \geq |V(G)| - 1$. Then by Propositions 4.3 and 4.4, Corollary 4.6, and Lemma 4.7, none of the forbidden structures in Figure 4.1 occurs in G.

If G has three or more components, then $\mathrm{M}(G) \geq 3$ and $q(G) \leq n-2$, which is impossible. Next, assume that $G = G_1 \sqcup G_2$. If both G_1 and G_2 have at least two vertices, then by scaling and shifting G allows a matrix with two eigenvalues of multiplicity two, which implies $q(G) \leq n-2$, a contradiction. Therefore, G is isomorphic to $G_1 \sqcup K_1$ with $q(G_1) = n-1$. Hence by Theorem 4.1, G is the disjoint union of a path and an isolated vertex.

Now assume G is connected. By Corollary 4.6, each cycle of G has length 3. Two 3-cycles that are not disjoint contain C_4 or the Campstool as a subgraph. If two 3-cycles are vertex disjoint, then G contains a $K_3 \sqcup K_3$ as a subgraph. Hence by Propositions 4.3 and 4.4, G contains at most one cycle, and it is of length 3.

Assume G contains a $K_3 = C_3$. If each vertex of the K_3 has degree at least 3 in G, then G has a 3-sun. Since 3-sun is forbidden, at least one of the three vertices on K_3 has degree 2. If any of the three vertices on K_3 has degree 4 or more, then G contains the Campstool, violating Proposition 4.4. If there is a vertex w of degree at least 3 that is not on K_3, then $G - w$ has at least 3 components with at least one not isomorphic to a path, violating Lemma 4.7. Therefore, G is a path with an extra edge joining two vertices of distance 2.

Finally assume G is connected and contains no cycle. Then G is a tree. By Lemma 4.7, G does not have a cut-vertex with 4 or more branches. Thus, the maximum degree of G is at most 3. If G has two adjacent vertices of degree 3, then G contains an H-tree. If G has two non-adjacent vertices of degree 3, then removing one of the two vertices leads to 3 components, where the one containing the other vertex is not a path, violating Lemma 4.7. Therefore, G has at most one vertex of degree 3. Since $S(2,2,2)$ is forbidden by Proposition 4.4, G is either a path or $S(1,a,b)$ with $a,b \geq 1$, which is a path with one leaf attached to an interior vertex.

Conversely, any of the listed graphs has a unique shortest path on $|V(G)| - 1$ vertices, so $q(G) \geq |V(G)| - 1$ by Theorem 4.8. □

4.2. Augmentation Lemma

Lemma 3.11 and Lemma 3.13 provide powerful tools to achieve some spectra from the spanning subgraphs of a graph. Unfortunately, there is a gap: Not all spectra that occur in $\mathcal{S}(G)$ can be achieved by its spanning subgraphs. Cycles are such examples. If H is a proper spanning subgraph of a cycle C_n, then H is a disjoint union of paths. A path only allows simple eigenvalues, while the spectra from different components of H have to be disjoint so that the matrix may have the SSP. This means each matrix with the SSP in $\mathcal{S}(H)$ has only simple eigenvalues. However, a cycle has maximum nullity 2, and there are SSP matrices in $\mathcal{S}(C_n)$ with several multiple eigenvalues.

One possible approach to fill this theoretical gap is to first find an $(n-1)\times(n-1)$ matrix and then try to find an appropriate last row and column. Suppose A is a matrix and λ one of its eigenvalues. If we want to append one row and one column to increase the multiplicity of λ by one, it is natural to think of $A \oplus [\lambda]$. However, since λ is in $\operatorname{spec}(A)$, the matrix $A \oplus [\lambda]$ does not have the SSP and Lemma 3.11 does not apply. A different technique, the Augmentation Lemma, to add nonzero entries to the last row and column and get the desired spectrum has been developed. Here the *support* $\operatorname{supp}(\mathbf{v})$ of a vector $\mathbf{v}$ is the set of indices where $\mathbf{v}$ is nonzero.

LEMMA 4.10 (Augmentation Lemma [**31**]). *Let G be a graph, $A \in \mathcal{S}(G)$ an matrix with the SSP, and λ an eigenvalue of A of multiplicity k. Suppose $\alpha \subseteq V(G)$ is a set of size $k+1$ such that $|\operatorname{supp}(\mathbf{v}) \cap \alpha| \geq 2$ for each eigenvector $\mathbf{v}$ of A corresponding to eigenvalue λ. Let H be the graph obtained from G by adding a new vertex joined to each vertex in α. Then there is a matrix $A' \in \mathcal{S}(H)$ with the SSP such that $\operatorname{spec}(A') = \operatorname{spec}(A) \cup \{\lambda\}$. Moreover, the matrix A' can be chosen arbitrarily close to $A \oplus [\lambda]$.*

COROLLARY 4.11. [**31**] *For any multiset $\Lambda = \{\lambda_1^{(2)}, \lambda_2, \dots, \lambda_{n-1}\}$ such that $\lambda_1, \dots, \lambda_{n-1}$ are distinct, there is a matrix in $\mathcal{S}(C_n)$ with the SSP whose spectrum is Λ.*

PROOF. Let $\Lambda' = \{\lambda_1, \ldots, \lambda_{n-1}\}$ be a set of $n-1$ distinct real numbers. By Example 3.12, there is an matrix $A \in \mathcal{S}(P_{n-1})$ with the SSP and with spectrum Λ, where P_{n-1} is the path 1–2– $\cdots$ –$(n-1)$. Let $\alpha = \{1, n-1\}$ and $\lambda = \lambda_1$ be an eigenvalue of A, whose multiplicity is 1. Suppose $\mathbf{v}$ is an eigenvector of A corresponding to eigenvalue λ, i.e., $\mathbf{v}$ is in the kernel of $A - \lambda I \in \mathcal{S}(P_{n-1})$. Since $\{1\}$ is a zero forcing set of P_{n-1}, the first entry of $\mathbf{v}$ is nonzero. (Alternatively, the same result comes from the fact that the first $n-1$ columns of $A - \lambda I$ always form an independent set regardless the choice of A.) Similarly, the $(n-1)$-entry of $\mathbf{v}$ is nonzero. Therefore, $|\operatorname{supp}(\mathbf{v}) \cap \alpha| \geq 2$ for each eigenvector $\mathbf{v}$ of A corresponding to λ. By Lemma 4.10, one may construct a graph by adding a new vertex n joining to two vertices 1 and $n-1$, which is the cycle graph C_n, and there is a matrix $A' \in \mathcal{S}(C_n)$ with the SSP whose spectrum is $\Lambda = \Lambda' \cup \{\lambda\}$. □

EXAMPLE 4.12. Let K_n be the complete graph and J be the $n \times n$ all-ones matrix in $\mathcal{S}(K_n)$. The spectrum of J is $\{n, 0^{(n-1)}\}$. The eigenvalue n has multiplicity $k = 1$, and the corresponding eigenvector is the all-ones vector $\mathbb{1}$. Pick α to consist of any two vertices of K_n. Then $|\alpha| = 2$ and $|\operatorname{supp}(\mathbf{v}) \cap \alpha| \geq 2$ for each eigenvector for n. Let H be the graph obtained from K_n by adding a new vertex joined to the two vertices in α. Then by Lemma 4.10 there is an matrix $A' \in \mathcal{S}(H)$ with the SSP such that $\operatorname{spec}(A') = \{n^{(2)}, 0^{(n-1)}\}$. Since A' has the SSP, any supergraph of H of the same order also allows this spectrum.

As we will see in Section 4.5, the Augmentation Lemma is a special case of the Matrix Liberation Lemma, so a proof will be deferred to Section 4.5. According to the proof, the matrix A' is obtained by perturbing the matrix $A \oplus [\lambda]$, so it can be chosen arbitrarily close to $A \oplus [\lambda]$. We emphasize that $A \oplus [\lambda]$ does not have the SSP but A' does have the SSP, which is an interesting feature of the Augmentation Lemma.

4.3. The IEP-G for Small Graphs

In this section, we illustrate how to use the SSP (and some classical results) to generate the solutions to the IEP-G for small graphs G with 4 or fewer vertices. First we focus on paths and complete graphs.

THEOREM 4.13. [**176, 33**] *A multiset of real numbers Λ is the spectrum of a matrix in $\mathcal{S}(P_n)$ if and only if all elements in Λ are distinct. Moreover, the matrix in $\mathcal{S}(P_n)$ with the spectrum Λ can be chosen to have the SSP.*

PROOF. Suppose Λ is the spectrum of a matrix in $\mathcal{S}(P_n)$. Since the zero forcing number of a path is 1, all elements in Λ are distinct.

Conversely, if all elements in Λ are distinct, then there is a matrix $A \in \mathcal{S}(P_n)$ with the SSP and $\operatorname{spec}(A) = \Lambda$ by Example 3.12. □

Stronger results are known for paths. Suppose $\Lambda = \{\lambda_1, \ldots, \lambda_n\}$ and $\Omega = \{\mu_1, \ldots, \mu_{n-1}\}$ such that

$$\lambda_1 < \mu_1 < \lambda_2 < \cdots < \lambda_{n-1} < \mu_{n-1} < \lambda_n.$$

Gary and Wilson [**162**] showed that there is a matrix $A \in \mathcal{S}(P_n)$ such that $\operatorname{spec}(A) = \Lambda$ and $\operatorname{spec}(A(1)) = \Omega$, where vertex 1 is one of the endpoints of P_n. The proof of Gary and Wilson is constructive and all off-diagonal entries are positive. Indeed, Hochstadt showed that such a matrix A with $\operatorname{spec}(A) = \Lambda$, $\operatorname{spec}(A(1)) = \Omega$, and all off-diagonal entries positive is unique for any given Λ and Ω that are strictly interlacing.

Suppose G is a tree on n vertices and v is any vertex of G. Duarte [**119**] proved that for each collection of n distinct real numbers Λ and $n-1$ real numbers Ω such that Λ and Ω are strictly interlacing, there is a matrix $A \in \mathcal{S}(G)$ such that $\operatorname{spec}(A) = \Lambda$ and $\operatorname{spec}(A(v)) = \Omega$. Shader and Monfared [**233**] then proved that the same result is true for any connected graph (see Theorem 3.1).

The IEP-G for complete graphs K_n was solved by Barrett et al. [**35**]. The proof is constructive and uses Givens rotations. Recall that a Givens rotation can be written as an $n \times n$ matrix $G(i,j,\theta)$ such that all entries are the same as the identity matrix except that the principal submatrix induced on the rows and columns indexed by i and j is the rotation matrix

$$\begin{bmatrix} \cos\theta & -\sin\theta \\ \sin\theta & \cos\theta \end{bmatrix}.$$

THEOREM 4.14. [**35**] *For $n \geq 2$, a multiset of real numbers Λ is the spectrum of a matrix in $\mathcal{S}(K_n)$ if and only if Λ contains at least two distinct elements.*

PROOF. Suppose $\Lambda = \{\lambda^{(n)}\}$ for some λ. Then a matrix with spectrum Λ has to be λI, which is not in $\mathcal{S}(K_n)$ whenever $n \geq 2$.

Conversely, suppose $\Lambda = \{\lambda_1, \ldots, \lambda_n\}$ is a multiset with $\lambda_1 \neq \lambda_2$. Then a matrix $A \in \mathcal{S}(K_n)$ with $\operatorname{spec}(K_n) = \Lambda$ can be constructed by the following steps.

(1) Let $A^{(1)} = \begin{bmatrix}\lambda_1\end{bmatrix}$.
(2) For $k = 2, \ldots, n$,
  (*a*) let $A^{(k)} = A^{(k-1)} \oplus \begin{bmatrix}\lambda_k\end{bmatrix}$, and
  (*b*) replace $A^{(k)}$ by $R^\top A^{(k)} R$ for some Givens rotation $R = G(k-1, k, \theta)$ such that the new $A^{(k)}$ is in $\mathcal{S}(K_k)$ and has the k,k-entry not equal to λ_{k+1} (ignore the condition on the k,k-entry when $k = n$).

The constructed $A^{(k)}$ has the spectrum $\{\lambda_1, \ldots, \lambda_k\}$ and $A^{(k)} \in \mathcal{S}(K_k)$. Also, the k,k-entry of $A^{(k)}$ is different from λ_{k+1} for any $k = 1, \ldots, n-1$. This condition is to ensure the existence of R in step (2b). Consequently, a matrix $A^{(n)}$ is constructed with the spectrum Λ. □

Note that every matrix $A \in \mathcal{S}(K_n)$ has the SSP since $A \circ X = O$ and $I \circ X = O$ are enough to guarantee $X = O$. In fact, the same phenomenon also happens for matrices in $\mathcal{S}(P_n)$; namely, every matrix in $\mathcal{S}(P_n)$ has the SSP. Section 4.7 will provide a reason for this phenomenon.

Now we are ready to see that the SSP along with the Supergraph Lemma (Lemma 3.15) is enough to solve the IEP-G for graphs with at most 4 vertices. Since the possible spectra of a disconnected graphs can be found by the union of the possible spectra of its connected components, it is enough to focus on connected graphs. Also, every connected graph with fewer than four vertices is either a path or a complete graph, which is solved by Theorems 4.13 and 4.14. Therefore, we start our discussion from connected graphs on 4 vertices. Table 4.1 lists all connected graphs on 4 vertices and summarizes the answers to the IEP-G for them. It turns out that for connected graphs on 4 vertices, every realizable ordered multiplicity list is spectrally arbitrary with a matrix satisfying the SSP. Also, if an ordered multiplicity list can be achieved, then each of its refinements can also be achieved. (The *refinement* of an ordered multiplicity list is an ordered multiplicity list obtained by replacing each multiplicity m_i with a list of positive integers whose sum is m_i. An ordered multiplicity list $\mathbf{m}$ of a graph G is *maximal* for G if $\mathbf{m}$ is not a refinement of another ordered multiplicity list of G.) More discussions about the ordered multiplicity list can be found in Chapter 6.

graph	maximal ordered multiplicity lists
K_4 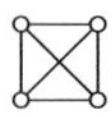	$(3,1)$, $(1,3)$, $(2,2)$
Dmnd 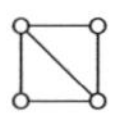	$(2,2)$, $(1,2,1)$
C_4 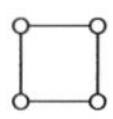	$(2,2)$, $(1,2,1)$
Paw 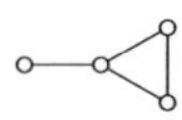	$(2,1,1)$, $(1,1,2)$, $(1,2,1)$
$K_{1,3}$ 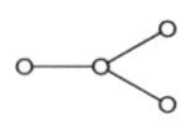	$(1,2,1)$
P_4 	$(1,1,1,1)$

TABLE 4.1. Connected graphs on 4 vertices with its maximal ordered multiplicity lists. For graphs on 4 vertices, if an ordered multiplicity list can be achieved, its refinements can also be achieved.

Case: $G = P_4$ and $G = K_4$.

Theorems 4.13 and 4.14 completely resolve the inverse eigenvalue problem for these graphs. Obviously, every matrix $A \in \mathcal{S}(K_4)$ has the SSP. In Section 4.7, we will see every matrix $A \in \mathcal{S}(P_4)$ has the SSP.

Case: $G = K_{1,3}$.

Since $\mathrm{Z}(K_{1,3}) = 2$ and $\mathrm{Z}_+(K_{1,3}) = 1$, the possible ordered multiplicity lists are $(1,1,1,1)$ and $(1,2,1)$, where the former one is achievable by Example 3.12.

The matrix

$$A = \begin{bmatrix} 2a & 1 & 1 & 1 \\ 1 & 0 & 0 & 0 \\ 1 & 0 & 0 & 0 \\ 1 & 0 & 0 & 0 \end{bmatrix}$$

is a matrix in $\mathcal{S}(K_{1,3})$. The spectrum of A is $\{a - \sqrt{a^2+3}, 0^{(2)}, a + \sqrt{a^2+3}\}$. Note that the ratio between $(a + \sqrt{a^2+3}) - 0$ and $0 - (a - \sqrt{a^2+3})$ can be any positive value. Therefore, $pA + qI$ with $p \neq 0$ can realize any spectrum with the ordered multiplicity list $(1,2,1)$. As we will see in Section 4.7, each matrix in $\mathcal{S}(K_{1,3})$ has the SSP.

Case: $G =$ Paw.

Since $\mathrm{Z}(\mathsf{Paw}) = 2$ and Paw has a unique shortest path on 3 vertices, the possible ordered multiplicity lists are $(1,1,1,1)$, $(1,2,1)$, $(2,1,1)$, and $(1,1,2)$. The first one is spectrally arbitrary and achievable by Example 3.12. Note that the lists $(1,2)$

and $(2,1)$ are achievable by SSP matrices in $\mathcal{S}(K_3)$ and are spectrally arbitrary. Therefore, $K_3 \sqcup K_1$ allows SSP matrices with spectrally arbitrary ordered multiplicity lists $(1,2,1)$, $(2,1,1)$ and $(1,1,2)$, and the same properties also occur for the Paw graph by Lemma 3.11 since $K_3 \sqcup K_1$ is a spanning subgraph of Paw.

Case: $G = C_4$.

Since $\mathrm{Z}(C_4) = 2$, the possible ordered multiplicity lists are $(1,1,1,1)$, $(1,2,1)$, $(2,1,1)$, $(1,1,2)$, and $(2,2)$. The case $(1,1,1,1)$ is handled by Example 3.12. The lists $(1,2,1)$, $(2,1,1)$, and $(1,1,2)$ are achievable for C_4 and spectrally arbitrary by Corollary 4.11.

The matrix

$$A = \begin{bmatrix} 0 & 0 & 1 & 1 \\ 0 & 0 & 1 & -1 \\ 1 & 1 & 0 & 0 \\ 1 & -1 & 0 & 0 \end{bmatrix}$$

has spectrum $\{\sqrt{2}^{(2)}, -\sqrt{2}^{(2)}\}$ and has the SSP (Example 4.20). Thus, $pA + qI$, $p \neq 0$, can realize all spectra with ordered multiplicity list $(2,2)$. As we will see in Proposition 4.15, $pA + qI$ with $p \neq 0$ has the SSP if A has the SSP. Therefore, $(2,2)$ is achievable and spectrally arbitrary for C_4.

Case: $G =$ Dmnd.

Since $\mathrm{Z}(\mathsf{Dmnd}) = 2$, the possible ordered multiplicity lists are $(1,1,1,1)$, $(1,2,1)$, $(2,1,1)$, $(1,1,2)$, and $(2,2)$. The case $(1,1,1,1)$ is handled by Example 3.12. By Lemma 3.11, the lists $(1,2,1)$, $(2,1,1)$, and $(1,1,2)$ are given by the spanning subgraph Paw, while the list $(2,2)$ is given by C_4.

As we have seen here, the SSP is powerful in implicitly proving the existence of matrices with given graph and spectrum. In particular, the discussion on the Dmnd graph only relies on Lemma 3.11 and some nice spanning subgraphs. The SSP reduces the effort significantly for constructing matrices explicitly and finding their spectra. The same ideas have been applied to solve the IEP-G for connected graphs on 5 vertices [**31**], and the Figure 1 of the same paper illustrates the possible ordered multiplicity lists. One might notice that for connected graphs on 4 vertices, every realizable spectrum is realizable by a matrix with the SSP. This is not the case for graphs on 5 vertices. For example, the graph $K_{1,4}$ allows multiplicity 3, but each matrix in $\mathcal{S}(K_{1,4})$ with multiplicity 3 is forced to lose the SSP since every such matrix has the form

$$A = \begin{bmatrix} d & a_1 & a_2 & a_3 & a_4 \\ a_1 & 0 & 0 & 0 & 0 \\ a_2 & 0 & 0 & 0 & 0 \\ a_3 & 0 & 0 & 0 & 0 \\ a_4 & 0 & 0 & 0 & 0 \end{bmatrix}$$

and the matrix

$$X = \begin{bmatrix} 0 & 0 & 0 & 0 & 0 \\ 0 & 0 & 0 & a_2a_4 & -a_2a_3 \\ 0 & 0 & 0 & -a_1a_4 & a_1a_3 \\ 0 & a_2a_4 & -a_1a_4 & 0 & 0 \\ 0 & -a_2a_3 & a_1a_3 & 0 & 0 \end{bmatrix}$$

witnesses the loss of the SAP of A.

If an ordered multiplicity list $\mathbf{m}$ is achievable by matrix A and contains only two distinct eigenvalues, then it is spectrally arbitrary since $pA+qI$ can realize all possible spectra governed by $\mathbf{m}$. This technique is called *scaling and shifting*.

PROPOSITION 4.15. [**31**] *For any $p \neq 0$, a matrix A has the SSP if and only if $pA+qI$ has the SSP.*

PROOF. The proposition follows from two observations. First, whenever $p \neq 0$, under the assumption $I \circ X = O$, the two conditions $A \circ X = O$ and $(pA+qI) \circ X = O$ are equivalent. Also,

$$[pA+qI, X] = p[A,X] + q[I,X] = p[A,X]$$

implies that $[A,X] = O$ is equivalent to $[pA+qI, X] = O$. Therefore, $A \circ X = O$, $I \circ X = O$, and $[A,X] = O$ if and only if $(pA+qI) \circ X = O$, $I \circ X = O$, and $[pA+qI, X] = O$. □

4.4. Verification Matrices

As we have seen in Example 2.52, checking the SAP for a given matrix is equivalent to checking whether a homogeneous system of linear equations has only the trivial solution. In this section, we describe the process in detail and introduce the verification matrices as a general strategy to verify the strong properties.

Our discussion starts with the SAP, as similar techniques apply to other strong properties. Let A be an $n \times n$ matrix in $\mathcal{S}(G)$. Recall that A has the SAP if the system of matrix equations $A \circ X = O$, $I \circ X = O$, and $AX = O$, where X is a symmetric matrix, only has the trivial solution $X = O$. Indeed, the system of matrix equations can be viewed as a system of linear equations in the entries of X. Since $A \circ X = O$ and $I \circ X = O$, we may assume $X = \begin{bmatrix} x_{ij} \end{bmatrix}$ such that $x_{ij} = x_{ji}$ and $x_{ij} = 0$ whenever $i = j$ or $\{i,j\} \in E(G)$. Thus, X contains $|E(\overline{G})|$ independent variables. We call X a (symmetric) *variable $\overline{G}$-matrix.*

In the context of variable matrices, the SAP asserts that the equation $AX = O$ on a variable $\overline{G}$-matrix X has only the trivial solution $X = O$. By comparing the entries on each side of $AX = O$, there are n^2 linear equations, namely, $(AX)_{ij} = 0$ for each $i,j \in \{1,\ldots,n\}$, on variables $\{x_{ij} : \{i,j\} \in E(\overline{G})\}$. (Note that for the case of the SAP, the equations $(AX)_{ij} = 0$ and $(AX)_{ji} = 0$ are not the same since AX is not necessarily a symmetric matrix. In contrast, for the case of the SSP, $(AX-XA)_{ij} = 0$ is equivalent to $(AX-XA)_{ji} = 0$.) As a consequence, a matrix A has the SAP if and only if the homogeneous system of linear equations $(AX)_{ij} = 0$, $i,j \in \{1,\ldots,n\}$, only has the trivial solution. Thus, one may verify the SAP with standard linear algebra techniques.

Let S_{ij} be the $n \times n$ symmetric matrix whose i,j-entry and j,i-entry are 1 while other entries are 0. Thus, the variable matrix X can be written as $X = \sum_{\{i,j\} \in E(\overline{G})} x_{ij} S_{ij}$. (Note that S_{ij} is a matrix while x_{ij} is the i,j-entry of X, which is a variable.) Then the equations for the strong properties can be rephrased as in Proposition 4.16.

PROPOSITION 4.16. [**31**] *Let G be a graph, $A \in \mathcal{S}(G)$, and $q = q(A)$. Then A has the SSP, SMP, and SAP if and only if the linear systems* (1), (2), *and* (3) *below only have the trivial solution $x_{ij} = 0$ for all $\{i,j\} \in E(\overline{G})$, respectively.*

$$(1) \quad \sum_{\{i,j\} \in E(\overline{G})} x_{ij} [A, S_{ij}] = O$$

(2) $\displaystyle\sum_{\{i,j\}\in E(\overline{G})} x_{ij}[A,S_{ij}] = O$ *and* $\displaystyle\sum_{\{i,j\}\in E(\overline{G})} x_{ij}(A^k)_{ij} = 0$ *for* $k = 0,1,\ldots,q-1$

(3) $\displaystyle\sum_{\{i,j\}\in E(\overline{G})} x_{ij}AS_{ij} = O$

EXAMPLE 4.17. Here we show that every matrix $A \in \mathcal{S}(K_{1,3})$ has the SSP. We may assume that

$$A = \begin{bmatrix} d_1 & a_{1,2} & a_{1,3} & a_{1,4} \\ a_{1,2} & d_2 & 0 & 0 \\ a_{1,3} & 0 & d_3 & 0 \\ a_{1,4} & 0 & 0 & d_4 \end{bmatrix} \text{ and } X = \begin{bmatrix} 0 & 0 & 0 & 0 \\ 0 & 0 & x_{2,3} & x_{2,4} \\ 0 & x_{2,3} & 0 & x_{3,4} \\ 0 & x_{2,4} & x_{3,4} & 0 \end{bmatrix},$$

where $a_{ij} \neq 0$ and the d_i are arbitrary real numbers. The equality $[A,X] = O$ is equivalent to

$$x_{2,3}[A,S_{2,3}] + x_{2,4}[A,S_{2,4}] + x_{3,4}[A,S_{3,4}] = O,$$

where

$$[A,S_{2,3}] = \begin{bmatrix} 0 & a_{1,3} & a_{1,2} & 0 \\ -a_{1,3} & 0 & d_2-d_3 & 0 \\ -a_{1,2} & d_3-d_2 & 0 & 0 \\ 0 & 0 & 0 & 0 \end{bmatrix},$$

$$[A,S_{2,4}] = \begin{bmatrix} 0 & a_{1,4} & 0 & a_{1,2} \\ -a_{1,4} & 0 & 0 & d_2-d_4 \\ 0 & 0 & 0 & 0 \\ -a_{1,2} & d_4-d_2 & 0 & 0 \end{bmatrix}, \text{ and}$$

$$[A,S_{3,4}] = \begin{bmatrix} 0 & 0 & a_{1,4} & a_{1,3} \\ 0 & 0 & 0 & 0 \\ -a_{1,4} & 0 & 0 & d_3-d_4 \\ -a_{1,3} & 0 & d_4-d_3 & 0 \end{bmatrix}.$$

By focusing on the first row of the equation, we get

$$\begin{bmatrix} x_{2,3} & x_{2,4} & x_{3,4} \end{bmatrix} \begin{bmatrix} 0 & a_{1,3} & a_{1,2} & 0 \\ 0 & a_{1,4} & 0 & a_{1,2} \\ 0 & 0 & a_{1,4} & a_{1,3} \end{bmatrix} = \mathbf{0}.$$

Since

$$\det \begin{bmatrix} a_{1,3} & a_{1,2} & 0 \\ a_{1,4} & 0 & a_{1,2} \\ 0 & a_{1,4} & a_{1,3} \end{bmatrix} = -2a_{1,3}a_{1,2}a_{1,4} \neq 0,$$

we know $x_{2,3} = x_{2,4} = x_{3,4} = 0$ and $X = O$ for any choice of $A \in \mathcal{S}(K_{1,3})$.

To make the idea of verifying the strong properties more concrete, the next step is to write down the matrix representation of the mentioned linear equations. For pairs (i,j) of integers, the *lexicographic ordering* is $(i_1,j_1) < (i_2,j_2)$ if and only $i_1 < i_2$ or ($i_1 = i_2$ and $j_1 < j_2$). Thus, we may order the entries of a matrix by the lexicographic ordering of its indices. Let $B = \begin{bmatrix} b_{ij} \end{bmatrix}$ be a matrix. Define $\operatorname{vec}_\square(B)$ as the vector in $\mathbb{R}^{n^2}$ that records all entries of B in lexicographic ordering. Define $\operatorname{vec}_\wedge(B)$ be the vector in $\mathbb{R}^{\binom{n}{2}}$ that records all the off-diagonal entries b_{ij} with $i < j$ in the lexicographic ordering. (The subscript $\square$ and $\wedge$ suggest that the vector records the whole and the strict upper-triangular parts of the matrix, respectively.) In general, when E is a set of pairs (i,j), $\mathrm{vec}_E(B)$ is defined as the vector in $\mathbb{R}^{|E|}$ that records the entries in E in the lexicographic ordering.

DEFINITION 4.18. [**31**] Let G be a graph on n vertices, $A \in \mathcal{S}(G)$, and $q = q(A)$. Let $\overline{E} = \{(i,j) : i < j, \{i,j\} \in E(\overline{G})\}$ and $p = |\overline{E}|$.

(1) The *SSP verification matrix* Ψ_S of A is the $p \times \binom{n}{2}$ matrix whose (i,j)-row is $\mathrm{vec}_\wedge([A, S_{ij}])$ for $(i,j) \in \overline{E}$.
(2) The *SMP verification matrix* Ψ_M of A is the $p \times \left(\binom{n}{2} + q\right)$ matrix obtained by augmenting Ψ_S with q columns $\mathrm{vec}_{\overline{E}}(A^k)$ for $k = 0, 1 \ldots, q-1$.
(3) The *SAP verification matrix* Ψ_A of A is the $p \times n^2$ matrix whose (i,j)-row is $\mathrm{vec}_\square(AS_{ij})$ for $(i,j) \in \overline{E}$.

For each of Ψ_S, Ψ_M, and Ψ_A, the rows are indexed by $\overline{E}$ in the lexicographic ordering.

PROPOSITION 4.19. [**31**] *Let G be a graph and $A \in \mathcal{S}(G)$. Then A has the SSP, SMP, and SAP respectively if and only if Ψ_S, Ψ_M, and Ψ_A have the full row-rank respectively.*

EXAMPLE 4.20. The matrix

$$A = \begin{bmatrix} 0 & 0 & 1 & 1 \\ 0 & 0 & 1 & -1 \\ 1 & 1 & 0 & 0 \\ 1 & -1 & 0 & 0 \end{bmatrix}$$

is a matrix in $\mathcal{S}(C_4)$ and was used to construct the spectrum $\{\sqrt{2}^{(2)}, -\sqrt{2}^{(2)}\}$ in Section 4.3. Here we verify that A has the SSP. Let X be the variable $\overline{C_4}$-matrix, which as two variables $x_{1,2}$ and $x_{3,4}$. Then $[A, X] = O$ is equivalent to

$$x_{1,2}[A, S_{1,2}] + x_{3,4}[A, S_{3,4}] = O.$$

Compute

$$[A, S_{1,2}] = \begin{bmatrix} 0 & 0 & -1 & 1 \\ 0 & 0 & -1 & -1 \\ 1 & 1 & 0 & 0 \\ -1 & 1 & 0 & 0 \end{bmatrix} \text{ and } [A, S_{3,4}] = \begin{bmatrix} 0 & 0 & 1 & 1 \\ 0 & 0 & -1 & 1 \\ -1 & 1 & 0 & 0 \\ -1 & -1 & 0 & 0 \end{bmatrix}.$$

By recording the (strictly) upper triangular part of these two matrices, the SSP verification matrix of A is

$$\Psi_S = \begin{bmatrix} 0 & -1 & 1 & -1 & -1 & 0 \\ 0 & 1 & 1 & -1 & 1 & 0 \end{bmatrix}.$$

Since Ψ_S has full row-rank, A has the SSP.

Calculating the verification matrices can be tedious when the matrix becomes large, but the calculation can be done by a computer in polynomial time. For example, one may use the Sage code, e.g., [**223**], to do this task.

As of now, we know that verifying the strong properties is equivalent to checking if the rank of the corresponding verification matrices is full, which can be easily done when a specified $A \in \mathcal{S}(G)$ is given. Once the SSP is verified, the matrix can be used to prove the existence of other matrices with the same spectrum and with more nonzero entries. This greatly accelerates the process of finding matrices for IEP-G. On the other hand, Example 4.17 shows the potential of treating entries in

$A \in \mathcal{S}(G)$ as unknown values, and the SSP of A is successfully verified regardless of the choice of $A \in \mathcal{S}(G)$. In Section 4.7, this idea will be formulated into theorems for granting the strong property to all matrices of some graph.

4.5. Matrix Liberation Lemma

If one looks back on the proof of Lemma 3.11, it is actually stronger than what it states—if A is a matrix with a strong property, then one may perturb "any" zero entries by an arbitrarily small value and make small adjustments to the nonzero or diagonal elements while preserving the desired spectral property. However, powerful results come with demanding assumptions, which do not always hold. For example, if one would like to increase the multiplicity of an eigenvalue λ of A, then one might consider using the matrix $A \oplus \begin{bmatrix}\lambda\end{bmatrix}$. Unfortunately, by Lemma 3.13, $A \oplus B$ has the SSP if and only if A and B have the SSP and do not share any common eigenvalue. From this point of view, the SSP seems less useful in increasing the multiplicity of an eigenvalue.

The Matrix Liberation Lemma can be viewed as an enhanced version of the Supergraph Lemma (Lemma 3.11). It gives a weaker condition (than the strong properties) so that "some" specified zero entries of a matrix can be perturbed while preserving the spectral properties. In fact, we have seen this kind of phenomenon before. In the statement of the Augmentation Lemma (Lemma 4.10), the set α depends on the original matrix and determines where we are allowed to add nonzero entries. In this section, we will introduce the Matrix Liberation Lemma and explain why the Augmentation Lemma is a special case of it.

Let G be a graph and $A \in \mathcal{S}(G)$. Recall that the rows of a verification matrix $\Psi(A)$ are indexed by the elements in $E(\overline{G})$, regardless the choice of Ψ_S, Ψ_M, or Ψ_A. Consequently, if $\mathbf{v}$ is a vector in the column space of $\Psi(A)$, then $\operatorname{supp}(\mathbf{v})$ can be viewed as a subset of the non-edges in G, and it is safe to say that $G + \operatorname{supp}(\mathbf{v})$ is the graph obtained from G by adding edges corresponding to $\operatorname{supp}(\mathbf{v})$.

LEMMA 4.21 (Matrix Liberation Lemma [**31**]). *Let G be a graph and $A \in \mathcal{S}(G)$. Let*

(1) $\Psi = \Psi_S(A)$,
(2) $\Psi = \Psi_M(A)$, *or*
(3) $\Psi = \Psi_A(A)$,

respectively. If $\mathbf{v}$ is a vector in the column space of Ψ such that the rows of Ψ corresponding to $E(\overline{G}) \setminus \operatorname{supp}(\mathbf{v})$ are linearly independent, then there is a matrix $A' \in \mathcal{S}(G + \operatorname{supp}(\mathbf{v}))$ such that

(1) $\operatorname{spec}(A') = \operatorname{spec}(A)$ *and A' has the SSP,*
(2) $\mathbf{m}(A') = \mathbf{m}(A)$ *and A' has the SMP, or*
(3) $\operatorname{rank} A' = \operatorname{rank} A$ *and A' has the SAP,*

respectively.

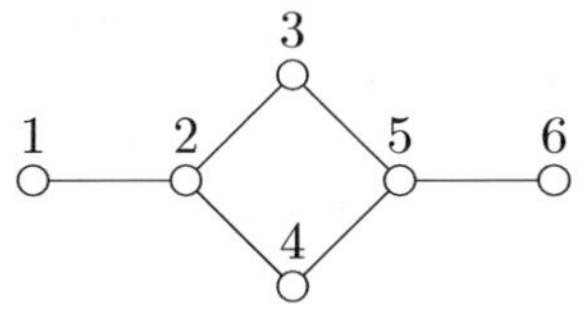

FIGURE 4.2. The graph G99

EXAMPLE 4.22. Let G be the graph in Figure 4.2. The matrix

$$A = \begin{bmatrix} 0 & 1 & 0 & 0 & 0 & 0 \\ 1 & 0 & -1 & 1 & 0 & 0 \\ 0 & -1 & 0 & 0 & 1 & 0 \\ 0 & 1 & 0 & 0 & 1 & 0 \\ 0 & 0 & 1 & 1 & 0 & 1 \\ 0 & 0 & 0 & 0 & 1 & 0 \end{bmatrix}$$

is in $\mathcal{S}(G)$ and has spectrum $\{1^{(2)}, \sqrt{3}^{(2)}, -\sqrt{3}^{(2)}\}$. Also, its SSP verification matrix $\Psi = \Psi_S(A)$ is

$$\begin{bmatrix} 1 & 0 & 0 & -1 & 0 & 1 & 0 & 0 & 0 & 0 & 0 & 0 & 0 & 0 & 0 \\ -1 & 0 & 0 & -1 & 0 & 0 & 1 & 0 & 0 & 0 & 0 & 0 & 0 & 0 & 0 \\ 0 & -1 & -1 & 0 & -1 & 0 & 0 & 1 & 0 & 0 & 0 & 0 & 0 & 0 & 0 \\ 0 & 0 & 0 & -1 & 0 & 0 & 0 & 0 & 1 & 0 & 0 & 0 & 0 & 0 & 0 \\ 0 & 0 & 0 & 1 & 0 & -1 & -1 & 0 & -1 & 0 & -1 & 0 & 1 & 0 & 0 \\ 0 & 0 & 0 & 0 & 1 & 0 & 0 & -1 & 0 & 0 & 0 & -1 & 0 & 1 & 0 \\ 0 & 0 & 0 & 0 & 0 & 1 & -1 & 0 & 0 & 0 & -1 & 0 & -1 & 0 & 0 \\ 0 & 0 & 0 & 0 & 0 & 0 & 0 & 0 & -1 & 0 & -1 & 0 & 0 & 0 & 1 \\ 0 & 0 & 0 & 0 & 0 & 0 & 0 & 0 & 1 & 0 & 0 & 0 & -1 & 0 & 1 \end{bmatrix},$$

where the rows are indexed by the non-edges $\{1,3\}$, $\{1,4\}$, $\{1,5\}$, $\{1,6\}$, $\{2,5\}$, $\{2,6\}$, $\{3,4\}$, $\{3,6\}$, and $\{4,6\}$ in order. Since Ψ only has rank 8, A does not have the SSP. The first column $\mathbf{v}$ of Ψ has $\operatorname{supp}(\mathbf{v}) = \{\{1,3\},\{1,4\}\}$. Meanwhile, the rows of Ψ corresponding to $E(\overline{G}) \setminus \{\{1,3\},\{1,4\}\}$, that is, all rows except for the first and the second, form a linearly independent set. Therefore, by Lemma 4.21 there is a matrix $A' \in \mathcal{S}(G + \operatorname{supp}(\mathbf{v}))$ with the same spectrum as A.

Indeed, not every supergraph of G with the same order allows the spectrum $\{1^{(2)}, \sqrt{3}^{(2)}, -\sqrt{3}^{(2)}\}$. For example, let H be the graph obtained from G by inserting the edge between 2 and 5. Then H has a unique shortest path on 4 vertices, which forbids any matrix in $\mathcal{S}(H)$ from having only three distinct eigenvalues.

Note that if a matrix A has the SSP, then $\Psi_S(A)$ has full row-rank. Consequently, the column space of $\Psi_S(A)$ occupies the whole space $\mathbb{R}^{\overline{m}}$ with $\overline{m} = |E(\overline{G})|$, and obviously the rows of $\Psi_S(A)$ corresponding to $E(\overline{G}) \setminus \operatorname{supp}(\mathbf{v})$ are linearly independent for any choice of $\mathbf{v}$. These observation is also true for the SMP and the SAP. Therefore, the Matrix Liberation Lemma is a generalization of Lemma 3.11.

EXAMPLE 4.23. Let G be a connected graph on $n \geq 3$ vertices and $A \in \mathcal{S}(G)$. Then $[A, S_{ij}]$ (respectively, AS_{ij}) is a nonzero matrix for each $\{i,j\} \in E(\overline{G})$. Therefore, for $\Psi = \Psi_S(A)$ (respectively, $\Psi = \Psi_A(A)$), there is no zero row in Ψ. Consequently, the column space of Ψ contains a nowhere-zero vector. By the Matrix

Liberation Lemma, there is a matrix $A' \in \mathcal{S}(K_n)$ with $\operatorname{spec}(A') = \operatorname{spec}(A)$ (respectively, $\operatorname{rank} A' = \operatorname{rank} A$).

One may examine the condition for Ψ having no zero row in more detail.

PROPOSITION 4.24. [**31**] *Let G be a graph and $A = [a_{ij}] \in \mathcal{S}(G)$.*

(1) *$\Psi_S(A)$ has no zero row if and only if $a_{ii} \neq a_{jj}$ for every pair of distinct isolated vertices i and j.*
(2) *$\Psi_M(A)$ has no zero row if and only if $\Psi_S(A)$ has no zero row.*
(3) *$\Psi_A(A)$ has no zero row if and only if A contains at most one zero column.*

The Matrix Liberation Lemma can be rephrased in terms of the left kernel of the verification matrix. Recall that the orthogonal complement of the left kernel of Ψ is the column space of Ψ for any matrix Ψ.

COROLLARY 4.25 (Matrix Liberation Lemma, kernel version [**31**]). *Let G be a graph and $A \in \mathcal{S}(G)$. Let*

(1) $\Psi = \Psi_S(A)$,
(2) $\Psi = \Psi_M(A)$, *or*
(3) $\Psi = \Psi_A(A)$,

respectively. Let $\ker(\Psi^\top)$ be the left kernel of Ψ. If $\mathbf{v}$ is a vector in $\ker(\Psi^\top)^\perp$ such that $\operatorname{supp}(\mathbf{x}) \cap \operatorname{supp}(\mathbf{v}) \neq \emptyset$ for any $\mathbf{x} \in \ker(\Psi^\top)$, then there is a matrix $A' \in \mathcal{S}(G + \operatorname{supp}(\mathbf{v}))$ such that

(1) $\operatorname{spec}(A') = \operatorname{spec}(A)$ *and A' has the SSP,*
(2) $\mathbf{m}(A') = \mathbf{m}(A)$ *and A' has the SMP, or*
(3) $\operatorname{rank} A' = \operatorname{rank} A$ *and A' has the SAP,*

respectively.

PROPOSITION 4.26. [**31**] *Let G be a graph, $\overline{E} = E(\overline{G})$, and $A \in \mathcal{S}(G)$. Let Ψ be the SSP, SMP, or SAP verification matrix of A, respectively. Then the left kernel $\ker(\Psi^\top)$ of Ψ is*

(1) $\{\operatorname{vec}_{\overline{E}}(X) : A \circ X = I \circ X = [A, X] = O\}$,
(2) $\{\operatorname{vec}_{\overline{E}}(X) : A \circ X = I \circ X = [A, X] = O, \operatorname{tr}(A^k X) = 0, k = 0, 1, \ldots, q(A)\}$, *or*
(3) $\{\operatorname{vec}_{\overline{E}}(X) : A \circ X = I \circ X = AX = O\}$,

respectively.

PROOF. These characterizations follow immediately from Proposition 4.16 and Definition 4.18. □

With Corollary 4.25 and Proposition 4.26, we are ready to show that the Augmentation Lemma follows from applying the Matrix Liberation Lemma to the matrix $A_\lambda = A \oplus [\lambda]$, where λ is an eigenvalue of A.

PROOF OF LEMMA 4.10. Let G be a graph on n vertices and $A \in \mathcal{S}(G)$ with the SSP. Suppose λ is an eigenvalue of A of multiplicity k and E_λ is the corresponding eigenspace. We will perturb the matrix $A_\lambda = A \oplus [\lambda]$ without changing its spectrum.

Let $\Psi = \Psi_S(A_\lambda)$. By Proposition 4.26, $\mathbf{x} \in \ker(\Psi^\top)$ if and only if $\mathbf{x} = \operatorname{vec}_{\overline{E}}(X)$ for some X that satisfies $A \circ X = O$, $I \circ X = O$, and $[A, X] = O$, where $\overline{E} = E(\overline{G})$.

Suppose X is a symmetric matrix that satisfies $A \circ X = O$, $I \circ X = O$, and $[A, X] = O$. We may write $X = \begin{bmatrix} Y & \mathbf{b} \\ \mathbf{b}^\top & 0 \end{bmatrix}$ and have

$$[A_\lambda, X] = \begin{bmatrix} AY - YA & A\mathbf{b} - \lambda\mathbf{b} \\ (A\mathbf{b} - \lambda\mathbf{b})^\top & 0 \end{bmatrix} = O.$$

Thus, $Y = O$ since A has the SSP. Therefore,

$$\ker(\Psi^\top) = \left\{ \mathrm{vec}_{\overline{E}}(X) : X = \begin{bmatrix} O & \mathbf{b} \\ \mathbf{b}^\top & 0 \end{bmatrix}, \mathbf{b} \in E_\lambda \right\}$$

and $\dim \ker(\Psi^\top) = k$.

Suppose $\alpha \subseteq V(G)$ is a set of cardinality $k + 1$ such that $|\operatorname{supp}(\mathbf{b}) \cap \alpha| \geq 2$ for each $\mathbf{b} \in E_\lambda$. Since $\{\mathbf{v} \in \mathbb{R}^n : \operatorname{supp}(\mathbf{v}) \subseteq \alpha\}$ has dimension $k + 1$ and $\dim E_\lambda = k$, there is a vector $\mathbf{b}_0$ in $E_\lambda^\perp$ with $\operatorname{supp}(\mathbf{b}_0) \subseteq \alpha$. Next we claim that $\operatorname{supp}(\mathbf{b}_0) = \alpha$. For each 2-subset β of α, since $\dim E_\lambda = k$ and $|\alpha \setminus \beta| = k - 1$, there is a nonzero vector $\mathbf{b}_\beta \in E_\lambda$ such that $\mathbf{b}_\beta$ vanishes on $\alpha \setminus \beta$. Then the assumption $|\operatorname{supp}(\mathbf{b}_\beta) \cap \alpha| = 2$ implies $\operatorname{supp}(\mathbf{b}_\beta) \cap \alpha = \beta$. As a consequence, $\mathbf{b}_\beta \cdot \mathbf{b}_0 = 0$ implies $\mathbf{b}_0$ is nonzero on each of the entries in β. Since the argument works for each 2-subset β of α, $\operatorname{supp}(\mathbf{b}_0) = \alpha$.

Let $X_0 = \begin{bmatrix} O & \mathbf{b}_0 \\ \mathbf{b}_0^\top & 0 \end{bmatrix}$ and $\mathbf{v}_0 = \mathrm{vec}_{\overline{E}}(X_0)$. Then $\mathbf{v}_0 \in \ker(\Psi^\top)^\perp$ and $\operatorname{supp}(\mathbf{v}_0) = \{(i, n+1) : i \in \alpha\} \subseteq \overline{E}$. Meanwhile, if a vector $\mathbf{v}$ is in $\ker(\Psi^\top)$, then $\mathbf{v} = \mathrm{vec}_{\overline{E}}(X)$ with $X = \begin{bmatrix} O & \mathbf{b} \\ \mathbf{b}^\top & 0 \end{bmatrix}$ for some $\mathbf{b} \in E_\lambda$. Therefore, $\operatorname{supp}(\mathbf{b}) \cap \operatorname{supp}(\mathbf{b}_0) = \operatorname{supp}(\mathbf{b}) \cap \alpha \neq \emptyset$, and $\operatorname{supp}(\mathbf{v}) \cap \operatorname{supp}(\mathbf{v}_0) \neq \emptyset$.

Let $H = (G \sqcup K_1) + \alpha$. According to Corollary 4.25, there is a matrix $A' \in \mathcal{S}(H)$ that is a perturbation of A_λ such that $\operatorname{spec}(A') = \operatorname{spec}(A_\lambda) = \operatorname{spec}(A) \cup \{\lambda\}$. □

4.6. Minor Monotonicity

One of the key factors that makes the Colin de Verdière number $\mu(G)$ so powerful is minor-monotonicity. It is natural to ask if there is some type of monotonicity for ordered multiplicity lists or for spectra. Obtaining a minor involves three graph operations, namely, edge deletion, isolated vertex deletion, and edge contraction. The case of edge deletion is handled by Lemma 3.11, which provides the results for SSP and SMP analogous to that of SAP. The case of isolated vertex deletion is addressed by Lemma 3.15, in which adding a new vertex is corresponding to adding a new eigenvalue of multiplicity one. It seems that inserting a new eigenvalue is a reasonable approach for building the minor-monotonicity.

If G is a subgraph of H, then SSP ensures that the spectra achievable by an SSP matrix in $\mathcal{S}(G)$ is also achievable in $\mathcal{S}(H)$, with possibly adding some new simple eigenvalues. Following the same thoughts, if G is obtained from H by contracting an edge, then we hope that a spectrum in $\mathcal{S}(G)$ can be extended to a spectrum in $\mathcal{S}(H)$ by inserting a new simple eigenvalue. This hope is met by the following result.

LEMMA 4.27 (Decontraction Lemma [**27, 31**]). *Let G and H be graphs such that G is obtained from H by contracting an edge.*

(1) *If $A \in \mathcal{S}(G)$ has the SSP, then there is a real number L such that for any λ with $|\lambda| \geq L$ there is a matrix $A' \in \mathcal{S}(H)$ with the SSP and* $\operatorname{spec}(A') = \operatorname{spec}(A) \cup \{\lambda\}$.

(2) *If* $A \in \mathcal{S}(G)$ *has the SMP, then there is a matrix* $A' \in \mathcal{S}(H)$ *such that* $\mathbf{m}(A')$ *is obtained from* $\mathbf{m}(A)$ *by appending a* 1 *at the beginning or at the end.*
(3) *If* $A \in \mathcal{S}(G)$ *has the SAP, then there is a matrix* $A' \in \mathcal{S}(G)$ *such that* $\operatorname{rank} A' = \operatorname{rank} A + 1$, *and equivalently,* $\operatorname{null} A' = \operatorname{null} A$.

There are two approaches to prove the Decontraction Lemma. For the case of SAP stated in Lemma 4.27(3), it relies on the transversality of three manifolds, which preserve the pattern, the rank, the rank of a submatrix, respectively. A comprehensive proof including the required background is given in [**27**], where the monotonicity of $\xi(G)$ was introduced. This proof was motivated by Colin de Verdière's proof [**88, 89**] on the minor monotonicity of $\mu(G)$, which can also be found in [**195**]. On the other hand, the proof of Lemma 4.27(1) and (2) uses the Givens rotation to perturb the matrix and applies Brouwer's Fixed Point Theorem to assert the existence of the desired matrices. This proof appears in [**31**]. Since the proofs of the Decontraction Lemma can be technical, we do not include them in this book.

The Decontraction Lemma enforces extra conditions to the new eigenvalue; that is, the new eigenvalue is sufficiently large. This condition seems annoying but is required. For example, C_4 allows a matrix A with $\mathbf{m}(A) = (2, 2)$, but C_5 does not allow any matrix with the ordered multiplicity list $(2, 1, 2)$. (This is because C_5 is a unicyclic graph with an odd cycle; see e.g., [**31**, Lemma 3.3].)

Now that the results for SSP and SMP have been established for edge deletion, isolated vertex deletion, and edge contraction, we are ready to introduce the Minor Monotonicity Theorem for SSP and SMP.

THEOREM 4.28 (Minor Monotonicity Theorem [**27, 31**]). *Let* G *and* H *be graphs such that* G *is a minor of* H. *We may assume that* G *is obtained from* H *by contracting* r *edges, deleting* s *isolated vertices, and deleting some edges.*

(1) *If* $A \in \mathcal{S}(G)$ *has the SSP, then there is a matrix* $A' \in \mathcal{S}(H)$ *such that* $\operatorname{spec}(A')$ *is obtained from* $\operatorname{spec}(A)$ *by inserting* $r + s$ *new simple eigenvalues, of which* r *simple eigenvalues should be large enough while the other* s *simple eigenvalues can be arbitrarily chosen.*
(2) *If* $A \in \mathcal{S}(G)$ *has the SMP, then there is a matrix* $A' \in \mathcal{S}(H)$ *such that* $\mathbf{m}(A')$ *is obtained from* $\mathbf{m}(A)$ *by inserting* $r + s$ *ones, of which at most* s *ones can be inserted between elements in* $\mathbf{m}(A)$.
(3) *If* $A \in \mathcal{S}(G)$ *has the SAP, then there is a matrix* $A' \in \mathcal{S}(G)$ *such that* $\operatorname{rank} A' = \operatorname{rank} A + r + s$, *and equivalently,* $\operatorname{null} A' = \operatorname{null} A$.

According to the Minor Monotonicity Theorem, if a minor G of a graph H allows two multiple eigenvalues with the strong property, then H also allows two multiple eigenvalues. Figure 4.3 lists some graphs that allows two multiple eigenvalues. As a consequence, a graph that allows at most one multiple eigenvalue does not have any graph in Figure 4.3 as a minor. Indeed, these graphs are the complete collection of forbidden minors for the graphs allowing at most one multiple eigenvalue [**31**].

To state the results, we need to define some graphs. Recall that, as defined in Appendix A, a generalized star $S(n_1, \dots, n_t)$, $t \geq 3$, $n_i \geq 1$ for $i = 1, \dots, t$, is the graph obtained from $P_{n_1} \sqcup \cdots \sqcup P_{n_t}$ by inserting a vertex that is joined to one endpoint to each of the paths. A *generalized* 3*-sun* is a graph obtained from

$K_3 \sqcup P_{n_1} \sqcup P_{n_2} \sqcup P_{n_3}$, $n_i \geq 1$ for $i = 1, 2, 3$, by adding an edge between each of the three vertices of the K_3 and one of the endpoints in each of the three paths.

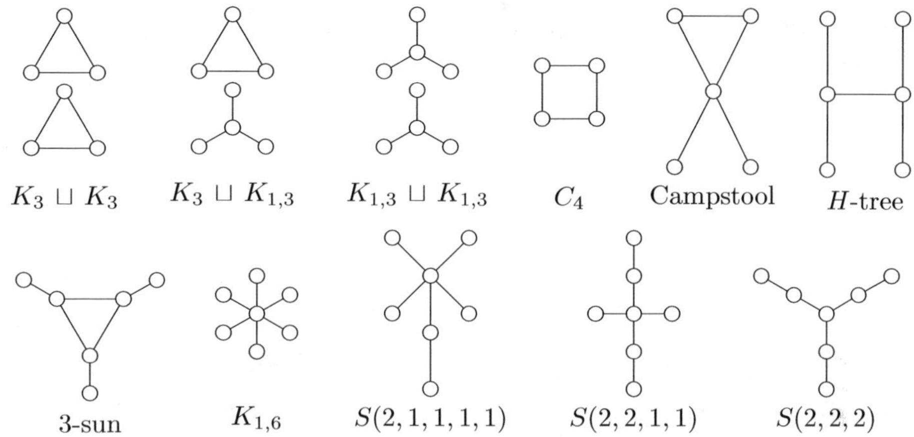

FIGURE 4.3. Forbidden minors for graphs allowing at most one multiple eigenvalue.

THEOREM 4.29. **[31]** *Let G be a connected graph. The following are equivalent.*

(1) *Every $A \in \mathcal{S}(G)$ has at most one multiple eigenvalue.*
(2) *Every $A \in \mathcal{S}(G)$ with the SMP has at most one multiple eigenvalue.*
(3) *Every $A \in \mathcal{S}(G)$ with the SSP has at most one multiple eigenvalue.*
(4) *None of the eleven graphs in Figure 4.3 is a minor of G.*
(5) *The graph G is either a path with an extra edge joining two vertices of distance 2, $K_{1,5}$, $S(1,1,1,a)$ with $a \geq 1$, $S(1,a,b)$ with $a, b \geq 1$, or a path.*

COROLLARY 4.30. **[31]** *Let G be a graph. The following are equivalent.*

(1) *Every $A \in \mathcal{S}(G)$ with the SMP has at most one multiple eigenvalue.*
(2) *Every $A \in \mathcal{S}(G)$ with the SSP has at most one multiple eigenvalue.*
(3) *None of the eleven graphs in Figure 4.3 is a minor of G.*
(4) *The graph G the disjoint union of G_1 and paths, where G_1 is either a path with an extra edge joining two vertices of distance 2, $K_{1,5}$, $S(1,1,1,a)$ with $a \geq 1$, $S(1,a,b)$ with $a, b \geq 1$, or a path.*

Among the eleven graphs in Figure 4.3, the two graphs $K_3 \sqcup K_{1,3}$ and $K_{1,3} \sqcup K_{1,3}$ are in fact not required for the connected case (Theorem 4.29), but they are required in the proof of Corollary 4.30.

Suppose A is a symmetric matrix with $\operatorname{spec}(A) = \{\lambda_1^{(m_1)}, \ldots, \lambda_q^{(m_q)}\}$ such that $\lambda_1 < \cdots < \lambda_q$. We say A has *consecutive multiple eigenvalues* if both λ_i and λ_{i+1} are multiple for some $i = 1, \ldots, q-1$. Suppose G is a minor of H and $A \in \mathcal{S}(G)$ has the SSP (or the SMP). The Matrix Decontraction Lemma (Lemma 4.27) allows us to insert new simple eigenvalues outside of the interval $[\min \Lambda, \max \Lambda]$, where $\Lambda = \operatorname{spec}(A)$. Therefore, if A has consecutive multiple eigenvalues, then H also allows a matrix with the SSP (or the SMP) that has consecutive multiple eigenvalues.

Let $\mathcal{F}_2' = \{K_3 \sqcup K_3, K_3 \sqcup K_{1,3}, K_{1,3} \sqcup K_{1,3}, C_4, \mathsf{Camp}, H\text{-graph}\}$, where Camp is the Campstool graph shown in Figure 4.3. Each graph in $\mathcal{F}_2'$ allows a matrix with consecutive multiple eigenvalues with the SMP. The results below show that $\mathcal{F}_2'$ is the complete list of forbidden minors.

THEOREM 4.31. [**31**] *Let G be a connected graph. The following are equivalent.*

(1) *Every $A \in \mathcal{S}(G)$ has no consecutive multiple eigenvalues.*
(2) *Every $A \in \mathcal{S}(G)$ with the SMP has no consecutive multiple eigenvalues.*
(3) *Every $A \in \mathcal{S}(G)$ with the SSP has no consecutive multiple eigenvalues.*
(4) *None of the six graphs in $\mathcal{F}_2'$ is a minor of G.*
(5) *The graph G is either a generalized star, a generalized 3-sun, a path with an extra edge joining two vertices of distance 2, or a path.*

COROLLARY 4.32. [**31**] *Let G be a graph. The following are equivalent.*

(1) *Every $A \in \mathcal{S}(G)$ with the SMP has no consecutive multiple eigenvalues.*
(2) *Every $A \in \mathcal{S}(G)$ with the SSP has no consecutive multiple eigenvalues.*
(3) *None of the six graphs in $\mathcal{F}_2'$ is a minor of G.*
(4) *The graph G the disjoint union of G_1 and paths, where G_1 is either is either a generalized star, a generalized 3-sun, a path with an extra edge joining two vertices of distance 2, or a path.*

Again, the graphs $K_3 \sqcup K_{1,3}$ and $K_{1,3} \sqcup K_{1,3}$ in $\mathcal{F}_2'$ are only required for Corollary 4.32 but not for Theorem 4.31.

We have seen that a parameter with minor monotonicity often has nice characterizations using forbidden minors. We remark that although many graph parameters are not minor monotone, one may take the minor monotone ceiling or the minor monotone floor of the parameter to generate associated parameters with minor monotonicity. This idea will be introduced in Section 9.8.

4.7. Guaranteed Strong Properties

By Example 4.17, we know that there are some graphs G such that every matrix $A \in \mathcal{S}(G)$ has the SSP. This kind of behavior also occurs among other strong properties. The key idea is to show that the verification matrix has full row-rank, regardless of the choice of $A \in \mathcal{S}(G)$. Here we start with an example on P_4.

EXAMPLE 4.33. Consider the graph P_4. One may write a matrix $A \in \mathcal{S}(P_4)$ as

$$\begin{bmatrix} d_1 & a_{1,2} & 0 & 0 \\ a_{1,2} & d_2 & a_{2,3} & 0 \\ 0 & a_{2,3} & d_3 & a_{3,4} \\ 0 & 0 & a_{3,4} & d_4 \end{bmatrix},$$

where a_{ij}'s are nonzero while d_i's are real numbers. Before knowing the values of the entries in A, we still can compute the SAP verification matrix $\Psi_A(A)$ as

$$\begin{bmatrix} 0 & 0 & d_1 & 0 & a_{2,3} & 0 & a_{1,2} & 0 & d_3 & 0 & 0 & 0 & a_{3,4} & 0 & 0 & 0 \\ 0 & 0 & 0 & d_1 & 0 & 0 & 0 & a_{1,2} & a_{3,4} & 0 & 0 & 0 & d_4 & 0 & 0 & 0 \\ 0 & 0 & 0 & a_{1,2} & 0 & 0 & 0 & d_2 & 0 & a_{3,4} & 0 & a_{2,3} & 0 & d_4 & 0 & 0 \end{bmatrix}.$$

It is enough to focus on the submatrix M of $\Psi_A(A)$ on the columns 5, 9, and 10. Thus,

$$M = \begin{bmatrix} a_{2,3} & d_3 & 0 \\ 0 & a_{3,4} & 0 \\ 0 & 0 & a_{3,4} \end{bmatrix}.$$

Since $a_{2,3}$ and $a_{3,4}$ are nonzero, both M and $\Psi_A(A)$ has full row-rank regardless the choices of $A \in \mathcal{S}(P_4)$. Therefore, every $A \in \mathcal{S}(P_4)$ has the SAP.

Let G be a graph on n vertices, $A = \begin{bmatrix} a_{ij} \end{bmatrix} \in \mathcal{S}(G)$, and $\overline{E} = E(\overline{G})$. Recall that the SAP verification matrix is simply a matrix representation of the equation $\sum_{\{u,v\} \in \overline{E}} x_{uv} AS_{uv} = O$, which can be written as the system of linear equations

$$\sum_{\{u,v\} \in \overline{E}} x_{uv} \left(AS_{uv}\right)_{ij} = 0$$

for all $i, j \in \{1, \ldots, n\}$. The rows of the verification matrix correspond to the variables x_{uv} for $\{u, v\} \in \overline{E}$, while each column represents an equation. Suppose for some i, j the corresponding equation has only one variable with nonzero coefficient. Then we know this variable must be zero. Once a variable is known to be zero, it may be possible to find other equation with the same property, which forces one more variable to be zero. Inductively, if all variables are zero, then we know the left kernel of the verification matrix is trivial and every $A \in \mathcal{S}(G)$ has the SAP. This idea is essentially playing the zero forcing game on the verification matrix.

Let G be a graph whose non-edges are colored blue or white. A *pendulum* on G (with respect to the current coloring on non-edges) is a triple (k, i, j) such that $\{i, j\}$ is an edge, $\{k, j\}$ is a white non-edge, and $\{k, i\}$ is either an edge or a blue non-edge; in particular, we say it is a pendulum starting from the tuple (k, i). As we will see immediately, if a pendulum occurs and some conditions hold, then the blue non-edge (or edge) ki will swing to the right and make kj blue, as shown in Figure 4.4.

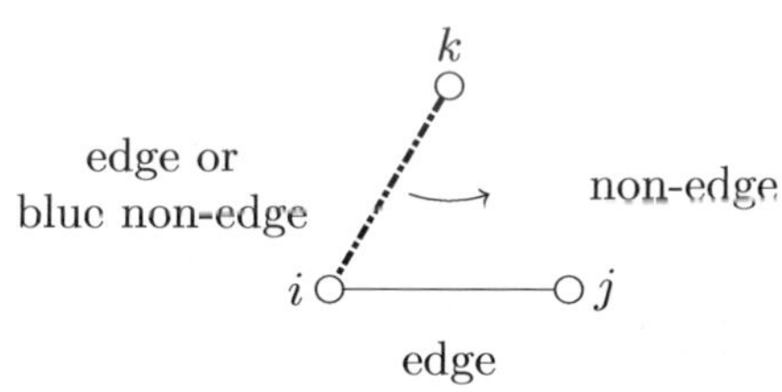

FIGURE 4.4. A pendulum

DEFINITION 4.34. The *SAP zero forcing game* on a graph G is a color change game where the non-edges $\overline{E} = E(\overline{G})$ are colored blue or white according to the Z_{SAP}*-color change rule*:

(1) If for some tuple (k, i) there is a unique j such that (k, i, j) is a pendulum, then the non-edge kj may turn blue. This action is denoted by $k \colon i \to j$.
(2) Let G_W be the graph on $V(G)$ whose edges are the white non-edges. If for some vertex k, the induced subgraph $G_W[N_G(k)]$ contains a component that is an odd cycle C, then all edges in $E(C)$ may turn blue. This action is denoted as $k \to C$.

A set of non-edges $B \subseteq \overline{E}$ is a Z_{SAP}*-zero forcing set* if starting with B as blue while other non-edges white, every non-edges in $\overline{E}$ turns blue by repeated applications of the Z_{SAP}-color change rule. The Z_{SAP}*-zero forcing number* $Z_{\text{SAP}}(G)$ is the cardinality of the minimum Z_{SAP}-zero forcing set.

THEOREM 4.35. [**221**] *If G is a graph with $Z_{SAP}(G) = 0$, then every matrix $A \in \mathcal{S}(G)$ has the SAP.*

COROLLARY 4.36. [**221**] *If G is a graph with $Z_{SAP}(G) = 0$, then $M(G) = \xi(G)$ and $M_+(G) = \nu(G)$.*

EXAMPLE 4.37. Let G be the graph in Figure 4.2. One may perform the following forces sequentially.

$$\begin{array}{ll} 2\colon 3 \to 5 & 3\colon 5 \to 6 \\ 2\colon 5 \to 6 & 4\colon 5 \to 6 \\ 5\colon 2 \to 1 & 1\colon 5 \to 6 \\ 2 \to (1,3,4) & \end{array}$$

Thus, every non-edge turns blue without setting any initial blue non-edges. Therefore, $Z_{SAP}(G) = 0$, and every $A \in \mathcal{S}(G)$ has the SAP. In contrast, one may verify that the matrix

$$\begin{bmatrix} 0 & 1 & 0 & 0 & 0 & 0 \\ 1 & 0 & 1 & 2 & 0 & 0 \\ 0 & 1 & 0 & 0 & 1 & 0 \\ 0 & 2 & 0 & 0 & 1 & 0 \\ 0 & 0 & 1 & 1 & 0 & 1 \\ 0 & 0 & 0 & 0 & 1 & 0 \end{bmatrix}$$

has neither the SMP nor the SSP.

Let G be a graph whose non-edges are colored blue or white. A tuple (k,i) is said to be *inactive* if there is no pendulum in G starting with (k,i). If ki is an edge or a blue non-edge, then the *reachable set* $R(ki)$ is the set of vertices j such that either (k,i,j) or (i,k,j) is a pendulum. Let Y_h, $h \geq 1$, be the generalized star $S(h,h,h)$. Recall that $G^{(r)}$ and G^r are the strong power and the power of G, respectively.

DEFINITION 4.38. The *SSP zero forcing game* on a graph G is a color change game where the non-edges $\overline{E} = E(\overline{G})$ are colored blue or white according to the Z_{SSP}*-color change rule*:

(1) If for some tuple (k,i) there is a unique j such that (k,i,j) is a pendulum and the tuple (i,k) is inactive, then the non-edge kj may turn blue. This action is denoted by $k\colon i \to j$.
(2) Let G_W be the graph on $V(G)$ whose edges are the white non-edges. If for some vertex k, the induced subgraph $G_W[N_G(k)]$ contains an odd cycle as a component, and $R(ki) \subseteq V(C)$ for every $i \in V(C)$, then all edges in $E(C)$ may turn blue. This action is denoted as $k \to C$.
(3) If Y_h is a subgraph of G such that $E(Y_h^h)$ contains edges or blue non-edges, $E(Y_h^{(h+1)})$ contains only white non-edges, and $R(e) \subseteq V(Y_h)$ for all $e \in E(Y_h^{(h)})$, then all edges in $E(Y_h^{(h+1)})$ may turn blue. This action is denoted as $Y_h^{(h)} \to Y_h^{(h+1)}$.

A set of non-edges $B \subseteq \overline{E}$ is a Z_{SSP}*-zero forcing set* if starting with B as blue while other non-edges white, every non-edges in $\overline{E}$ turns blue by repeated applications of the Z_{SSP}-color change rule. The Z_{SSP}*-zero forcing number* $Z_{SSP}(G)$ is the cardinality of the minimum Z_{SSP}-zero forcing set.

THEOREM 4.39. [**224**] *If G is a graph with* $\mathrm{Z}_{\mathrm{SSP}}(G) = 0$*, then every matrix* $A \in \mathcal{S}(G)$ *has the SSP.*

Let $\mathcal{G}^{\mathrm{SAP}}$, $\mathcal{G}^{\mathrm{SMP}}$, and $\mathcal{G}^{\mathrm{SSP}}$, respectively, be the family of graphs G such that every $A \in \mathcal{S}(G)$ has the SAP, SMP, and SSP, respectively. What we had is $\mathrm{Z}_{\mathrm{SAP}}(G) = 0 \implies G \in \mathcal{G}^{\mathrm{SAP}}$ and $\mathrm{Z}_{\mathrm{SSP}}(G) = 0 \implies G \in \mathcal{G}^{\mathrm{SSP}}$. There are examples G in $\mathcal{G}^{\mathrm{SSP}}$ with $\mathrm{Z}_{\mathrm{SSP}}(G) \neq 0$ [**224**]. However, there is no example known for $G \in \mathcal{G}^{\mathrm{SAP}}$ and $\mathrm{Z}_{\mathrm{SAP}}(G) \neq 0$.

Also, $\mathcal{G}^{\mathrm{SSP}} \subseteq \mathcal{G}^{\mathrm{SMP}} \subseteq \mathcal{G}^{\mathrm{SAP}}$ by their definitions. Example 4.37 gives a graph in $\mathcal{G}^{\mathrm{SAP}}$ but not in $\mathcal{G}^{\mathrm{SMP}}$. One may ask if there are some graphs that is in $\mathcal{G}^{\mathrm{SMP}}$ but not in $\mathcal{G}^{\mathrm{SSP}}$. Indeed, C_4 and $\overline{P_6}$ are graphs in $\mathcal{G}^{\mathrm{SMP}}$ but not in $\mathcal{G}^{\mathrm{SSP}}$. The argument for $\overline{P_6} \in \mathcal{G}^{\mathrm{SMP}} \setminus \mathcal{G}^{\mathrm{SSP}}$ can be found in [**224**], while the similar argument also applies to C_4.

This section has discussed some graphs G such that every matrix in $\mathcal{S}(G)$ has some strong property. This kind of behavior had been observed for another family of matrices associated with a graph. A generalized Laplacian matrix of G is a matrix in $\mathcal{S}(G)$ whose off-diagonal entries are either zero or negative. If G is a path, a 2-connected outer-planar graph, a 3-connected planar graph, or a 4-connected linklessly embeddable graph, then it is known that every generalized Laplacian matrix of G with exactly one negative eigenvalue has the SAP [**190, 252**]. One may expect that every $\mu(G)$-connected graph will have the same property, but not much is known beyond $\mu(G) = 4$.

CHAPTER 5

Theoretical Underpinnings of the Strong Properties

In this chapter, we prove many results mentioned in the earlier chapters, such as the Implicit Function Theorems (Theorems 3.2 and 3.3), the Supergraph Lemma (Lemma 3.11), and the Matrix Liberation Lemma (Lemma 4.21). Beyond that, we will prove the Bifurcation Lemma (Lemma 5.31) in Section 5.5. A goal here is to illustrate the proof techniques that may prove useful in developing additional strong properties.

We remark that the alternative argument for the Supergraph Lemma in Section 5.4 and the proof of the Bifurcation Lemma in Section 5.5 only rely on the perturbations defined in Section 5.2 and Theorem 5.18 and can be read without further understanding of the manifold theory. Therefore, Theorems 5.19 and 5.20 are not required for this purpose.

For people who wish to follow the manifold approach, we prove in Section 5.2 that the sets $\mathcal{E}_A$, $\mathcal{U}_A$, and $\mathcal{R}_A$ nearby A are submanifolds locally in the space of symmetric matrices. Although we do not prove they are manifolds globally, being manifolds locally is enough for the use of Theorems 3.2 and 3.3.

5.1. Spaces of Matrices

We start with introducing various vector spaces composed of matrices of interest and set up the standard basis and the inner product to be used for each of them throughout the chapter. The matrices considered here are over $\mathbb{R}$.

Before getting into the these spaces, we recall some standard results to clarify the meaning of our notations. Let U be a finite-dimensional vector space and $\alpha = \{\mathbf{u}_1, \ldots, \mathbf{u}_d\}$ a basis of U. Then the *vector representation* of the vector $\mathbf{u} \in U$ is

$$[\mathbf{u}]_\alpha = \begin{bmatrix} c_1 \\ \vdots \\ c_d \end{bmatrix},$$

where $\mathbf{u} = c_1\mathbf{u}_1 + \cdots + c_d\mathbf{u}_d$. Also, let W be a finite-dimensional vector space with a basis β. Let $f\colon U \to W$ be a linear transformation. Then the *matrix representation* of f is

$$[f]_\alpha^\beta = \begin{bmatrix} | & & | \\ [f(\mathbf{u}_1)]_\beta & \cdots & [f(\mathbf{u}_d)]_\beta \\ | & & | \end{bmatrix}.$$

Note $[f]_\alpha^\beta[\mathbf{u}]_\alpha = [f(\mathbf{u})]_\beta$ for all $\mathbf{u} \in U$.

The space of $n \times n$ matrices is denoted by $\mathbb{R}^{n\times n}$. Its dimension is n^2. Let E_{ij} be the 0,1-matrix whose i,j-entry is 1 while all other entries are zero. The standard basis of $\mathbb{R}^{n\times n}$ as $\beta = \{E_{ij}\}_{1\leq i,j\leq n}$, where the vectors are ordered by the lexicographical order of the indices. The bilinear form $\langle A, B\rangle_{\mathbb{R}^{n\times n}} = \operatorname{tr}(B^\top A)$ is an inner product on $\mathbb{R}^{n\times n}$. Note the dot product $[A]_\beta \cdot [B]_\beta$ equals the $\langle A, B\rangle_{\mathbb{R}^{n\times n}}$ for all $A, B \in \mathbb{R}^{n\times n}$.

The space of $n \times n$ symmetric matrices is denoted by $S_n(\mathbb{R})$. Its dimension is $\binom{n+1}{2}$. Let $S_{ij} = E_{ij} + E_{ji}$ if $i \neq j$ and $S_{ii} = \sqrt{2}E_{ii}$. We set the standard basis of $S_n(\mathbb{R})$ to be $\beta = \{S_{ij}\}_{1\leq i\leq j\leq n}$, following the lexicographical order. The inner product of $A, B \in S_n(\mathbb{R})$ is defined as $\langle A, B\rangle_{S_n(\mathbb{R})} = \frac{1}{2}\operatorname{tr}(B^\top A)$. Therefore, $[A]_\beta \cdot [B]_\beta = \langle A, B\rangle_{S_n(\mathbb{R})}$. (This justifies why we set $S_{ii} = \sqrt{2}E_{ii}$.) Note that the inner product of two symmetric matrices and the inner product of two general matrices are off by a scalar, but they are equivalent in terms of orthogonality. We take this setting is because we only focus on the upper triangular part of the symmetric matrices.

The space of $n\times n$ skew symmetric matrices is denoted by $S_n^-(R)$. Its dimension is $\binom{n}{2}$. Let $S_{ij}^- = E_{ij} - E_{ji}$ (so $S_{ii}^- = O$). We set the standard basis of $S_n^-(R)$ as $\beta = \{S_{ij}^-\}_{1\leq i<j\leq n}$, following the lexicographical order. The inner product of $A, B \in S_n^-(R)$ is defined as $\langle A, B\rangle_{S_n^-(R)} = \frac{1}{2}\operatorname{tr}(B^\top A)$. The reason for the scalar $\frac{1}{2}$ is again because we only focus on the (strict) upper triangular part.

Naturally, if α is a subset of any of these three standard basis, then it is called the standard basis of $\operatorname{span}(\alpha)$. For example, $\{S_{ij} : i = j \text{ or } \{i,j\} \in E(G)\}$ is the standard basis of $\mathcal{S}^{\mathrm{cl}}(G)$.

5.2. Manifolds and Tangent Spaces

Let $U \subseteq \mathbb{R}^n$ and $W \subseteq \mathbb{R}^m$ be open subsets. Recall that when $F\colon U \to W$ is a differentiable function defined by $F(\mathbf{x}) = (f_1(\mathbf{x}), \ldots, f_m(\mathbf{x})) \in \mathbb{R}^m$ for $\mathbf{x} = (x_1, \ldots, x_n) \in \mathbb{R}^n$, the derivative of F at a given point $\mathbf{x}_0$ is

$$\dot{F}\Big|_{\mathbf{x}=\mathbf{x}_0} = \left[\frac{\partial f_j}{\partial x_i}\right]\Big|_{\mathbf{x}=\mathbf{x}_0},$$

which is also known as the Jacobian matrix of F at $\mathbf{x}_0$. (When the context is clear, sometimes we just write $\dot{F}$ without specifying $\mathbf{x} = \mathbf{x}_0$.) Note that

$$\dot{F} \cdot \mathbf{v} = \frac{d}{dt}F(\mathbf{x}_0 + t\mathbf{v})\Big|_{t=0}$$

is the directional derivative along the direction $\mathbf{v}$ at $\mathbf{x}_0$. Therefore, $\dot{F}$ can be viewed as a linear operator from the tangent space of U, which is $\mathbb{R}^n$, to the tangent space of V, which is $\mathbb{R}^m$, that takes a direction $\mathbf{v}$ and outputs the directional derivative.

For example, let $\mathbf{u} = (1, 2, 3)^\top$ and $F(\mathbf{x}) = \mathbf{u} \cdot \mathbf{x} = x_1 + 2x_2 + 3x_3$. Thus, we have

$$\dot{F} = \begin{bmatrix} 1 & 2 & 3 \end{bmatrix}$$

is constant at all points. On the other hand, it can also be found by

$$\frac{d}{dt}F(\mathbf{x}_0 + t\mathbf{v})\Big|_{t=0} = \lim_{t\to 0}\frac{\mathbf{u}\cdot(\mathbf{x}_0 + t\mathbf{v}) - \mathbf{u}\cdot\mathbf{x}_0}{t} = \mathbf{u}\cdot\mathbf{v},$$

so $\dot{F}$ is the linear operator from $\mathbb{R}^3$ to $\mathbb{R}$ that sends $\mathbf{v}$ to $\mathbf{u}\cdot\mathbf{v}$.

The same idea can be applied when U and V are subsets of some finite dimensional vector spaces over $\mathbb{R}$, such as the space of matrices. Suppose F is a function whose input and output are matrices. We will again view the derivative of F as a linear operator sending a direction to its directional derivative. When $\dot{F}$ is such a linear operator and A a direction (possibly in the vector space of matrices), we use the notation $\dot{F} \cdot A$ for the output of the linear operator $\dot{F}$.

EXAMPLE 5.1. Let

$$A = \begin{bmatrix} 1 & 2 \\ 3 & 4 \end{bmatrix} \text{ and } X = \begin{bmatrix} x_1 & x_2 \\ x_3 & x_4 \end{bmatrix}.$$

Consider the function $F(X) = AX$. Then we may compute

$$\dot{F}\Big|_{X=O} \cdot Y = \lim_{t \to 0} \frac{A(O + tY) - A(O)}{t} = AY$$

for any 2×2 matrix Y.

On the other hand, one may consider F as

$$(x_1, x_2, x_3, x_4) \mapsto (x_1 + 2x_3, x_2 + 2x_4, 3x_1 + 4x_3, 3x_2 + 4x_4),$$

which sends the four entries in X to the four entries in AX in order. In this sense, the Jacobian matrix of F is

$$J = \begin{bmatrix} 1 & 0 & 2 & 0 \\ 0 & 1 & 0 & 2 \\ 3 & 0 & 4 & 0 \\ 0 & 3 & 0 & 4 \end{bmatrix}.$$

Indeed, if $[M]$ is the vector representation of a matrix M that records the entries of M in the row-major order, then

$$J[Y] = [\dot{F} \cdot Y].$$

That is, J is the matrix representation of $\dot{F}$ with respect to the standard basis of $\mathbb{R}^{n \times n}$. The benefit of the matrix representation is one may use standard linear algebra techniques to check whether $\dot{F}$ is injective or surjective or not. In contrast, when the basis is not critical, viewing $\dot{F}$ as a linear operator is clean and meaningful.

Recall that a function is *smooth* if it is infinitely differentiable. Proposition 5.2 is a well known fact.

PROPOSITION 5.2. *The function $F\colon S_n^-(R) \to \mathbb{R}^{n \times n}$ defined by $F(K) = e^K$ is a smooth function such that the derivative at O has $\dot{F} \cdot K = K$.*

PROOF. Recall that by definition

$$e^K = \frac{1}{0!}I + \frac{1}{1!}K + \frac{1}{2!}K^2 + \cdots,$$

so each entry of e^K is a convergent power series and e^K is smooth.

Let $\dot{F}$ be the derivative at O. Since K is skew symmetric, there are diagonal matrix D with pure imaginary numbers on the diagonal and orthogonal matrix Q such that $K = QDQ^\top$. Thus,

$$e^K = Qe^DQ^\top \text{ and } e^{tK} = Qe^{tD}Q^\top.$$

Therefore,

$$\begin{aligned}\dot{F}\cdot K &= \lim_{t\to 0}\frac{Qe^{O+tD}Q^\top - Qe^{O}Q^\top}{t}\\ &= Q\left(\lim_{t\to 0}\frac{e^{tD}-I}{t}\right)Q^\top\\ &= Q\left(\lim_{t\to 0}\frac{\frac{1}{0!}I+\frac{1}{1!}Dt+\frac{1}{2!}D^2t^2+\cdots-I}{t}\right)Q^\top\\ &= QDQ^\top = K.\end{aligned}$$

This completes the proof. □

Let $U\subseteq\mathbb{R}^n$ and $W\subseteq\mathbb{R}^m$ be two open subsets. A function $F\colon U\to W$ is a *diffeomorphism* if F is smooth and bijective (so invertible), and $F^{-1}\colon W\to U$ is also smooth. Necessarily, if there is a diffeomorphism between U and W, then $n=m$. Any diffeomorphism on U can be viewed as a new coordinate system, providing a different point of view of the same open subset.

One way to verify the diffeomorphism is by the Inverse Function Theorem, which can be found, e.g., in [**146**].

THEOREM 5.3 (Inverse Function Theorem). *Let U and V be two open subsets in $\mathbb{R}^m$. Let $F\colon U\to W$ be a smooth function with $F(\mathbf{u}_0)=\mathbf{w}_0$. Let $\dot{F}$ be the derivative of F at $\mathbf{u}_0$. If $\dot{F}$ is surjective (or equivalently, invertible), then there is a open subset $U'\subseteq U$ containing $\mathbf{u}_0$ such that the restriction of F on U' is a diffeomorphism between U' and $F(U')$. Moreover, the inverse of the diffeomorphism has derivative $(\dot{F})^{-1}$ at $\mathbf{v}_0$.*

The following definitions are required for setting up a diffeomorphism around a matrix, which can be viewed as a new coordinate system. Let $D=\operatorname{diag}(d_1,\ldots,d_n)$ be an $n\times n$ diagonal matrix. A pair of indices (i,j) is said to be *inner* with respect to D if $d_i=d_j$ and *outer* otherwise. We use $\operatorname{in}(D)$ and $\operatorname{out}(D)$ for the set of inner index pairs and the set of outer index pairs with respect to D. Suppose the diagonal entries of D contains m_i copies of λ_i for $i=1,\ldots,q$. Then up to some permutation similarity one may write $D=\oplus_{i=1}^q\lambda I_{m_i}$ and the inner and outer index pairs are illustrated in Figure 5.1.

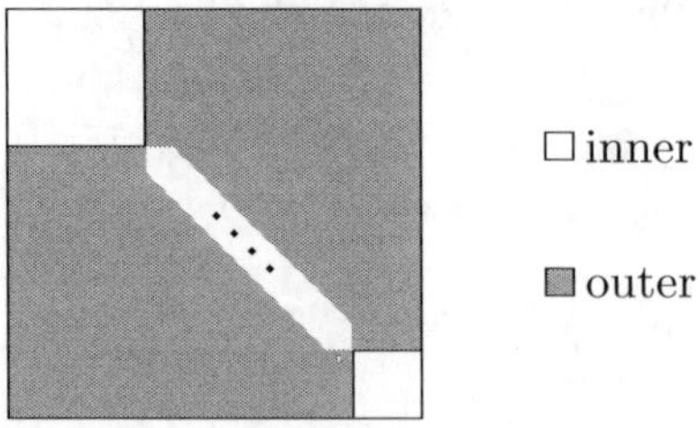

FIGURE 5.1. Illustration of inner and outer index pairs of a diagonal matrix.

The support of matrix M is the set of indices on which the entry is nonzero, denoted by $\operatorname{supp}(M)$. We consider the vector space $S_{\text{in}}(D)$ composed of all $n\times n$ symmetric matrices M with $\operatorname{supp}(M)\subseteq\operatorname{in}(D)$ and the vector space $K_{\text{out}}(D)$ composed of all $n\times n$ skew symmetric matrices M with $\operatorname{supp}(M)\subseteq\operatorname{out}(D)$. Thus, the *D-perturbation space* is the Cartesian product $S_{\text{in}}(D) \times K_{\text{out}}(D)$. Note that the dimension of the D-perturbation space is always $\binom{n+1}{2}$ for any $n \times n$ diagonal matrix D.

Next we show that there is a diffeomorphism from the D-perturbation space to the space of symmetric matrices.

THEOREM 5.4. *Let A be an $n \times n$ symmetric matrix and $A = Q^\top DQ$ for some orthogonal matrix Q and diagonal matrix D. Let*

$$\begin{array}{rccc} F\colon & S_{\text{in}}(D) \times K_{\text{out}}(D) & \to & S_n(\mathbb{R}) \\ & (S, K) & \mapsto & Q^\top e^{-K}(D + S)e^K Q \end{array}$$

be a function. Then there is an open set U in $S_{\text{in}}(D) \times K_{\text{out}}(D)$ around (O, O) and an open set W around A such that $F\big|_U$ is a diffeomorphism between them.

PROOF. Observe that

$$F(O, O) = Q^\top e^{-O}(D + O)e^O Q = Q^\top DQ = A.$$

Since F is smooth and the domain and codomain of F have the same dimension, it is enough to show that the derivative of F at (O, O) is injective.

Let $\dot{F}$ be the derivative of F at (O, O). By direct computation,

$$\begin{aligned} \dot{F} \cdot (S, K) &= Q^\top(-K)DQ + Q^\top SQ + Q^\top DKQ, \\ &= Q^\top(K^\top D + S + DK)Q. \end{aligned}$$

Suppose $\dot{F} \cdot (S, K) = O$. Then $K^\top D + S + DK = O$. Since $\operatorname{supp}(S) \subseteq \operatorname{in}(D)$, the i, j-entry of $K^\top D + DK$ is zero for any $(i, j) \in \operatorname{out}(D)$. Suppose $D = \operatorname{diag}(d_1, \ldots, d_n)$ and $K = \begin{bmatrix} k_{ij} \end{bmatrix}$. For any $(i, j) \in \operatorname{out}(D)$, we know $d_i \neq d_j$ and the i, j-entry of $K^\top D + DK$ is $-k_{ij}d_j + d_i k_{ij} = 0$. This means $k_{ij} = 0$. Since the choice of (i, j) is arbitrary in $\operatorname{out}(D)$ and $\operatorname{supp}(K) \subseteq \operatorname{out}(D)$, $K = O$. Therefore, $\dot{F}$ is injective and invertible.

By the inverse eigenvalue theorem (Theorem 5.3), the desired open sets and diffeomorphism exist. □

DEFINITION 5.5. Let A be a symmetric matrix. The diffeomorphism in Theorem 5.4 is called the *spectral diffeomorphism* at A.

The spectral diffeomorphism provides a new coordinate system around a matrix A such that S_{in} controls the eigenvalues and K_{out} controls the eigenspaces. With this notion, we may analyze the manifolds around A that preserve its spectrum, ordered multiplicity list, and rank.

5.2.1. Isospectral manifold.

DEFINITION 5.6 (Isospectral perturbation). Let A be an $n \times n$ symmetric matrix. The *isospectral perturbation* of A is the function $F_{\mathcal{E}_A, A}\colon S_n^-(R) \to S_n(\mathbb{R})$ defined by $K \mapsto e^{-K}Ae^K$. We let $\dot{F}_{\mathcal{E}_A, A}$ denote the derivative of $F_{\mathcal{E}_A, A}$ at O.

COROLLARY 5.7. *Let A be an $n \times n$ symmetric matrix such that* $\operatorname{spec}(A) = \{\lambda_1^{(m_1)}, \ldots, \lambda_q^{(m_q)}\}$. *There is a neighborhood N_A of A such that $N_A \cap \mathcal{E}_A$ is a submanifold in $S_n(\mathbb{R})$. Its dimension is*

$$\binom{n+1}{2} - \binom{m_1 + 1}{2} - \cdots - \binom{m_q + 1}{2},$$

and its tangent space at A is

$$\text{range}(\dot{F}_{\mathcal{E}_A,A}) = \{K^\top A + AK : K \in S_n^-(R)\}.$$

PROOF. Let $F\colon U \to W$ be the spectral diffeomorphism at A. Let Q and D be the orthogonal matrix and diagonal matrix used in the spectral diffeomorphism. Let U' be the intersection of U and the subspace of $S_{\text{in}}(D) \times K_{\text{out}}(D)$ composed of (S, K) with $S = O$. Since U' is a submanifold and F is a diffeomorphism, $F(U')$ is a submanifold in $S_n(\mathbb{R})$. Since the spectrum of $F(S, K)$ is different from A whenever $S \neq O$, $F(U')$ is indeed all the matrices around A with the same spectrum. Since F is a diffeomorphism, the dimension of $F(U')$ is the dimension of U', which is the number of elements in $\{(i, j) \in K_{\text{out}}(D) : i \leq j\}$.

Finally, consider the isospectral perturbation $F_{\mathcal{E}_A,A}$ at A. Since $F_{\mathcal{E}_A,A}(K)$ has the same spectrum as A, $\text{range}(\dot{F}_{\mathcal{E}_A,A})$ is contained in the tangent space of $F(U')$. However,

$$\text{range}(\dot{F}_{\mathcal{E}_A,A}) = \{K^\top A + AK : K \in S_n^-(R)\}$$

contains

$$\{\dot{F} \cdot (O, K) : K \in K_{\text{out}}(D)\} = \{Q^\top (K^\top D + DK)Q : K \in K_{\text{out}}(D)\},$$

so $\text{range}(\dot{F}_{\mathcal{E}_A,A})$ is exactly the tangent space of $F(U')$. □

COROLLARY 5.8. **[33]** *Let A be a symmetric matrix. The normal space of $\mathcal{E}_A$ at A is*

$$\{X \in S_n(\mathbb{R}) : AX - XA = O\}.$$

PROOF. By Corollary 5.7, the tangent space of $\mathcal{E}_A$ at A is

$$\{K^\top A + AK : K \in S_n^-(R)\}.$$

Suppose $\langle X, K^\top A + AK\rangle = 0$ for all $K \in S_n^-(R)$. Then

$$\begin{aligned}
\langle X, K^\top A + AK\rangle &= \text{tr}((K^\top A + AK)^\top X) \\
&= \text{tr}(K^\top AX) + \text{tr}(AKX) \\
&= \text{tr}(K^\top AX) + \text{tr}(KXA) \\
&= \text{tr}(K^\top AX) - \text{tr}(K^\top XA) \\
&= \text{tr}(K^\top (AX - XA)) \\
&= \langle AX - XA, K\rangle = 0
\end{aligned}$$

for any $K \in S_n^-(R)$. Since $AX-XA$ is a skew symmetric matrix, $\langle AX-XA, K\rangle = 0$ for all $K \in S_n^-(R)$ if and only if $AX - XA = O$. □

5.2.2. Iso-oml manifold.

DEFINITION 5.9 (Iso-oml perturbation). Let A be an $n \times n$ symmetric matrix with $q(A) = q$. The *iso-oml perturbation* of A is the function $F_{\mathcal{U}_A,A}\colon S_n^-(R) \times \mathbb{R}^q \to S_n(\mathbb{R})$ defined by

$$(K, (c_0, \ldots, c_{q-1})) \mapsto e^{-K} p(A) e^K,$$

where $p(x) = x + \sum_{k=0}^{q-1} c_k x^k$. We let $\dot{F}_{\mathcal{U}_A,A}$ denote the derivative of $F_{\mathcal{U}_A,A}$ at $(O, \mathbf{0})$.

COROLLARY 5.10. *Let A be an $n \times n$ symmetric matrix such that* $\operatorname{spec}(A) = \{\lambda_1^{(m_1)}, \ldots, \lambda_q^{(m_q)}\}$. *There is a neighborhood N_A of A such that $N_A \cap \mathcal{U}_A$ is a submanifold in $S_n(\mathbb{R})$. Its dimension is*

$$\binom{n+1}{2} - \binom{m_1+1}{2} - \cdots - \binom{m_q+1}{2} + q,$$

and its tangent space at A is

$$\operatorname{range}(\dot{F}_{\mathcal{U}_A,A}) = \{K^\top A + AK + \sum_{k=0}^{q-1} c_k A^k : (c_0, \ldots, c_{q-1}) \in \mathbb{R}^q, K \in S_n^-(R)\}.$$

PROOF. Let $F\colon U \to W$ be the spectral diffeomorphism at A. Let Q and D be the orthogonal matrix and diagonal matrix used in the spectral diffeomorphism. Let U' be the intersection of U and the subspace of $S_{\text{in}}(D) \times K_{\text{out}}(D)$ composed of (S, K) with $S = \oplus_{k=1}^q d_k I_{m_k}$ for some d_k. Since U' is a submanifold and F is a diffeomorphism, $F(U')$ is a submanifold in $S_n(\mathbb{R})$. Since the ordered multiplicity list of $F(S, K)$ is different from A whenever S is not of the form $S = \oplus_{k=1}^q d_k I_{m_k}$, $F(U')$ is indeed all the matrices around A with the same ordered multiplicity list. Since F is a diffeomorphism, the dimension of $F(U')$ is the dimension of U', which is the number of elements in $\{(i,j) \in K_{\text{out}}(D) : i \leq j\}$ plus q.

Finally, consider the iso-oml perturbation $F_{\mathcal{U}_A,A}$ at A. Since $F_{\mathcal{U}_A,A}(K)$ has the same ordered multiplicity list as A, $\operatorname{range}(\dot{F}_{\mathcal{U}_A,A})$ is contained in the tangent space of $F(U')$. However,

$$\operatorname{range}(\dot{F}_{\mathcal{U}_A,A}) = \{K^\top A + AK + \sum_{k=0}^{q-1} c_k A^k : (c_0, \ldots, c_{q-1}) \in \mathbb{R}^q, K \in S_n^-(R)\}$$

contains

$$\begin{aligned} &\{\dot{F} \cdot (\oplus_{k=1}^q d_k I_{m_k}, K) : (d_1, \ldots, d_q) \in \mathbb{R}^q, K \in K_{\text{out}}(D)\} \\ &= \{Q^\top (K^\top D + DK + \oplus_{k=1}^q d_k I_{m_k}) Q : (d_1, \ldots, d_q) \in \mathbb{R}^q, K \in K_{\text{out}}(D)\}, \end{aligned}$$

so $\operatorname{range}(\dot{F}_{\mathcal{U}_A,A})$ is exactly the tangent space of $F(U')$. □

COROLLARY 5.11. **[33]** *Let A be an $n \times n$ symmetric matrix with $q(A) = q$. The normal space of $\mathcal{U}_A$ at A is*

$$\{X \in S_n(\mathbb{R}) : AX - XA = O, \operatorname{tr}(A^k X) = 0 \text{ for } k = 0, \ldots, q-1\}.$$

PROOF. By Corollary 5.7, the tangent space of $\mathcal{U}_A$ at A is

$$T + \operatorname{span}(\{I, \ldots, A^{q-1}\}),$$

where T is the tangent space of $\mathcal{E}_A$ at A. Thus,

$$T^\perp \cap \operatorname{span}(\{I, \ldots, A^{q-1}\})^\perp$$

is the normal space. □

5.2.3. Isorank manifold. In the space $\mathbb{R}^{n\times n}$, let $\mathcal{B}_n$ be the set of matrices A with $\|A\| < 0.5$. Thus, $I + L$ is invertible for any $L \in \mathcal{B}_n$.

DEFINITION 5.12 (Isorank perturbation). Let A be an $n \times n$ symmetric matrix. The *isorank perturbation* of A is the function $F_{\mathcal{R}_A,A}\colon \mathcal{B}_n \to S_n(\mathbb{R})$ defined by

$$L \mapsto (I+L)^\top A (I+L).$$

We let $\dot{F}_{\mathcal{R}_A,A}$ denote the derivative of $F_{\mathcal{R}_A,A}$ at O.

COROLLARY 5.13. *Let A be an $n \times n$ symmetric matrix with nullity m. There is a neighborhood N_A of A such that $N_A \cap \mathcal{R}_A$ is a submanifold in $S_n(\mathbb{R})$. Its dimension is*

$$\binom{n+1}{2} - \binom{m+1}{2},$$

and its tangent space at A is

$$\operatorname{range}(\dot{F}_{\mathcal{R}_A,A}) = \{L^\top A + AL : L \in \mathbb{R}^{n\times n}\}.$$

PROOF. Let $F\colon U \to W$ be the spectral diffeomorphism at A. Let Q and $D = \operatorname{diag}(d_1, \ldots, d_n)$ be the orthogonal matrix and diagonal matrix used in the spectral diffeomorphism. Let V be the subspace of $S_{\text{in}}(D) \times K_{\text{out}}(D)$ composed of (S, K) such that the i, j-entry of S is zero whenever $d_i = d_j = 0$. Let $U' = V \cap U$. Since U' is a submanifold and F is a diffeomorphism, $F(U')$ is a submanifold in $S_n(\mathbb{R})$. Since the rank of $F(S, K)$ is different from A whenever an i, j-entry of S with $d_i = d_j = 0$ is nonzero, $F(U')$ is indeed all the matrices around A with the same rank. Since F is a diffeomorphism, the dimension of $F(U')$ is the dimension of U', which is the dimension of $S_{\text{in}}(D) \times K_{\text{out}}(D)$ minus the number of elements in $\{(i, j) : d_i = d_j = 0, i \le j\}$.

Finally, consider the isorank perturbation $F_{\mathcal{R}_A,A}$ at A. Since $F_{\mathcal{R}_A,A}(K)$ has the same rank as A, $\operatorname{range}(\dot{F}_{\mathcal{R}_A,A})$ is contained in the tangent space of $F(U')$. However,

$$\operatorname{range}(\dot{F}_{\mathcal{R}_A,A}) = \{L^\top A + AL : L \in \mathbb{R}^{n\times n}\}$$

contains

$$\{\dot{F} \cdot (S, K) : (S, K) \in V\} = \{Q^\top (K^\top D + DK + S)Q : (S, K) \in V\},$$

so $\operatorname{range}(\dot{F}_{\mathcal{R}_A,A})$ is exactly the tangent space of $F(U')$. □

COROLLARY 5.14. **[33]** *Let A be an $n \times n$ symmetric matrix. The normal space of $\mathcal{R}_A$ at A is*

$$\{X \in S_n(\mathbb{R}) : AX = O\}.$$

PROOF. By Corollary 5.13, the tangent space of $\mathcal{R}_A$ at A is

$$\{L^\top A + AL : L \in \mathbb{R}^{n\times n}\}.$$

Suppose $\langle X, L^\top A + AL\rangle = 0$ for all $L \in \mathbb{R}^{n\times n}$. Then

$$\begin{aligned}
\langle X, L^\top A + AL\rangle &= \operatorname{tr}((L^\top A + AL)^\top X) \\
&= \operatorname{tr}(L^\top AX) + \operatorname{tr}(ALX) \\
&= \operatorname{tr}(L^\top AX) + \operatorname{tr}(XAL) \\
&= \operatorname{tr}(L^\top AX) + \operatorname{tr}(L^\top AX) \\
&= 2\operatorname{tr}(L^\top (AX)) \\
&= 2\langle AX, L\rangle = 0
\end{aligned}$$

for any $L \in \mathbb{R}^{n\times n}$, which is equivalent to the condition $AX = O$. □

5.2.4. Pattern manifold. Let G be a graph. The set $\mathcal{S}(G)$ is an open set in $S_n(\mathbb{R})$, so naturally it is a manifold. Indeed, it is a subspace if we include the limit points. Recall that $\mathcal{S}^{\mathrm{cl}}(G)$ is the topological closure of $\mathcal{S}(G)$. That is,

$$\mathcal{S}^{\mathrm{cl}}(G) = \{A = \begin{bmatrix} a_{ij} \end{bmatrix} \in S_n(\mathbb{R}) : a_{ij} = 0 \text{ if } \{i,j\} \in E(\overline{G})\}.$$

For convenience, we let $\mathcal{S}_0^{\mathrm{cl}}(G)$ be the set of matrices in $\mathcal{S}^{\mathrm{cl}}(G)$ with zero diagonal. Note that with $A \in \mathcal{S}(G)$, we have $A \in \mathcal{S}^{\mathrm{cl}}(H)$ if and only if G is a spanning subgraph of H.

DEFINITION 5.15 (Pattern perturbation). Let H be a graph and $A \in \mathcal{S}^{\mathrm{cl}}(H)$. The *pattern perturbation* of A with respect to H is the function $F_{\mathcal{S}(H),A} \colon \mathcal{S}^{\mathrm{cl}}(H) \to S_n(\mathbb{R})$ defined by

$$B \mapsto A + B.$$

We let $\dot{F}_{\mathcal{S}(H),A}$ denote the derivative of $F_{\mathcal{S}(H),A}$ at O.

It is easy to see the following.

PROPOSITION 5.16. *Let H be a graph with n vertices and m edges. Then both $\mathcal{S}(H)$ and $\mathcal{S}^{\mathrm{cl}}(H)$ are submanifolds in $S_n(\mathbb{R})$ of dimension $m+n$. For both of them, the tangent space at any point is $\mathcal{S}^{\mathrm{cl}}(H)$, while the normal space is $\mathcal{S}_0^{\mathrm{cl}}(\overline{H})$.*

In this section, we went through several perturbations and showed that the local structures of $\mathcal{E}_A$, $\mathcal{U}_A$, and $\mathcal{R}_A$ are the range of these perturbations, respectively. Although we did not prove the sets $\mathcal{E}_A$, $\mathcal{U}_A$, and $\mathcal{R}_A$ are manifolds globally, being a manifold locally is all we need for Theorems 3.2 and 3.3. These two theorems will be proved in the next section to support the historical arguments, such as the proofs of the Supergraph Lemma (Lemma 3.11) and the Matrix Liberation Lemma (Lemma 5.41). However, we provide an alternative treatment for the Supergraph Lemma in Section 5.4 and a proof of the Bifurcation Lemma in Section 5.5 that relies on an appropriate version of the Implicit Function Theorem (Theorem 5.18) without using the manifold language.

5.3. Implicit Function Theorem Revisited

In this section, we establish the theoretical background of Theorems 3.2 and 3.3, the manifold versions of the Implicit Function Theorem.

We start with a seemingly more general, but actually equivalent, statement of the Inverse Function Theorem. Here we use id_U for the identity map from U to U. The classical Inverse Function Theorem says F is locally invertible if the derivative $\dot{F}$ is invertible. Theorem 5.17 states that F is locally surjective if the derivative $\dot{F}$ is surjective. Here we use $F \circ T$ for the composition of two functions.

THEOREM 5.17. *Let $U \subseteq \mathbb{R}^n$ and $W \subseteq \mathbb{R}^m$ be open subsets. Let $F \colon U \to W$ be a smooth function with $F(\mathbf{u}_0) = \mathbf{w}_0$. Let $\dot{F}$ be the derivative of F at $\mathbf{u}_0$. If $\dot{F}$ is surjective, then there exist an open subset $W' \subseteq W$ containing $\mathbf{w}_0$ and a smooth function T on W' such that $T(\mathbf{w}_0) = \mathbf{u}_0$ and $F \circ T = \mathrm{id}_{W'}$. Moreover, for any linear operator E with $\dot{F} \circ E = \mathrm{id}_{\mathbb{R}^m}$, the function T can be chosen such that the derivative of T at $\mathbf{w}_0$ is E.*

PROOF. Let E be a linear operator such that $\dot{F} \circ E = \mathrm{id}_{\mathbb{R}^m}$. Let $U_x = \mathrm{range}(E)$ and $U_y = U_x^{\perp}$. Thus, $\mathbb{R}^n = U_x \oplus U_y$ and every $\mathbf{u} \in U$ near $\mathbf{u}_0$ can be uniquely written as $\mathbf{u} = \mathbf{u}_0 + \mathbf{u}_x + \mathbf{u}_y$ for some $\mathbf{u}_x \in U_x$ and $\mathbf{u}_y \in U_y$. Since $\dot{F} \circ E = \mathrm{id}_{\mathbb{R}^m}$, $\dot{F}\Big|_{U_x}$ is invertible.

Let $\widetilde{U}$ be the intersection of U and the affine subspace $\mathbf{u}_0 + U_x$. Let $\widetilde{F}$ be the restriction of F on $\widetilde{U}$. Then the derivative $\dot{\widetilde{F}}$ of $\widetilde{F}$ at $\mathbf{u}_0$ is the same as $\dot{F}\Big|_{U_x}$, which is invertible. By the Inverse Function Theorem (Theorem 5.3), there is an open set $W' \subset W$ containing $\mathbf{w}_0$ and a smooth function $\widetilde{T}$ such that $\widetilde{T}(\mathbf{w}_0) = \mathbf{u}_0$ and $\widetilde{F} \circ \widetilde{T} = \mathrm{id}_{W'}$. Moreover, $\dot{\widetilde{F}} \circ \dot{\widetilde{T}} = \mathrm{id}_{\mathbb{R}^m}$, where $\dot{\widetilde{T}} = \dot{\widetilde{T}}\Big|_{\mathbf{w}=\mathbf{w}_0}$.

Let $T\colon W' \to U$ be the function such that $T(\mathbf{v}) = \widetilde{T}(\mathbf{v})$. Then T is smooth, $T(\mathbf{w}_0) = \mathbf{u}_0$, and $F \circ T = \mathrm{id}_{W'}$. Finally, let $\dot{T}$ be the derivative of T at $\mathbf{w}_0$. Then $\dot{T}$ and $\dot{\widetilde{T}}$ are the same except that the codomain of $\dot{T}$ is $\mathbb{R}^n$ and the codomain of $\dot{\widetilde{T}}$ is $U_x \subseteq \mathbb{R}^n$. Therefore, $\dot{F} \circ \dot{T} = \dot{\widetilde{F}} \circ \dot{\widetilde{T}} = \mathrm{id}_{\mathbb{R}^m}$. Since both $\mathrm{range}(\dot{T}) = \mathrm{range}(E) = U_x$ and $\dot{F}$ is nonsingular on U_x, we conclude that $\dot{T} = E$. □

Similarly, here we provide a surjective version of the Implicit Function Theorem.

THEOREM 5.18 (Implicit Function Theorem). *Let $U \subseteq \mathbb{R}^{n+k}$ and $W \subseteq \mathbb{R}^m$ be open subsets. Let $F\colon U \to W$ be a smooth function $F(\mathbf{x}, \mathbf{y})$ with $F(\mathbf{x}_0, \mathbf{y}_0) = \mathbf{0}$ for some $\mathbf{x}_0 \in \mathbb{R}^n$ and $\mathbf{y}_0 \in \mathbb{R}^k$. Let $\dot{F}_{\mathbf{x}}$ and $\dot{F}_{\mathbf{y}}$ be the partial derivatives of F with respect to $\mathbf{x}$ and $\mathbf{y}$ at $(\mathbf{x}_0, \mathbf{y}_0)$. If $\dot{F}_{\mathbf{x}}$ is surjective, then there are an open set $U_y \subseteq \mathbb{R}^k$ containing $\mathbf{y}_0$ and a smooth function T on U_y such that $T(\mathbf{y}_0) = \mathbf{x}_0$ and $F(T(\mathbf{y}), \mathbf{y}) = \mathbf{0}$ for all $\mathbf{y} \in U_y$. Moreover, for any linear operator E with $\dot{F}_{\mathbf{x}} \circ E = \mathrm{id}_{\mathbb{R}^m}$, the function T can be chosen such that the derivative of T at $\mathbf{y}_0$ is $-E \circ \dot{F}_{\mathbf{y}}$.*

PROOF. Based on F, define a new function $\widetilde{F}$ on U by $\widetilde{F}(\mathbf{x}, \mathbf{y}) = (F(\mathbf{x}, \mathbf{y}), \mathbf{y})$. Let $\dot{\widetilde{F}}$ be the derivative of $\widetilde{F}$ at $(\mathbf{x}_0, \mathbf{y}_0)$. Then

$$\dot{\widetilde{F}} \cdot (\mathbf{x}, \mathbf{y}) = (\dot{F}_{\mathbf{x}} \cdot \mathbf{x} + \dot{F}_{\mathbf{y}} \cdot \mathbf{y}, \mathbf{y}),$$

which is surjective since $\dot{F}_{\mathbf{x}}$ is surjective. By the Inverse Function Theorem (Theorem 5.17), there is an open set $\widetilde{W} \subseteq \mathbb{R}^{m+k}$ containing $(\mathbf{0}, \mathbf{y}_0)$ and a smooth function $\widetilde{T}$ on $\widetilde{W}$ such that $\widetilde{T}(\mathbf{0}, \mathbf{y}_0) = (\mathbf{x}_0, \mathbf{y}_0)$ and $\widetilde{F} \circ \widetilde{T} = \mathrm{id}_{\widetilde{W}}$.

Let U_y be an open subset such that $(\mathbf{0}, \mathbf{y}) \in \widetilde{W}$ whenever $\mathbf{y} \in U_y$. Define the a map T on U_y by

$$\widetilde{T}(\mathbf{0}, \mathbf{y}) = (T(\mathbf{y}), \mathbf{y}).$$

Following the properties of $\widetilde{T}$, we have $T(\mathbf{y}_0) = \mathbf{x}_0$ and $F(T(\mathbf{y}), \mathbf{y})$ for all $\mathbf{y} \in U_y$.

Finally, for a given linear operator $E\colon \mathbb{R}^m \to \mathbb{R}^n$ such that $\dot{F}_{\mathbf{x}} \circ E = \mathrm{id}_{\mathbb{R}^m}$, we may define the linear operator $\widetilde{E}\colon \mathbb{R}^{m+k} \to \mathbb{R}^{n+k}$ by $\widetilde{E} \cdot (\mathbf{v}, \mathbf{y}) = (E \cdot (\mathbf{v} - \dot{F}_{\mathbf{y}} \cdot \mathbf{y}), \mathbf{y})$ so that $\dot{\widetilde{F}} \circ \widetilde{E} = \mathrm{id}_{\mathbb{R}^{m+k}}$. Again, by the Inverse Function Theorem (Theorem 5.17), we may choose $\widetilde{T}$ such that its derivative at $(\mathbf{0}, \mathbf{y}_0)$ is $\widetilde{E}$. Consequently, the derivative of T at $\mathbf{y}_0$ is

$$\dot{T} \cdot \mathbf{y} = E \cdot (\mathbf{v} - \dot{F}_{\mathbf{y}} \cdot \mathbf{y})\Big|_{\mathbf{v}=\mathbf{0}} = -E \circ \dot{F}_{\mathbf{y}} \cdot \mathbf{y}.$$

Thus, we have found the function T with the desired properties. □

Now we are able to prove Theorems 3.2 and 3.3, which we restate here for convenience.

THEOREM 5.19. [**195**] *Let $\mathcal{M}_1(s)$ and $\mathcal{M}_2(t)$ be smooth families of manifolds in $\mathbb{R}^d$ for $s \in (-1,1)$ and $t \in (-1,1)$. Suppose $\mathcal{M}_1(0)$ and $\mathcal{M}_2(0)$ intersect transversally at a point $\mathbf{y}_0$. Then there exist a neighborhood $W \subseteq \mathbb{R}^2$ of the origin and a continuous function $f\colon W \to \mathbb{R}^d$ such that $f(0,0) = \mathbf{y}_0$ and $\mathcal{M}_1(s)$ and $\mathcal{M}_2(t)$ intersect transversally at $f(s,t)$ for all $(s,t) \in W$.*

PROOF. Let n_1 and n_2 be the dimensions of $\mathcal{M}_1(s)$ and $\mathcal{M}_2(t)$, respectively. Since $\mathcal{M}_1(s)$ is a smooth family of manifolds, there is a smooth function $F_1(\mathbf{x}_1, s)$ from an open set in $\mathbb{R}^{n_1+1}$ containing $(\mathbf{0}, 0)$ to $\mathcal{M}(s)$ such that $F_1(\mathbf{0}, 0) = \mathbf{y}_0$ and the range of the partial derivative $(\dot{F}_1)_{\mathbf{x}_1} = (\dot{F}_1)_{\mathbf{x}_1}\Big|_{(\mathbf{0},0)}$ is the tangent space of $\mathcal{M}_1(0)$ at $\mathbf{y}_0$. Similarly, there is another function $F_2(\mathbf{x}_2, t)$ from an open set in $\mathbb{R}^{n_2+1}$ to $\mathcal{M}_2(t)$ such that $F_2(\mathbf{0}, 0) = \mathbf{y}_0$ and the range of the partial derivative $(\dot{F}_2)_{\mathbf{x}_2} = (\dot{F}_2)_{\mathbf{x}_2}\Big|_{(\mathbf{0},0)}$ is the tangent space of $\mathcal{M}_2(0)$ at $\mathbf{y}_0$. Since $\mathcal{M}_1(0)$ and $\mathcal{M}_2(0)$ intersect transversally at $\mathbf{y}_0$, $\text{range}((\dot{F}_1)_{\mathbf{x}_1}) + \text{range}((\dot{F}_2)_{\mathbf{x}_2}) = \mathbb{R}^d$.

Define the function

$$\widetilde{F}(\mathbf{x}_1, \mathbf{x}_2, s, t) = F_1(\mathbf{x}_1, s) - F_2(\mathbf{x}_2, t).$$

Thus, $\widetilde{F}(\mathbf{0}, \mathbf{0}, 0, 0) = \mathbf{y}_0 - \mathbf{y}_0 = \mathbf{0}$. Let $\dot{\widetilde{F}}_{(\mathbf{x}_1,\mathbf{x}_2)}$ be the partial derivative of $\widetilde{F}$ at $(\mathbf{0}, \mathbf{0}, 0, 0)$. Thus,

$$\dot{\widetilde{F}}_{(\mathbf{x}_1,\mathbf{x}_2)} \cdot (\mathbf{x}_1, \mathbf{x}_2) = (\dot{F}_1)_{\mathbf{x}_1} \cdot \mathbf{x}_1 - (\dot{F}_2)_{\mathbf{x}_2} \cdot \mathbf{x}_2,$$

and by the transversality, $\dot{\widetilde{F}}_{(\mathbf{x}_1,\mathbf{x}_2)}$ is surjective. By the Implicit Function Theorem (Theorem 5.18), $\mathbf{x}_1$ and $\mathbf{x}_2$ can be written as smooth functions of (s,t) and

$$\widetilde{F}(\mathbf{x}_1(s,t), \mathbf{x}_2(s,t), s, t) = F_1(\mathbf{x}_1(s,t), s) - F_2(\mathbf{x}_2(s,t), t) = \mathbf{0}.$$

Therefore, $\mathcal{M}_1(s)$ and $\mathcal{M}_2(t)$ intersect at $F_1(\mathbf{x}_1(s,t), s) = F_2(\mathbf{x}_2(s,t), t)$. When s and t are small enough, the tangent spaces change only slightly, so the intersection remains transversal. □

THEOREM 5.20. [**195**] *Assume $\mathcal{M}_1$ and $\mathcal{M}_2$ are smooth manifolds that intersect transversally at $\mathbf{y}$ and let $\mathbf{v}$ be a common tangent to each of $\mathcal{M}_1$ and $\mathcal{M}_2$ with $\|\mathbf{v}\| = 1$. Then for every $\epsilon > 0$ there exists a point $\mathbf{y}' \neq \mathbf{y}$ such that $\mathcal{M}_1$ and $\mathcal{M}_2$ intersect transversally at $\mathbf{y}'$, and*

$$\left\| \frac{1}{\|\mathbf{y} - \mathbf{y}'\|}(\mathbf{y} - \mathbf{y}') - \mathbf{v} \right\| < \epsilon.$$

For Theorem 3.3, we prove a stronger statement as in Theorem 5.21.

THEOREM 5.21. *Let $\mathcal{M}_1$ and $\mathcal{M}_2$ be submanifolds of $\mathbb{R}^d$ such that they intersect transversally at $\mathbf{y}_0$. Suppose $\mathcal{M}_1$ and $\mathcal{M}_2$ have a common tangent vector $\mathbf{v}$ at $\mathbf{y}_0$. Then there is a function $f(t)$ such that $f(0) = \mathbf{y}_0$, $f'(0) = \mathbf{v}$, and $\mathcal{M}_1$ intersects with $\mathcal{M}_2$ at $f(t)$ for small $t \geq 0$.*

PROOF. Let n_1 and n_2 be the dimensions of $\mathcal{M}_1(s)$ and $\mathcal{M}_2(t)$, respectively. Since $\mathcal{M}_1$ is a manifold, there is a function F_1 from $\mathbb{R}^{n_1}$ to $\mathcal{M}_1$ such that the range of its derivative is the tangent space of $\mathcal{M}_1$ at $\mathbf{y}_0$. Since $\mathbf{v}$ is a tangent vector, we may further assume that F_1 can be written as $F_1(\mathbf{x}_1, s)$ such that $\mathbf{x}_1 \in \mathbb{R}^{n_1-1}$ and the partial derivative of F_1 with respect to s gives the tangent vector $\mathbf{v}$. Similarly, let $F_2(\mathbf{x}_2, t)$ be a function for $\mathcal{M}_2$ such that $\mathbf{x}_2 \in \mathbb{R}^{n_2-1}$ and the partial derivative of F_2 with respect to t leads to $\mathbf{v}$.

Define

$$\widetilde{F}(\mathbf{x}_1, \mathbf{x}_2, s, t) = F_1(\mathbf{x}_1, s) - F_2(\mathbf{x}_2, t).$$

Thus, $\widetilde{F}(\mathbf{0}, \mathbf{0}, 0, 0) = \mathbf{y}_0 - \mathbf{y}_0 = \mathbf{0}$. Since $\mathcal{M}_1$ and $\mathcal{M}_2$ intersect transversally and both s and t lead to the same tangent vector $\mathbf{v}$, the partial derivative $\dot{\widetilde{F}}_{(\mathbf{x}_1,\mathbf{x}_2,s)} = \dot{\widetilde{F}}_{(\mathbf{x}_1,\mathbf{x}_2,s)}\Big|_{(\mathbf{0},\mathbf{0},0)}$ with respect to $(\mathbf{x}_1, \mathbf{x}_2, s)$ has the same range as the derivative of $\widetilde{F}$, and both of them are surjective. Let E be a linear operator such that $E \cdot \mathbf{v} = (\mathbf{0}, \mathbf{0}, 1)^\top$ and $\dot{\widetilde{F}}_{(\mathbf{x}_1,\mathbf{x}_2,s)} E = \mathrm{id}_{\mathbb{R}^d}$. By the Implicit Function Theorem (Theorem 5.18), there is a smooth function $T(t) = (\mathbf{x}_1(t), \mathbf{x}_2(t), s(t))^\top$ such that

$$F_1(\mathbf{x}_1(t), s(t)) = F_2(\mathbf{x}_2(t), t)$$

for small $t \geq 0$ and the derivative of T at 0 is $-E \cdot (-\mathbf{v}) = (\mathbf{0}, \mathbf{0}, 1)^\top$. Therefore, we may let $f(t) = F_2(\mathbf{x}_2(t), t) \in \mathcal{M}_1 \cap \mathcal{M}_2$. Thus, $f(0) = F_2(\mathbf{0}, 0) = \mathbf{y}_0$ and

$$f'(0) = (\dot{F}_2)_{\mathbf{x}_2}\Big|_{\mathbf{x}_2=\mathbf{0}} \cdot \mathbf{0} + (\dot{F}_2)_t\Big|_{t=0} \cdot 1 = \mathbf{v}.$$

Because the transversality is equivalent to the derivative of $\widetilde{F}$ being surjective, the transversality remains when t is small. □

5.4. Strong Properties and Supergraph Lemma Revisited

Recall that two manifolds intersect transversally at a point A if their tangent spaces at A span the whole space, or, equivalently, their normal spaces have trivial intersection. In this section, we revisit the strong properties and see they are equivalent to the transversality.

Suppose A is a matrix in $\mathcal{S}(G)$ for some G. Then A is also an element in $\mathcal{S}^{\mathrm{cl}}(H)$ for any supergraph of G with $V(G) = V(H)$. Here we first make some more general definitions for strong properties by considering the matrix A as an element in $\mathcal{S}^{\mathrm{cl}}(H)$.

Definition 5.22. Let H be a graph and $A \in \mathcal{S}^{\mathrm{cl}}(H)$.

- The matrix A has the *Strong Spectral Property with respect to H* if $X = O$ is the only symmetric matrix satisfying $X \in \mathcal{S}^{\mathrm{cl}}_0(\overline{H})$ and $[A, X] = O$.
- The matrix A has the *Strong Multiplicity Property with respect to H* if $X = O$ is the only symmetric matrix satisfying $X \in \mathcal{S}^{\mathrm{cl}}_0(\overline{H})$, $[A, X] = O$, and $\mathrm{tr}(A^k X) = 0$ for $k = 0, \ldots, q-1$.
- The matrix A has the *Strong Arnold Property with respect to H* if $X = O$ is the only symmetric matrix satisfying $X \in \mathcal{S}^{\mathrm{cl}}_0(\overline{H})$ and $AX = O$.

Remark 5.23. If G is a graph and $A \in \mathcal{S}(G)$, then A has the strong property (one of SSP, SMP, or SAP) if and only if A has the strong property with respect to G itself.

By the discussions in Section 5.2, we have the following equivalence conditions.

Proposition 5.24. *Let H be a graph and $A \in \mathcal{S}^{\mathrm{cl}}(H)$. Then the following are equivalent.*

- *A has the SSP with respect to H.*
- *$\mathcal{E}_A$ and $\mathcal{S}^{\mathrm{cl}}(H)$ intersect transversally at A.*

- *The union of the range of* $\dot{F}_{\mathcal{E}_A,A}$ *and the range of* $\dot{F}_{\mathcal{S}(H),A}$ *spans* $S_n(\mathbb{R})$.

PROPOSITION 5.25. *Let* H *be a graph and* $A \in \mathcal{S}^{\mathrm{cl}}(H)$. *Then the following are equivalent.*

- A *has the SMP with respect to* H.
- $\mathcal{U}_A$ *and* $\mathcal{S}^{\mathrm{cl}}(H)$ *intersect transversally at* A.
- *The union of the range of* $\dot{F}_{\mathcal{U}_A,A}$ *and the range of* $\dot{F}_{\mathcal{S}(H),A}$ *spans* $S_n(\mathbb{R})$.

PROPOSITION 5.26. *Let* H *be a graph and* $A \in \mathcal{S}^{\mathrm{cl}}(H)$. *Then the following are equivalent.*

- A *has the SAP with respect to* H.
- $\mathcal{R}_A$ *and* $\mathcal{S}^{\mathrm{cl}}(H)$ *intersect transversally at* A.
- *The union of the range of* $\dot{F}_{\mathcal{R}_A,A}$ *and the range of* $\dot{F}_{\mathcal{S}(H),A}$ *spans* $S_n(\mathbb{R})$.

Note that two subspaces can possibly span the whose space $S_n(\mathbb{R})$ only when the sum of their dimensions is at least $\binom{n+1}{2}$. This leads to the following inequalities.

COROLLARY 5.27. [**33**] *Let* G *be a graph and* $A \in \mathcal{S}(G)$. *Suppose* $\mathbf{m}(A) = (m_1, \ldots, m_q)$ *and* $\operatorname{null} A = m$. *Then the following inequalities hold.*

(1) *If* A *has the SSP, then*

$$|V(G)| + |E(G)| \geq \binom{m_1+1}{2} + \cdots + \binom{m_q+1}{2}.$$

(2) *If* A *has the SMP, then*

$$|V(G)| + |E(G)| \geq \binom{m_1+1}{2} + \cdots + \binom{m_q+1}{2} - q.$$

(3) *If* A *has the SAP, then*

$$|V(G)| + |E(G)| \geq \binom{m+1}{2}.$$

By examining the intersection of the two subspaces in more detail, one may obtain better bounds; see, e.g., [**170**, Theorem 6.5]. On the other hand, one may use these inequalities to construct a graph such that for some particular ordered multiplicity, the SMP is possible while no matrix with this ordered multiplicity has the SSP [**31**].

Now we are ready for a more general version of the Supergraph Lemma, where Lemma 3.11 is a special case when $G = H$.

LEMMA 5.28. [**225**] *Let* G, H, *and* H' *be graphs such that* $V(G) = V(H) = V(H')$ *and* $E(G) \subseteq E(H) \subseteq E(H')$. *If* $A \in \mathcal{S}(G)$ *is a matrix with the*

(1) *SSP,*
(2) *SMP, or*
(3) *SAP,*

with respect to H, *respectively, then there is a matrix* $A' \in \mathcal{S}^{\mathrm{cl}}(H')$ *such that the* i,j*-entry of* A' *is nonzero whenever* $\{i,j\} \in E(G) \cup (E(H' - H))$ *and*

(1) $\operatorname{spec}(A') = \operatorname{spec}(A)$ *and* A' *has the SSP with respect to* H',
(2) $\mathbf{m}(A') = \mathbf{m}(A)$ *and* A' *has the SMP with respect to* H', *or*
(3) $\operatorname{rank} A' = \operatorname{rank} A$ *and* A' *has the SAP with respect to* H',

respectively.

PROOF. We will prove the case for the SSP, since other cases are similar. Let P be the adjacency matrix of $H' - H$. Consider the function

$$F(Q,B,t) = F_{\mathcal{E}_A,A}(Q) - F_{\mathcal{S}(H),A}(B) - tP.$$

Note that $F(O,O,0) = O$. Since A has the SSP with respect to H, the partial derivative of F with respect to (Q,B) at $(I,O,0)$ is surjective. By the Implicit Function Theorem (Theorem 5.18), for any small choice of t, there are smooth functions $Q(t)$ and $B(t)$ such that

$$F_{\mathcal{E}_A,A}(Q(t)) = F_{\mathcal{S}(H),A}(B(t)) + tP.$$

When t is small enough, $A' = F_{\mathcal{S}(H),A}(B(t)) + tP$ has the desired properties. □

Using this extended Supergraph Lemma, one may prove the Clique Appending Lemma introduced in [**225**].

THEOREM 5.29 (Clique Appending Lemma [**225**]). *Let G be a graph, $A \in \mathcal{S}(G)$ with the SSP, $v \in V(G)$, and s a positive integer. Suppose λ is a real number such that $\lambda \notin \operatorname{spec}(A) \cup \operatorname{spec}(A(v))$. Then there is a matrix $A' \in \mathcal{S}(G \oplus_v K_{s+1})$ with the SSP such that*

$$\operatorname{spec}(A') = \operatorname{spec}(A) \cup \{\lambda^{(s)}\},$$

where $G \oplus_v K_{s+1}$ is obtained from G and K_{s+1} by identifying the vertex v of G with a vertex of K_{s+1}.

PROOF. The matrix λI_s has the SSP with respect to K_s. Since A has the SSP, one may check that $A \oplus \lambda I_s$ has the SSP with respect to $H = G \sqcup K_s$. Let $H' = G \oplus_v K_{s+1}$. Then by the extended Supergraph Lemma (Lemma 5.28), there is a matrix $A' \in \mathcal{S}^{\mathrm{cl}}(H')$ such that A' has the SSP with respect to H' and $\operatorname{spec}(A') = \operatorname{spec}(A) \cup \{\lambda^{(s)}\}$. Moreover, the i,j-entry of A' is nonzero if $\{i,j\}$ is an edge of G or an edge of $H' - H$.

Now it is enough to show that $A' \in \mathcal{S}(H')$. Equivalently, we have to show that the i,j-entry of A' is nonzero for all distinct $i,j \in V(K_s)$. Suppose, for the purpose of yielding a contradiction, that the i,j-entry a'_{ij} of A' is zero for some distinct $i,j \in V(K_s)$. Then the matrix $A' - \lambda I$ induced on the rows of $V(G) \cup \{i\}$ and the columns of $V(G) \cup \{j\}$ has the form

$$\begin{bmatrix} A'[V(G)] - \lambda I & \mathbf{a} & \mathbf{0} \\ \mathbf{a}^\top & a'_{vv} - \lambda & b \\ \mathbf{0}^\top & c & a'_{ij} \end{bmatrix},$$

for some vector $\mathbf{a}$ and some numbers a'_{vv}, b, and c. Since $\lambda \notin \operatorname{spec}(A(v))$, this submatrix of matrix $A' - \lambda I$ has rank $|V(G)| + 1$ if $a'_{ij} = 0$. This is a contradiction to the fact rank $A' = |V(G)|$. Therefore, $A' \in \mathcal{S}(H')$ and A' has the SSP. □

EXAMPLE 5.30. Let $G = K_n$, $v \in V(G)$, and $A \in \mathcal{S}(G)$ the all-ones matrix. Thus, A has the SSP, $\operatorname{spec}(A) = \{0^{(n-1)}, n\}$, and $\operatorname{spec}(A(v)) = \{0^{(n-2)}, n-1\}$. If λ is not one of 0, n, $n-1$, then by the Clique Appending Lemma, there is a matrix $A' \in G \oplus_v K_{s+1}$ with the SSP and

$$\operatorname{spec}(A') = \{0^{(n-1)}, n, \lambda^{(s)}\}$$

for any positive integer s.

5.5. Bifurcation Lemma

In general, it seems hard to achieve an eigenvalue with high multiplicity, while making a high multiplicity lower seems easier. For example, if G is a graph and $A \in \mathcal{S}(G)$ with an eigenvalue $\lambda = 0$ of multiplicity m, then we may construct matrices $A_k \in \mathcal{S}(G)$ with an eigenvalue $\lambda = 0$ of multiplicity k for each $k = 1, \ldots, m$ in the following manner. Pick a small $\epsilon > 0$ such that $A - \epsilon I$ is nonsingular. We may write

$$\epsilon I = \sum_{i=1}^{n} \epsilon E_{ii},$$

where E_{ii} is the $n \times n$ 0,1-matrix with a single 1 at the i, i-entry. Since E_{ii} is a matrix of rank one, the nullities of any symmetric matrix B and $B + \epsilon E_{ii}$ are off by at most one. Therefore, the sequence

$$A, A + \epsilon E_{1,1}, A + \epsilon E_{1,1} + \epsilon E_{2,2}, \ldots, A + \sum_{i=1}^{n} \epsilon E_{ii}$$

contains all the desired A_k. The argument may apply to other nonzero eigenvalues by replacing A with $A - \lambda I$. However, in this process, we lose the control of other eigenvalues. Therefore, given a graph G on n vertices and a matrix $A \in \mathcal{S}(G)$ with spectrum $\{\lambda_1^{(m_1)}, \ldots, \lambda_q^{(m_q)}\}$, some natural questions are the following:

- Inside $\mathcal{S}(G)$, can we decrease the multiplicity of one eigenvalue while preserving all other eigenvalues?
- Inside $\mathcal{S}(G)$, can we perturb each of the n eigenvalues independently?
- Inside $\mathcal{S}(G)$, can we find a matrix whose ordered multiplicity list is a refinement of $\mathbf{m}(A)$?

The following results answer these questions in the affirmative subject to the size of the perturbation.

LEMMA 5.31 (Bifurcation Lemma [**126**]). *Let G be a graph and $A \in \mathcal{S}(G)$. If A has the*

(1) *SSP,*
(2) *SMP, or*
(3) *SAP,*

respectively, then there is an $\epsilon > 0$ such that for any C with $\|C - A\| < \epsilon$, there is a matrix $A' \in \mathcal{S}(G)$ such that

(1) $\operatorname{spec}(A') = \operatorname{spec}(C)$ *and A' has the SSP,*
(2) $\mathbf{m}(A') = \mathbf{m}(C)$ *and A' has the SMP, or*
(3) $\operatorname{rank} A' = \operatorname{rank} C$ *and A' has the SAP,*

respectively.

PROOF. We prove the case for the SSP, while the other cases are similar. Consider the function $F\colon S_n^-(R) \times \mathcal{S}^{\mathrm{cl}}(G) \to S_n(\mathbb{R})$ defined by

$$F(K, B) = e^{-K}(A + B)e^{K}.$$

Thus, the derivative of F at (O, O) satisfies

$$\dot{F} \cdot (K, B) = \dot{F}_{\mathcal{E}_A, A} \cdot K + \dot{F}_{\mathcal{S}(G), A} \cdot B.$$

Thus, $\dot{F}$ is surjective since A has the SSP. Note that $F(O, O) = A$. By the Inverse Function Theorem (Theorem 5.17), there is $\epsilon > 0$ such that for any C with

$\|C - A\| < \epsilon$ there are corresponding values of K and B satisfying $F(K, B) = C$. Therefore, $A' = A + B$ has the same spectrum as C. Since the perturbation is small, the pattern and the transversality are preserved. □

COROLLARY 5.32 (SSP Bifurcation Corollary [**126**]). *Let G be a graph and $A \in \mathcal{S}(G)$ a matrix with the SSP and* $\operatorname{spec}(A) = \{\lambda_1, \ldots, \lambda_n\}$. *Then there is $\epsilon > 0$ such that for any $\mu_1, \ldots, \mu_n$ with $|\mu_i - \lambda_i| < \epsilon$ for all $i = 1, \ldots, n$, a matrix $A' \in \mathcal{S}(G)$ exists with the SSP and* $\operatorname{spec}(A') = \{\mu_1, \ldots, \mu_n\}$.

COROLLARY 5.33 (SMP Bifurcation Corollary [**126**]). *Let G be a graph and $A \in \mathcal{S}(G)$ a matrix with the SMP and* $\mathbf{m}(A) = \mathbf{m}$. *Then for any refinement* $\mathbf{m}'$ *of* $\mathbf{m}$, *there is a matrix $A' \in \mathcal{S}(G)$ with the SMP and* $\mathbf{m}(B) = \mathbf{m}'$.

COROLLARY 5.34 (SAP Bifurcation Corollary [**126**]). *Let G be a graph on n vertices and $A \in \mathcal{S}(G)$ a matrix with the SAP and* $\operatorname{rank} A = r$. *Then for any r' with $r \le r' \le n$ there is a matrix $A' \in \mathcal{S}(G)$ with the SAP and* $\operatorname{rank} B = r'$.

More applications of the Bifurcation Lemma and a version for non-symmetric matrices can be found in [**126**].

5.6. Tangent Space Matrix

We have seen that for a given symmetric matrix A, the tangent spaces of $\mathcal{E}_A$, $\mathcal{U}_A$, and $\mathcal{R}_A$ are the ranges of the linear operators $\dot{F}_{\mathcal{E}_A,A}$, $\dot{F}_{\mathcal{U}_A,A}$, and $\dot{F}_{\mathcal{R}_A,A}$, respectively. Sometimes, it is convenient to have their matrix representations in order to apply some standard linear algebra techniques. For example, we may use the row operations to determine if the tangent space contains a vector with a specific support. In this section, we introduce the tangent space matrices of each of the three manifolds whose column space corresponds to each of the tangent space.

Recall that the standard bases of $\mathbb{R}^{n\times n}$, $S_n(\mathbb{R})$, and $S_n^-(R)$ have been set up in Section 5.1. Here we further assume the standard basis of $S_n^-(R) \times \mathbb{R}^q$ as $\alpha \cup \beta$, where α is the standard basis of $S_n^-(R)$ and β is the standard basis of $\mathbb{R}^q$ that is composed of the column vectors of the identity matrix I_q. The elements of $\alpha \cup \beta$ are ordered by the orders of α and β, respectively, such that elements in α are prior to elements in β. The inner product of $(A, \mathbf{x})$ and $(B, \mathbf{y})$ in $S_n^-(R) \times \mathbb{R}^q$ is $\langle A, B\rangle_{S_n^-(R)} + \mathbf{y}^\top \mathbf{x}$.

DEFINITION 5.35. Let A be an $n \times n$ symmetric matrix with q distinct eigenvalues.

(1) The *SSP tangent space matrix* of A is $\mathrm{TS_S}(A) = [\dot{F}_{\mathcal{E}_A,A}]_\alpha^\beta$, where α and β are the standard bases of $S_n^-(R)$ and $S_n(\mathbb{R})$.
(2) The *SMP tangent space matrix* of A is $\mathrm{TS_M}(A) = [\dot{F}_{\mathcal{U}_A,A}]_\alpha^\beta$, where α and β are the standard bases of $S_n^-(R) \times \mathbb{R}^q$ and $S_n(\mathbb{R})$.
(3) The *SAP tangent space matrix* of A is $\mathrm{TS_A}(A) = [\dot{F}_{\mathcal{R}_A,A}]_\alpha^\beta$, where α and β are the standard bases of $\mathbb{R}^{n\times n}$ and $S_n(\mathbb{R})$.

REMARK 5.36. Note that the height of each tangent space matrix in Definition 5.35 is $\binom{n+1}{2}$ while the tangent space matrices defined in [**31**] has height $\binom{n}{2}$. This is because the tangent space matrix in [**31**] only extracts the entries in the strict upper triangular part. However, the codomain of each derivative is $S_n(\mathbb{R})$, so we include the entries on the diagonal. Under this treatment, we may safely say the derivative is surjective if and only if the tangent space matrix has full column-rank.

Our next goal is to make a connection between the tangent space matrices and the verification matrices defined in Section 4.4. Before that, let us recall a standard result in linear algebra.

REMARK 5.37. Let U and W be inner product spaces over $\mathbb{R}$ and $f\colon U \to W$ a linear transformation. If $f^*\colon W \to U$ is a linear transformation such that

$$\langle f(\mathbf{u}), \mathbf{w}\rangle_W = \langle \mathbf{u}, f^*(\mathbf{w})\rangle_U$$

for any $\mathbf{u} \in U$ and $\mathbf{w} \in W$, then f^* is the *adjoint operator* of f. Let α be a basis of U such that

$$\langle \mathbf{u}_1, \mathbf{u}_2\rangle_U = [\mathbf{u}_1]_\alpha \cdot [\mathbf{u}_2]_\alpha$$

for any $\mathbf{u}_1, \mathbf{u}_2 \in U$, where the right hand side is the standard dot product of two vectors. Similarly, let β be a basis W with the same property. Then $[f^*]^\alpha_\beta = ([f]^\beta_\alpha)^\top$.

Now, for the derivative of each perturbations, we will find their adjoint operators.

PROPOSITION 5.38. *Let A be a symmetric matrix with q distinct eigenvalues.*

(1) *Let $\dot{F}^*\colon S_n(\mathbb{R}) \to S_n^-(R)$ defined by $\dot{F}^*(X) = [A, X]$. Then $\dot{F}^*$ is the adjoint operator of $\dot{F}_{\mathcal{E}_A,A}$.*
(2) *Let $\dot{F}^*\colon S_n(\mathbb{R}) \to S_n^-(R)\times\mathbb{R}^q$ defined by $\dot{F}^*(X) = ([A, X], (t_0, \dots, t_{q-1}))$, where $t_k = \frac{1}{2}\operatorname{tr}(A^kX)$. Then $\dot{F}^*$ is the adjoint operator of $\dot{F}_{\mathcal{U}_A,A}$.*
(3) *Let $\dot{F}^*\colon S_n(\mathbb{R}) \to \mathbb{R}^{n\times n}$ defined by $\dot{F}^*(X) = AX$. Then $\dot{F}^*$ is the adjoint operator of $\dot{F}_{\mathcal{R}_A,A}$.*

PROOF. The proofs of Corollaries 5.8, 5.11, and 5.14 actually proved these statements. □

COROLLARY 5.39. **[31]** *Let G be a graph, $A \in \mathcal{S}(G)$, and $\overline{E} = \{(i,j) : i < j, \{i,j\} \in E(\overline{G})\}$. Let T be the SSP (SMP, or SAP, respectively) tangent space matrix of A. Let Ψ be the SSP (SMP, or SAP, respectively) verification matrix of A. Then $\Psi = T[\overline{E}, :]$.*

PROOF. By Proposition 5.38, the matrix representation of $\dot{F}^*$ is the transpose of the matrix representation of the corresponding derivatives. Indeed, the verification matrices are the matrix representation of the corresponding $\dot{F}^*$ restricted on $\overline{E}$. □

COROLLARY 5.40. *Let H be a graph, $A \in \mathcal{S}^{\rm cl}(H)$, and Ψ its SSP (SMP, or SAP, respectively) verification matrix. Let $\overline{E} = \{(i,j) : i < j, \{i,j\} \in E(\overline{H})\}$. Then A has the SSP (SMP, or SAP, respectively) with respect to H if and only if $\Psi[\overline{E}, :]$ has full row-rank.*

5.7. Matrix Liberation Lemma Revisited

In this section, we will prove the Matrix Liberation Lemma (Lemma 4.21). With the notion of the strong properties with respect to a supergraph, we can restate the Matrix Liberation Lemma as follows.

LEMMA 5.41. *Let G be a graph and $A \in \mathcal{S}(G)$. Let*

(1) $\Psi = \Psi_S(A)$,
(2) $\Psi = \Psi_M(A)$, *or*

(3) $\Psi = \Psi_A(A)$,

respectively. If $\mathbf{v}$ *is a vector in the column space of* Ψ *such that* A *has the SSP (SMP, or SAP, respectively) with respect to* $H = G + \operatorname{supp}(\mathbf{v})$, *then there is a matrix* $A' \in \mathcal{S}(H)$ *such that*

(1) $\operatorname{spec}(A') = \operatorname{spec}(A)$ *and* B *has the SSP,*
(2) $\mathbf{m}(A') = \mathbf{m}(A)$ *and* B *has the SMP, or*
(3) $\operatorname{rank} A' = \operatorname{rank} A$ *and* B *has the SAP,*

respectively.

Proof. We focus on the case of the SSP, since the cases for other strong properties are similar. Suppose A has the SSP with respect to H. Then $\mathcal{E}_A$ and $\mathcal{S}^{\text{cl}}(H)$ intersect transversally at A. On the other hand, $\mathbf{v}$ is a column vector of Ψ, which is a submatrix of $\text{TS}_{\text{S}}(A)$, so there is a matrix $B \in \mathcal{S}^{\text{cl}}(H)$ such that

$$\operatorname{supp}(B) \cap E(\overline{G}) = \operatorname{supp}(\mathbf{v}) \cap E(\overline{G}).$$

Thus, B is a common tangent vector of $\mathcal{E}_A$ and $\mathcal{S}^{\text{cl}}(H)$. By the Implicit Function Theorem (Theorem 5.21), there is a smooth function $f(t)$ such that $f(0) = A$ and $f'(0) = B$. For $\epsilon > 0$ small enough, $f(\epsilon)$ is the desired matrix. $\square$

5.8. Future Work

The strong properties have provided fruitful results for the IEP-G, yet there are still many directions to be explored. For example, the Augmentation Lemma applies the Matrix Liberation Lemma to a matrix of the form $A \oplus [\lambda]$ where λ is an eigenvalue of A. More generally, what can we achieve if we apply the Matrix Liberation Lemma to matrices of the form $A \oplus B$, where A and B have common eigenvalues? Also, the Matrix Liberation Lemma requires a vector in the columns space of the verification matrix and with some extra properties. Is there an efficient way to find such vectors? So far, many applications of the Matrix Liberation Lemma focus on the SSP. Using the SAP, what are its applications to the Colin de Verdière parameter?

By Proposition 5.24, we have seen that the strong properties can be characterized as the nice interaction between two perturbation functions. These techniques are not limited to the perturbations preserving the spectrum, the ordered multiplicity list, or the rank. The same idea should work for many other perturbations preserving different spectral properties. For example, the SIPP is enforced for preserving orthogonality in Section 3.3. One may also focus on the perturbation preserving both $\operatorname{spec}(A)$ and $\operatorname{spec}(A(v))$ for a specific index v, or the perturbations preserving other spectral properties of the matrix and its principal submatrix.

The perturbations are not limited to the symmetric matrices either. Section 3.3 provides an application of the Implicit Function to the sign pattern problem. Naturally, one may consider the perturbations a non-symmetric matrix without changing its spectrum, rank, inertia, and so on.

In summary, the use of the Implicit Function Theorem and the Inverse Function Theorem for constructing matrices with desired properties has a long history. For example, in 1971, Arnol'd used them to study versal deformation [**16**], and in 1998 Laffey applied similar techniques on the nonnegative inverse eigenvalue problem [**216**]. The strong properties and the perturbation techniques have become a powerful toolkit for building new matrices. And the studies of the existing strong properties and the discovery of new strong properties have been growing rapidly. Therefore, it is foreseeable that the study along this line will provide significant impacts on the IEP-G and more.

Part 3

Further Discussion of Ancillary Problems

Due to the challenging nature of the IEP-G, historically it has been approached through the study of graph parameters that provide partial solutions, called ancillary problems. Chapter 2 introduced ancillary problems and focused on the maximum eigenvalue multiplicity and zero forcing. In this part we examine two additional ancillary problems in detail.

Chapter 8 discusses the minimum possible number of distinct eigenvalues. This is a relatively recent area of research but currently a very active one. For a matrix A, the number of distinct eigenvalues of A is denoted by $q(A)$ and the minimum possible number of distinct eigenvalues of a graph of G is

$$q(G) = \min\{q(A) : A \in \mathcal{S}(G)\}.$$

Ordered multiplicity lists and (unordered) multiplicity lists are discussed in Chapter 6. Let A be a real symmetric matrix with distinct eigenvalues $\lambda_1 < \cdots < \lambda_q$ having multiplicities $m_1, \ldots, m_q$, respectively. The *ordered multiplicity list* of A is $\mathbf{m}(A) = (m_1, \ldots, m_q)$ and the *multiplicity list* of A is $\mathbf{mult}(A) = (n_1, \ldots, n_q)$, where $n_1, \ldots, n_q$ is the result from arranging $m_1, \ldots, m_q$ in non-increasing order. The *ordered multiplicity list* and *multiplicity list* a graph of G are

$$\mathbf{m}(G) = \{\mathbf{m}(A) : A \in \mathcal{S}(G)\} \text{ and } \mathbf{mult}(G) = \{\mathbf{mult}(A) : A \in \mathcal{S}(G)\}.$$

Rigid linkages, which are defined and discussed in Chapter 7, provide additional information about possible ordered multiplicity lists of matrices described by a graph and are useful for studying $q(G)$.

CHAPTER 6

Ordered Multiplicity Lists of a Graph

The inverse eigenvalue problem for graphs appears in general to be a difficult problem. It has been completely solved for graphs of order less than 4 in [**36**] and for graphs of order $n = 5$ in [**31**]. A survey of these results is given in Section 4.3. The subtleties of some realizable spectra, such as those with many multiple eigenvalues, as well as the infinitude of possible spectra contribute to the difficulty.

To gain some insight, it has proven useful to study related but less specific parameters associated with the IEP-G. Historically, one way this has been approached is to focus on the possible lists (or ordered lists) of multiplicities among matrices with a given graph.

6.1. Multiplicity Lists for Special Families of Graphs

A *refinement* of an ordered list $\mathbf{m} = (m_1, \ldots, m_q)$ of positive integers is an ordered list obtained from $\mathbf{m}$ by replacing each m_i by an ordered list of positive integers whose sum is m_i; e.g., the refinements of $(2, 1, 3)$ are

$$(2,1,3), (1,1,1,3), (2,1,2,1), (1,1,1,2,1),$$
$$(2,1,1,2), (1,1,1,1,2), (2,1,1,1,1), (1,1,1,1,1,1).$$

Let G be a graph. An ordered list $\mathbf{m} = (m_1, \ldots, m_q) \in \mathbf{m}(G)$ is *spectrally arbitrary* provided for each ordered list $\lambda_1 < \cdots < \lambda_q$ of real numbers there exists a matrix in $\mathcal{S}(G)$ with spectrum $\{\lambda_1^{(m_1)}, \ldots, \lambda_q^{(m_q)}\}$.[1] Some basic properties of the multiplicity lists and ordered multiplicity lists of a graph are given in the next result.

THEOREM 6.1. *The following statements hold for the multiplicity and ordered multiplicity lists.*

(1) *For each n, $\mathbf{mult}(P_n) = \{(1, 1, \ldots, 1)\} = \mathbf{m}(P_n)$.*
(2) *If G is a graph, then the first entry of each element of $\mathbf{mult}(G)$ is bounded above by $M(G)$, and the first and last entry of each element of $\mathbf{m}(G)$ are bounded above by $\mathrm{M}_+(G)$.*
(3) *If T is a tree, then the first and last entry of each element of $\mathbf{m}(T)$ equal 1.*
(4) *If G is an odd unicyclic graph, then each element of $\mathbf{m}(G)$ has its first or last entry equal to 1.*

PROOF. By Example 2.4, the eigenvalues of each $A \in \mathcal{S}(P_n)$ are distinct. Hence (1) holds.

By definition, the first entry of each $\mathbf{n} \in \mathbf{mult}(G)$ is at most $\mathrm{M}(G)$. If the least (respectively, greatest) eigenvalue of $A \in \mathcal{S}(G)$ is λ_1 (respectively, λ_q), then $A - \lambda_1 I$ (respectively, $\lambda_q I - A$) is positive semidefinite and $\mathrm{mult}_A(\lambda_1) = \mathrm{null}(A - \lambda_1 I)$

[1] We caution that the notions of a spectrally arbitrary sign-pattern and a spectrally arbitrary ordered multiplicity list are distinct.

(respectively, $\operatorname{mult}_A(\lambda_q) = \operatorname{null}(\lambda_q I - A)$). The upper bound for the first and last entries of each element of $\mathbf{m}(G)$ now follows from the definition of $\mathrm{M}_+(G)$. Thus (2) holds.

Theorem 1.14 implies (3).

To establish (4), let $A \in \mathcal{S}(G)$. Then either A or $-A$ has an even number of negative entries above the main diagonal corresponding to edges of the odd cycle in G. Using this and Lemma 1.13 we see that either A or $-A$ is diagonally similar to a matrix whose off-diagonal entries are nonnegative. Hence, as in the proof of Theorem 1.14, either the least or greatest eigenvalue of A is simple. This yields (4). □

The set of multiplicity lists for a star is also known. The number of simple eigenvalues of a matrix A is denoted by $\operatorname{sim}(A)$, and the minimum number $\operatorname{sim}(A)$ over $A \in \mathcal{S}(G)$ is denoted by $\operatorname{sim}(G)$. For a non-increasing sequence $\mathbf{n} = (n_1, n_2, \ldots, n_{q-1}, n_q)$ of positive integers, the number of ones in $\mathbf{n}$ is denoted by $s(\mathbf{n})$; that is, $n_{q-s(\mathbf{n})} > 1$ and $n_{q-s(\mathbf{n})+1} = 1$ when $s(\mathbf{n}) < q$.

THEOREM 6.2. *Let $n \geq 3$. Then*

(1) *For each $A \in \mathcal{S}(K_{1,n-1})$, $q(A) \geq 3$ and $\operatorname{sim}(A) \geq \frac{q(A)+1}{2}$.*
(2) *$\mathbf{mult}(K_{1,n-1})$ is the set of all non-increasing sequences $\mathbf{n} = (n_1, \ldots, n_q)$ of $q \geq 3$ positive integers with $s(\mathbf{n}) \geq \frac{q+1}{2}$.*
(3) *$s(K_{1,n-1}) = 2$ for $n \geq 4$, and $s(K_{1,2}) = 3$.*

PROOF. Without loss of generality assume that vertex n is the center of $K_{1,n-1}$. First, consider $A \in \mathcal{S}(K_{1,n-1})$ with $\mathbf{mult}(A) = (n_1, \ldots, n_{q(A)})$. Let $q = q(A)$ and $s = \operatorname{sim}(A)$. By Theorem 6.1(3), $s \geq 2$. Since $n \geq 3$, $q \geq 3$. Let R be the set of $q - s$ eigenvalues of A having multiplicity at least 2. Note that vertex n is the only vertex of $K_{1,n-1}$ of degree at least 3. For each $\mu \in R$, Theorem 2.19 implies that $\operatorname{mult}_{A(n)}(\mu) = 1 + \operatorname{mult}_A(\mu)$. Thus

$$n - 1 \geq \sum_{\mu \in R} \operatorname{mult}_{A(1)}(\mu) = \sum_{\mu \in R} (1 + \operatorname{mult}_A(\mu)) = (q - s) + (n - s),$$

where the last equality uses the fact that $n = s + \sum_{\mu \in R} \operatorname{mult}_A(\mu)$. It now follows that $2s \geq q+1$, and thus $s \geq \frac{q+1}{2}$ is necessary. This establishes the first statement, which in turn implies that $q \geq 3$ and $s(\mathbf{n}) \geq \frac{q+1}{2}$ for each $\mathbf{n} = (n_1, \ldots, n_q) \in \mathcal{S}(K_{1,n-1})$.

To argue the sufficiency of the given conditions in (2), consider a non-increasing sequence of positive integers $\mathbf{n} = (n_1, \ldots, n_q)$ such that $q \geq 3$ and $s = s(\mathbf{n})$ satisfies $s \geq \frac{q+1}{2}$. Let $\lambda_1 < \cdots < \lambda_{q-s}$ be real numbers. The assumption $s \geq \frac{q+1}{2}$ implies that

$$n - 1 \geq n - (2s - q) = (q - s) + (n - s) = \sum_{i=1}^{q-s} (1 + n_i).$$

Thus, there is an $(n-1) \times (n-1)$ diagonal matrix $D = [d_{ij}]$ having $1 + n_i$ diagonal entries equal to λ_i for $i = 1, \ldots, q - s$ and whose remaining diagonal entries, if any, are distinct real numbers not in $\{\lambda_1, \ldots, \lambda_{q-s}\}$. Now let $A = [a_{ij}]$ be an $n \times n$ matrix in $\mathcal{S}(K_{1,n-1})$ such that $a_{ii} = d_{ii}$ for $i = 1, \ldots, n-1$. Then by Theorem 2.19, $\lambda_1, \ldots, \lambda_{q-s}$ are the multiple eigenvalues of A and $\operatorname{mult}_A(\lambda_i) = \operatorname{mult}_{A(n)}(\lambda_i) - 1 = n_i$ for $i = 1, \ldots, q - s$. As $\sum_{i=1}^{q-s} n_i = n - s$, A has $q - s$ multiple eigenvalues and s simple eigenvalues. Thus, $q(A) = q$ and $\mathbf{mult}(A) = \mathbf{n}$.

Statement (3) holds, since the adjacency matrix of $K_{1,n-1}$ has exactly two simple eigenvalues for $n \geq 4$, and $K_{1,2} = P_3$. $\square$

Theorem 6.1 suggests some questions about ordered multiplicity lists of graphs. One concerns the possible spectra for a graph and a given ordered multiplicity list. If the possible ordered multiplicity lists for a graph G have been determined and every possible ordered multiplicity list is spectrally arbitrary, then the IEP-G has been solved for G.

We now identify some families of trees for which every ordered multiplicity list is known to be spectrally arbitrary. A *generalized star* or *spider* is obtained from a star by subdividing zero or more edges possibly many times. A *double path* is obtained from two disjoint paths of order at least three by adding an edge between two vertices of degree two (one in each path). A *double generalized star* is obtained from two disjoint generalized stars, each of which has a high degree vertex, by inserting an edge between the two high degree vertices. A *linear tree* is a tree for which there is a path that contains each high degree vertex. The family of linear trees is the most general, as it contains the families of paths, double paths, stars, generalized stars, and double generalized stars. The resolution of the IEP for linear trees in [**206**] represents a significant breakthrough for the IEP for trees. Its statement and proof are quite technical and specific to the setting, and thus we don't give these here. We note that for all of these famiies except linear trees, the ordered multiplicity lists can be explicitly described. For linear trees, the order multiplicity lists are characterized by certain combinatorial conditions that don't lead to an easily describable set of ordered lists.

THEOREM 6.3. [**23, 138, 201, 202, 206**] *For the following trees, the possible ordered multiplicity lists have been characterized and every possible ordered multiplicity list is spectrally arbitrary:*

- *paths,*
- *double paths,*
- *stars,*
- *generalized stars,*
- *double generalized stars, and*
- *linear trees.*

Not every ordered multiplicity list of every tree is spectrally arbitrary. The graph T_{BF} in Figure 6.1 is known as the *Barioli-Fallat Tree* (although their original counterexample used the graph obtained from Figure 6.1 by adding a leaf to all but one of the pendant vertices). The Barioli-Fallat tree and variants of it seem to play roles for the IEP-G analogous to that played by the Petersen Graph for graph theory, by serving as convenient counterexamples to many questions. We show in Example 6.26 that T_{BF} has the ordered multiplicity list $(1, 2, 4, 2, 1)$ and that list is not spectrally arbitrary. This was originally established by applying the Parter-Wiener Theorem (Theorem 2.19) repeatedly, but we take another approach to illustrate path removal techniques.

Like the IEP for the path, the IEP for the cycle C_n has its historical roots of the study of vibrations, modeling a coupled system of masses in a ring (see Figure 6.2).

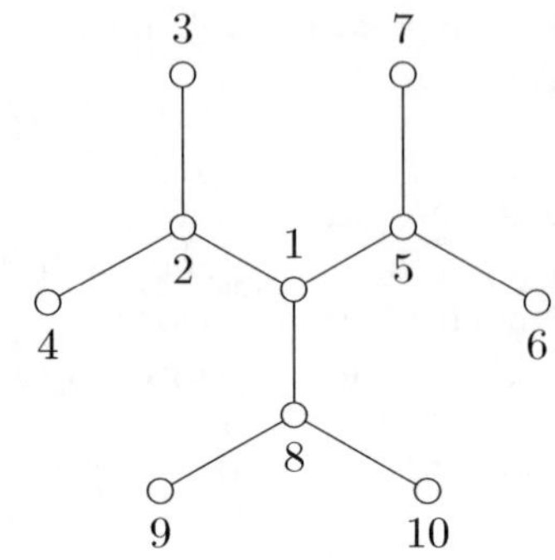

FIGURE 6.1. The Barioli-Fallat Tree T_{BF}.

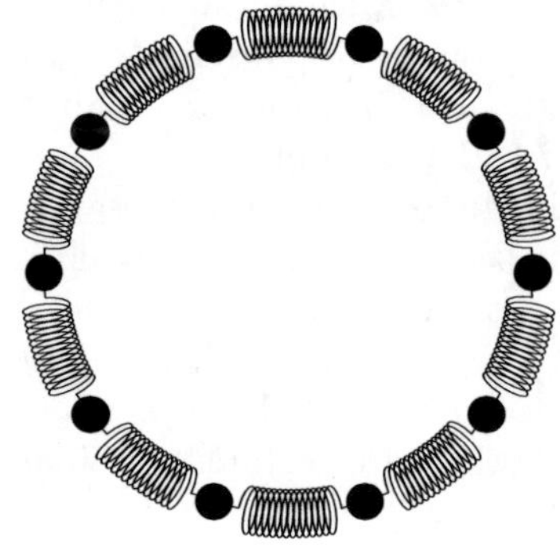

FIGURE 6.2. A coupled system of $n = 10$ masses in a ring.

A *periodic Jacobi matrix* is a real symmetric matrix of order $n \geq 3$ having the form

$$\begin{bmatrix} a_1 & b_1 & 0 & \cdots & b_n \\ b_1 & a_2 & b_2 & \ddots & \vdots \\ 0 & \ddots & \ddots & \ddots & 0 \\ \vdots & \ddots & b_{n-2} & a_{n-1} & b_{n-1} \\ b_n & \cdots & 0 & b_{n-1} & a_n \end{bmatrix},$$

where $b_i > 0$ for $i = 1, 2, \ldots, n$. If one has the same form except $b_i > 0$ for $i = 1, 2, \ldots, n-1$ and $b_n < 0$, then A is a *nearly periodic Jacobi matrix*. Thus the graph of an $n \times n$ periodic or aperiodic Jacobi matrix is the cycle C_n.

The matrix A is *sign-similar* to B provided there is a diagonal matrix D each of whose diagonal entries are in $\{1, -1\}$ with $D^{-1}AD = B$.

OBSERVATION 6.4. *Some simple facts about a matrix whose graph is a cycle are the following.*

(1) *By Lemma* 1.13, *each $A \in \mathcal{S}(C_n)$ is sign-similar to a periodic, or nearly periodic Jacobi matrix.*
(2) *No $A \in \mathcal{S}(C_n)$ is sign-similar to both a periodic and a nearly periodic Jacobi matrix.*

(3) *If n is odd and $A \in \mathcal{S}(C_n)$, then exactly one of A or $-A$ is sign-similar to a periodic Jacobi matrix.*
(4) *If n is even and $A \in \mathcal{S}(C_n)$, then either both A and $-A$ are sign-similar to a periodic Jacobi matrix, or both are sign-similar to a nearly periodic Jacobi matrix.*

The work of Ferguson [**132**] and Fernandes and da Fonseca [**134**] has solved the IEP-G for cycles (and thus determined the possible ordered multiplicity lists of cycles). Their proofs for the inverse eigenvalue problem of C_n are (necessarily) technical and analytic in nature. Here we provide a simple proof, based upon Theorem 1.18 and the Bifurcation Lemma for the SMP (Corollary 5.33), of the characterization of the multiplicity and ordered multiplicity lists. We note that statement (1) in the following theorem is equivalent to the assertion that $\mathbf{m}(C_n)$ consists of the refinements of $(2, 2, \dots, 2, 2)$ if n even; and the refinements of $(2, \dots, 2, 1)$ and $(1, 2, \dots, 2)$ if n is odd, where in each case the sum of the entries in the multiplicity list is n.

First we present some well known results about the spectra of a symmetric matrix and one of its principal submatrices.

LEMMA 6.5 (Cauchy interlacing (see [**196**])). *Let A be an $n \times n$ symmetric matrix whose eigenvalues are $\lambda_1 \le \lambda_2 \le \cdots \le \lambda_n$, and let the $(n-1) \times (n-1)$ principal submatrix $A(i)$ have eigenvalues $\mu_1 \le \mu_2 \le \cdots \le \mu_{n-1}$. Then*

$$\lambda_1 \le \mu_1 \le \lambda_2 \le \mu_2 \le \cdots \le \mu_{n-1} \le \lambda_n.$$

Moreover, μ is an eigenvalue of both A and $A(i)$ if and only if A has an eigenvector corresponding to μ whose i-th entry is zero.

LEMMA 6.6. [**139**] *Let T be a tree with $A \in \mathcal{S}(T)$ and i be a vertex of T. Then neither the largest nor smallest eigenvalue of A is an eigenvalue of $A(i)$.*

PROOF. By Lemma 1.13, A is diagonally similarly to a matrix whose off-diagonal entries are nonnegative, and hence its largest eigenvalue of A is simple and its (unique up to scalar multiplication) is nowhere zero. Thus, by Lemma 6.5, the largest eigenvalue of A is not an eigenvalue of $A(i)$.

The analogous result for the smallest eigenvalue follows by applying the previous argument to $-A$ □

THEOREM 6.7. [**132, 134**]
Let $n \ge 3$ be an integer. Then

(1) $\mathbf{m}(C_n)$ *consists of all $(m_1, m_2, \dots, m_q)$ with $m_i \in \{1, 2\}$ for $i = 1, \dots, q$, $\sum_{i=1}^{q} m_i = n$, and if $m_i = m_{i+k} = 2$ and $m_{i+j} = 1$ for $j = 1, \dots, k-1$, then k is odd.*
(2) $\mathbf{mult}(C_n) = \{(2, 2, \dots, 2, 1, \dots, 1) :$ *the sum of the entries is n*$\}$.
(3) *Each $\mathbf{m} \in \mathbf{m}(C_n)$ is spectrally arbitrary.*

PROOF. We note that (1) implies (2), and the combined results of [**132, 134**] give (3).

We first prove that the conditions in (1) are necessary for an ordered list to be an element of $\mathbf{m}(C_n)$. Let $A = [a_{ij}] \in \mathcal{S}(C_n)$, and without loss of generality assume that A is periodic or A is aperiodic (that is, $a_{i,i+1} > 0$ for $i = 1, \dots, n-1$). As $M(C_n) = 2$, each ordered list in $\mathcal{S}(G)$ is a list of ones and twos. It suffices to show that for each multiple eigenvalue λ of A the number of simple eigenvalues of A greater than λ is even (respectively, odd) if A is nearly periodic (respectively, periodic).

Since λ is a multiple eigenvalue, there exists a nonzero eigenvector $\mathbf{x} = [x_i]$ of A corresponding to λ whose first coordinate is zero. Hence $\mathbf{x}(1)$ is an eigenvector of $A(1)$ corresponding to λ.

As a pendant vertex, or a pair of consecutive vertices, of a path is a zero-forcing set of the path (see Example 2.10), both x_2 and x_n are nonzero, and for $i \in \{3, \dots, n-2\}$ at least one of x_i and x_{i+1} is nonzero. Also for $i \in \{3, \dots, n-1\}$, if $x_i = 0$ then $A\mathbf{x} = \lambda\mathbf{x}$ implies that $x_{i-1}x_{i+1} < 0$. Let

$$s = |\{i \in \{2, \dots, n-1\} : x_i x_{i+1} < 0 \text{ or } x_i = 0\}|.$$

Then $(-1)^s x_2$ and x_n have the same sign. Equating the first entries of $A\mathbf{x} = \lambda\mathbf{x}$ we see that x_2 and x_n have the same (respectively, opposite) sign when $a_{1,n} < 0$ (respectively, $a_{1,n} > 0$). Thus, s is even (respectively, odd) when A is nearly periodic (respectively, periodic).

By Theorem 1.18, there are exactly s eigenvalues of $A(1)$ (counting multiplicity) greater than λ. This, the fact that $\operatorname{mult}_A(\lambda) = 2$, and the Cauchy Interlacing inequalities (Lemma 6.5), imply that A has exactly s (counting multiplicity) eigenvalues greater than λ. Since each multiple eigenvalue of A has multiplicity 2, the number of simple eigenvalues of A greater than λ and s have the same parity. So the number of simple eigenvalues greater than λ is odd if A is a periodic Jacobian matrix and is even if A is a nearly periodic Jacobian matrix. This establishes the necessity of the conditions in (1).

By Theorem 4.5, if C is the $n \times n$ flipped-cycle matrix, then $C + C^\top$ has the SMP and ordered multiplicity list $(2, 2, \dots, 2)$ (respectively $(1, 2, \dots, 2)$ if n is even (respectively, odd). Furthermore, when n is odd $-C$ has the SMP with ordered multiplicity list $(2, 2, \dots, 2, 1)$. Thus by the Bifurcation Lemma for the SMP (Corollary 5.33), if n is even then each refinement of $(2, \dots, 2)$ and each of its refinements is an element of $\mathbf{m}(C_n)$, and if n is odd then each refinement of $(2, \dots, 2, 1)$ or of $(1, 2, \dots, 2)$ is an element of of $\mathbf{m}(C_n)$. Hence (1) holds. ☐

6.2. Constructive Techniques

This section discusses a variety of constructive techniques that have been used to study the IEP-G and multiplicity lists. The most recently developed techniques, the strong spectral property and the strong multiplicity property, are discussed in Chapters 3-5 are inherently non-constructive.

Recall (see Section 4.3) that a *Givens* rotation is an $n \times n$ orthogonal matrix $G(i, j, \theta)$ representing the transformation that rotates the plane spanned by the standard basis vectors $\mathbf{e}_i$ and $\mathbf{e}_j$ by an angle of θ radians, and sends each other standard basis vector to itself. For example, for $n = 5$,

$$G(2, 4, \theta) = \begin{bmatrix} 1 & 0 & 0 & 0 & 0 \\ 0 & \cos\theta & 0 & -\sin\theta & 0 \\ 0 & 0 & 1 & 0 & 0 \\ 0 & \sin\theta & 0 & \cos\theta & 0 \\ 0 & 0 & 0 & 0 & 1 \end{bmatrix}$$

The conjugate of the square matrix A by the invertible matrix P is the matrix $P^{-1}AP$. Each conjugate of A has the same spectrum as A; and if A is symmetric and P is orthogonal then the conjugate of A by P is symmetric. The graph of the conjugate $G(i,j,\theta)^\top AG(i,j,\theta)$ of A is readily obtained from $\mathcal{G}(A)$ when θ is sufficiently small and nonzero.

LEMMA 6.8. *Let A be an $n \times n$ matrix, and let i and j be distinct elements of $\{1,\dots,n\}$. Let $B = G(i,j,\theta)^\top AG(i,j,\theta)$. Then for θ nonzero and sufficiently small in magnitude the following hold:*

(1) *If $a_{ii} \neq a_{jj}$ or $a_{ij} \neq 0$, then $\mathcal{G}(B)$ is obtained from $\mathcal{G}(A)$ by inserting each edge of the form jk where $k \in N[i]$, and each edge of the form $i\ell$ where $\ell \in N[j]$.*
(2) *If $a_{ii} = a_{jj}$ and $a_{ij} = 0$, then $\mathcal{G}(B)$ is obtained from $\mathcal{G}(A)$ by inserting each edge of the form jk where $k \in N(i)$, and each edge of the form $i\ell$ where $\ell \in N(j)$.*

PROOF. Without loss of generality, we assume that $i = 1$ and $j = 1$, Thus,

$$A = \left[\begin{array}{cc|c} a_{11} & a_{12} & \mathbf{x}^\top \\ a_{12} & a_{22} & \mathbf{y}^\top \\ \hline \mathbf{x} & \mathbf{y} & A(\{1,2\}) \end{array}\right].$$

Calculations show that

$$B = \left[\begin{array}{cc|c} b_{11} & b_{12} & \mathbf{x}^\top \cos\theta + \mathbf{y}^\top \sin\theta \\ b_{12} & b_{22} & -\mathbf{x}^\top \sin\theta + \mathbf{y}^\top \cos\theta \\ \hline \mathbf{x}\cos\theta + \mathbf{y}\sin\theta & -\mathbf{x}\sin\theta + \mathbf{y}\cos\theta & A(\{1,2\}) \end{array}\right],$$

where $b_{12} = (\cos^2\theta - \sin^2\theta)a_{12} + (a_{22} - a_{11})\cos\theta\sin\theta$. Thus, for all but finitely many θ in $[0, 2\pi]$ the kth entries of $\mathbf{x}\cos\theta + \mathbf{y}\sin\theta$ and $-\mathbf{x}\sin\theta + \mathbf{y}\cos\theta$ will be nonzero if and only if the kth entry of x or the kth entry of y is nonzero. Thus for $k \notin \{1,2\}$ and θ sufficiently small in magnitude, k is adjacent to 1 in $\mathcal{G}(B)$ if and only if k is adjacent to 1 or 2 in $\mathcal{G}(A)$. If $a_{11} \neq a_{22}$ or $a_{12} \neq 0$, then $b_{12} \neq 0$ for θ sufficiently small in magnitude, and thus $\{1,2\}$ is an edge in $\mathcal{G}(B)$. If $a_{11} = a_{22}$ and $a_{12} = 0$, then $b_{12} = 0$ and $\{1,2\}$ is neither an edge of $\mathcal{G}(A)$ nor $\mathcal{G}(B)$. $\square$

REMARK 6.9. One can also show that for A and B as in the proof of Lemma 6.8, if $a_{ii} \neq a_{jj}$ and $a_{ij} = 0$, then as θ varies over all small in magnitude values, b_{jj} takes on infinitely many values.

Vertices v and w of a graph are *twins* provided they have the same set of neighbors (except possibly v and w). If the twins v and w are not adjacent, then they are *independent twins*; if v and w are adjacent, then v and w are *adjacent twins.* Define an *independent twinning* (respectively, *adjacent twinning*) of the graph at vertex $v \in V(G)$ as the graph obtained from G by appending a new vertex x such that $N(v) = N(x)$ (respectively, $N[v] = N[x]$).

COROLLARY 6.10. *Let $A = [a_{ij}]$ be an $n \times n$ matrix and λ be a real number. Then the following hold:*

(1) *If $a_{n,n} \neq \lambda$, then* $\operatorname{spec}(A) \cup \{\lambda\}$ *is a realizable spectrum for the adjacent twinning of $\mathcal{G}(A)$ at n. Additionally, there is such a realization whose last diagonal entry is not one of its eigenvalues.*
(2) *If $a_{n,n} = \lambda$, then* $\operatorname{spec}(A) \cup \{\lambda\}$ *is a realizable spectrum for the independent twinning of $\mathcal{G}(A)$ at n.*

PROOF. This follows from Lemma 6.8 and Remark 6.9 applied to the matrix $A \oplus [\lambda]$ with $i = n$ and $j = n + 1$. □

The solution to the IEP for complete graphs given in Theorem 4.14 due to [**35**] uses Givens rotations and (implicitly) Corollary 6.10. Using the terminology introduced in this chapter this becomes the following corollary.

COROLLARY 6.11. *For $n \geq 2$, $\mathbf{m}(K_n)$ is the set of ordered lists $\mathbf{m} = (m_1, \ldots, m_q)$ of positive integers where $q \geq 2$ and $\sum_{i=1}^{q} m_i = n$. Moreover, each element of $\mathbf{m}(K_n)$ is spectrally arbitrary.*

Barrett et al [**35**] also use Givens rotations to constructively solve the IEP for several families of graphs that are almost complete. To illustrate their techniques, we show how to use Corollary 6.10 to characterize the multiplicity lists for such graphs. We note these characterizations do follow from their resolution of the IEP. We also show how the SMP can be used to non-constructively arrive at these characterizations.

COROLLARY 6.12. [**35**] *Let $G = K_n - e$ be the complete graph on $n \geq 3$ vertices missing one edge. Then $\mathbf{mult}(G)$ is the set of all non-increasing sequences $\mathbf{n} = (n_1, \ldots, n_q)$ of positive integers with $n - 2 \geq n_1$ and $\sum_{i=1}^{q} n_i = n$.*

PROOF. Without loss of generality e is the edge between 1 and 2. Since G is not complete, $\mathrm{M}(G) \leq n - 2$ and the given conditions are necessary.

To prove the conditions are sufficient, we prove by induction on n that for each non-increasing sequence $\mathbf{n} = (n_1, \ldots, n_q)$ of positive integers with $n - 2 \geq n_1$ and $\sum_{i=1}^{q} m_i = n$, there is a matrix $A \in \mathcal{S}(G)$ with $\mathbf{mult}(A) = \mathbf{n}$ and $a_{nn} \notin \operatorname{spec}(A)$.

For $n = 3$, the only possible multiplicity list is $(1, 1, 1)$. This can be achieved by

$$A = \begin{bmatrix} 0 & 0 & 1 \\ 0 & 1 & 1 \\ 1 & 1 & 0 \end{bmatrix} \tag{6.1}$$

and its last diagonal entry is not one of its eigenvalues.

Now assume $n \geq 3$ and any list satisfying the conditions for n has a realization in $\mathcal{S}(G)$ whose last diagonal entry is not one of its eigenvalues. Let $\mathbf{n} = (n_1, \ldots, n_q)$ be a non-increasing sequence of positive integers with $n - 1 \geq n_1$ and $\sum_{i=1}^{q} n_i = n + 1$.

First consider the case that $n - 2 \geq n_1$. Let $\mathbf{n}' = (n_1, n_2, \ldots, n_{q-1}, n_q - 1)$ if $n_q > 1$, and $\mathbf{n}' = (n_1, n_2, \ldots, n_{q-1})$ if $n_q = 1$. Then by induction $\mathbf{n}'$ is the multiplicity list of some matrix $B \in \mathcal{S}(G)$ whose last diagonal entry is not in $\operatorname{spec}(B)$. In the case $n_q > 1$, apply Corollary 6.10(1) to B and the eigenvalue of B corresponding to $n_q - 1$. In the case $n_q = 1$, apply Corollary 6.10 (1) to B and any real number different from the last diagonal entry of B and each eigenvalue of B. In each case we get a matrix whose graph is $K_{n+1} - e$ having multiplicity list $\mathbf{n}$ whose last diagonal entry is not one its eigenvalues.

The remaining case is $n_1 = n - 1$. Now let $\mathbf{n}' = (n_1 - 1, n_2, \ldots, n_q)$. Since $n \geq 3$, $n_1 - 1 \geq n_2$ or $\mathbf{n} = (2, 2)$. In the former case, by induction, $\mathbf{n}' \in \mathbf{mult}(G)$, and as before we can use Corollary 6.10(1) to get a realization in $\mathcal{S}(K_{n+1} - e)$ with multiplicity list $\mathbf{n}$ whose last diagonal entry is not one of its eigenvalues. In the latter case we use the matrix

$$
\begin{bmatrix} 1 & 0 & 1 & 1 \\ 0 & -1 & 1 & -1 \\ 1 & 1 & 0 & -1 \\ 1 & -1 & -1 & 0 \end{bmatrix}. \tag{6.2}
$$

Thus the result follows from induction. □

We note that one cannot realize the multiplicity $(2,2)$ list for a matrix in $\mathcal{S}(K_4 - e)$ by utilizing Corollary 6.10(1) as this would require doing an adjacent twinning of P_3, and a P_3 doesn't allow multiplicity list $(2,1)$.

REMARK 6.13. Assume $n \geq 5$ is an integer. Let $\mathbf{u}$ the $n \times 1$ all-ones vector, and $\mathbf{v}$ be the $n \times 1$ vector $\begin{bmatrix} 1 & -1 & \sqrt{\frac{n-2}{2}} & -\sqrt{\frac{n-2}{2}} & 0 & \cdots & 0 \end{bmatrix}^\top$. Note that $\mathbf{u}$ and $\mathbf{v}$ are orthogonal vectors of length $\sqrt{n}$. This implies that the matrix $A = \mathbf{u}\mathbf{u}^\top + \mathbf{v}\mathbf{v}^\top$ has spectrum $\{n^{(2)}, 0^{(n-2)}\}$, and multiplicity list $(n-2, 2)$. Since $n \geq 5$ it is easy to verify that $\mathcal{G}(A)$ is K_n minus the edge joining 1 and 2, and that A has the SSP. Corollary 5.33 can now be used to (non-constructively) prove the existence of matrices with graph $K_n - e$ for each possible list in Corollary 6.12 for $n \geq 5$.

COROLLARY 6.14. **[35]** *Let $n \geq 4$ be an integer and $G = K_n - P_3$ denote the graph obtained from K_n by deleting the edges on the path 1—2—3. Then the non-increasing sequence of positive integers $(n_1, \ldots, n_q)$ is in $\mathbf{mult}(K_n - P_3)$ if and only if $n - 2 \geq n_1$, $\sum_{i=1}^q n_i = n$ and $(n_1, \ldots, n_q) \neq (2,2)$.*

PROOF. We first show necessity of the conditions. Since G is not complete, $\mathrm{M}(G) \leq n - 2$. If $A \in \mathcal{S}(G)$ has $\mathbf{mult}(A) = (2,2)$, then (by scaling and shifting) there is a symmetric orthogonal matrix B having graph G. However for each $B \in \mathcal{S}(G)$ the first two rows are not orthogonal. Hence $(2,2) \notin \mathbf{mult}(G)$, and the stated conditions are necessary.

We now prove that by induction on n that for each non-increasing sequence $\mathbf{n} = (n_1, \ldots, n_q)$ of positive integers satisfying the stated conditions there is a matrix $A \in \mathcal{S}(G)$ with $\mathbf{mult}(A) = \mathbf{n}$ whose last diagonal entry is not an eigenvalue of A.

For $n = 4$ the possible multiplicity lists are $(1,1,1,1)$ and $(2,1,1)$. These can be achieved by

$$
\begin{bmatrix} 0 & 0 & 1 & 1 \\ 0 & 4 & 0 & 1 \\ 1 & 0 & 8 & 1 \\ 1 & 1 & 1 & 14 \end{bmatrix} \text{ and } \begin{bmatrix} 1 & 0 & 1 & 1 \\ 0 & 2 & 0 & 1 \\ 1 & 0 & 1 & 1 \\ 1 & 1 & 1 & 3/2 \end{bmatrix}, \tag{6.3}
$$

respectively. For the first of these, the Gersgorin disks are mutually disjoint and hence the eigenvalues are distinct. One can readily verify that none of the diagonal entries is an eigenvalue. For the latter, the eigenvalues are $7/2, 2, 0, 0$.

For $n = 5$ the possible multiplicity lists are $(3,2)$, $(3,1,1)$, $(2,2,1)$, $(2,1,1,1)$ and $(1,1,1,1,1)$. For the last four of these one can use Corollary 6.10(1) and one of (6.3) to give a matrix in $\mathcal{S}(K_5 - P_3)$ with the desired multiplicity list whose last diagonal entry is not one of its eigenvalues. The matrix

$$\begin{bmatrix} 0 & 0 & 1/\sqrt{3} & 1/\sqrt{3} & 1/\sqrt{3} \\ 0 & 0 & 0 & 1/\sqrt{2} & -1/\sqrt{2} \\ 1/\sqrt{3} & 0 & 2/3 & -1/3 & -1/3 \\ 1/\sqrt{3} & 1/\sqrt{2} & -1/3 & 1/6 & 1/6 \\ 1/\sqrt{3} & -1/\sqrt{2} & -1/3 & 1/6 & 1/6 \end{bmatrix} \tag{6.4}$$

is an orthogonal matrix (and hence has eigenvalues ± 1) with multiplicity list $(3, 2)$. Note its last diagonal entry is not one of its eigenvalues. Thus we have established the claim for $n = 4$ and $n = 5$.

Now assume the claim holds for some $n \geq 5$, and $\mathbf{n} = (n_1, \ldots, n_q)$ is a list satisfying the conditions for $K_{n+1} - P_3$. Consider the list $\mathbf{n}' = (n_1 - 1, \ldots, n_q)$. If $n_1 > n_2$, then $\mathbf{n}'$ is a non-increasing sequence of positive integers satisfying the criteria to be in $\mathbf{mult}(G)$. Hence by induction there is $B \in \mathcal{S}(G)$ having multiplicity list $\mathbf{n}'$ whose last diagonal entry is not one of its eigenvalues. As before, we apply Corollary 6.10(1) to give the existence of $A \in \mathcal{S}(K_{n+1} - e)$ with multiplicity list $\mathbf{n}$ and whose last diagonal entry is not one of its eigenvalues. Otherwise $n_1 = n_2$. Now let $\mathbf{n}' = (n_1, \ldots, n_{q-1}, n_q - 1)$ if $n_q \geq 2$, and $\mathbf{n}' = (n_1, \ldots, n_{q-1}, n_q - 1)$. Since $n \geq 5$ and $n_1 = n_2$, $n_1 \leq n - 2$. Thus, in either case $\mathbf{n}'$ is a non-increasing sequence of positive integers satisfying the criteria to be in $\mathbf{mult}(G)$, and by induction there is a matrix $A \in \mathcal{S}(G)$ with multiplicity list $\mathbf{n}'$ whose last diagonal entry is not one of it eigenvalues. We now apply Corollary 6.10 as we did in the proof of Corollary 6.12 to conclude that $\mathbf{n}$ is the multiplicity list of some matrix $B \in \mathcal{S}(K_{n+1} - P_3)$ whose last diagonal entry is not in $\operatorname{spec}(B)$. The result now follows from induction. □

REMARK 6.15. As in the previous remark, one can construct a matrix with multiplicity list $(n - 2, 2)$, graph $K_n - P_3$ and the SMP, and then use Corollary 5.33 to prove the existence of a matrix with graph $K_n - P_3$ for each multiplicity list given in Corollary 6.14, for $n \geq 5$.

A proof similar to those of Corollaries 6.12 and 6.14 that uses the known results for the IEP-G of small graphs (see [**31**]) can be given to show the following.

COROLLARY 6.16. *Let $G = K_{2,2,\ldots,2,1,1,\ldots,1}$ be the complete multipartite graph with $r \geq 1$ parts of size 2 and $t \geq 0$ parts of size 1, and let $n = 2s + t$. Let q, $n_1, \ldots, n_q$ be positive integers with $n_1 \geq \cdots \geq n_1 \geq 1$ and $n = n_1 + \cdots + n_q$. Then $(n_1, \ldots, n_q) \in \mathbf{mult}(G)$ if and only if $n - 2 \geq n_1$ and $n = n_1 + \cdots + n_q$.*

The spectral theorem for symmetric matrices asserts that if A is a (real) symmetric $n \times n$ matrix, then there exists an orthogonal matrix Q and a diagonal matrix $D = \operatorname{diag}(\lambda_1, \ldots, \lambda_n)$ such that $A = QDQ^\top$. This implies that the jth column $\mathbf{q_j}$ of Q is an eigenvector of A corresponding to the eigenvalue λ_j, and $A = \sum_{j=1}^n \lambda_j \mathbf{q_j}\mathbf{q_j}^\top$. We refer to either $A = QDQ^\top$ or $A = \sum_{j=1}^n \lambda_j \mathbf{q_j}\mathbf{q_j}^\top$ as a *spectral decomposition of* A. We now briefly discuss a technique for constructing certain spectra using information about spectral decompositions of associated matrices.

LEMMA 6.17. *Let A be an $m \times m$ symmetric matrix with spectral decomposition $A = \sum_{i=1}^m \lambda_i \mathbf{q}_i \mathbf{q}_i^\top$, B be an $n \times n$ symmetric matrix having spectral decomposition*

$B = \sum_{j=1}^{n} \lambda_j \mathbf{r}_j \mathbf{r}_j^\top$, *and* $C = [c_{ij}]$ *be an* $m \times n$ *matrix. Then the matrix*

$$M = \left[\begin{array}{c|c} A & \sum_{i=1}^{m}\sum_{j=1}^{n} c_{ij}\mathbf{q}_i\mathbf{r}_j^\top \\ \hline \sum_{i=1}^{m}\sum_{j=1}^{n} c_{ij}\mathbf{r}_j\mathbf{q}_i^\top & B \end{array}\right]$$

is similar to the matrix

$$\left[\begin{array}{c|c} \mathrm{diag}(\lambda_1,\ldots,\lambda_m) & C \\ \hline C^\top & \mathrm{diag}(\mu_1,\ldots,\mu_n) \end{array}\right] \tag{6.5}$$

PROOF. Let $Q = \begin{bmatrix} \mathbf{q}_1 & \cdots & \mathbf{q}_m \end{bmatrix}$ and $R = \begin{bmatrix} \mathbf{r}_1 & \cdots & \mathbf{r}_n \end{bmatrix}$. Then

$$(Q^\top \oplus R^\top) M (Q \oplus R)$$

is the matrix in (6.5) and is similar to M. □

A vector is *nowhere zero* provided each of its entries is nonzero.

COROLLARY 6.18. *Let* $A \in S(G)$ *have a nowhere zero eigenvector* $\mathbf{v}$ *corresponding to the eigenvalue* λ. *Then for each pair* (μ, τ) *of real numbers with* $\mu < \lambda < \tau$ *there is a matrix* $B \in S(G \vee K_1)$ *whose spectrum is* $(\mathrm{spec}(A) \setminus \{\lambda\}) \cup \{\mu, \tau\}$.

PROOF. Without loss of generality we may assume that $\mathbf{v}$ has length 1, and the spectral decomposition $A = \sum_{j=1}^{n} \lambda_j \mathbf{q}_j \mathbf{q}_j^\top$ has $\mathbf{q_n} = \mathbf{v}$. Let a and b be real numbers with $b \neq 0$. Then the matrix

$$\left[\begin{array}{c|c} A & b\mathbf{v} \\ \hline b\mathbf{v}^\top & a \end{array}\right]$$

has graph $G \vee K_1$, and by Lemma 6.17 its spectrum is $(\mathrm{spec}(A) setminus \{\lambda\}) \cup \mathrm{spec}(L)$, where L is the 2×2 matrix

$$L = \begin{bmatrix} \lambda & b \\ b & a \end{bmatrix}.$$

It can be verified that the eigenvalues μ_1 and μ_2 of L satisfy

$$\begin{aligned} a &= \mu_1 + \mu_2 - \lambda \\ -b^2 &= (\lambda - \mu_1)(\lambda - \mu_2). \end{aligned}$$

This implies that for each pair of distinct real numbers with $\mu_1 < \lambda < \mu_2$, there exist real numbers a and b with $b \neq 0$ such that L has spectrum μ_1 and μ_2. □

Another application of Lemma 6.17 is the following upper bound on $q(G \vee G)$ for an arbitrary graph G. In Chapter 8, we show that $q(G \vee G) = 2$, for connected graphs of order at least 2 .

COROLLARY 6.19. *Let* G *be a graph. Then* $q(G) \leq 4$.

PROOF. If G has no edges, then $q(G) = 1 < 4$. Otherwise, let L be the Laplacian matrix of G, and let $\sum_{j=1}^{n} \lambda_j \mathbf{q_j} \mathbf{q_j}^\top$ be a spectral decomposition of L. We may assume that $\lambda_1 = 0$ and $\mathbf{q_1} = \frac{1}{\sqrt{n}}\mathbb{1}$.

For t sufficiently small and $s \in \mathbb{R}$, let

$$B(s,t) = \left[\begin{array}{c|c} tA & \sum_{j=2}^{n} \sqrt{1 - t^2\lambda_j^2}\,\mathbf{q_j}\mathbf{q_j}^\top + s\mathbf{q_1}\mathbf{q_1}^\top \\ \hline \sum_{j=2}^{n} \sqrt{1 - t^2\lambda_j^2}\,\mathbf{q_j}\mathbf{q_j}^\top + s\mathbf{q_1}\mathbf{q_1}^\top & -tA \end{array}\right].$$

Since $\mathbf{q_1}$ is nowhere zero, $B(s,t) \in \mathcal{S}(G \vee G)$ for s sufficiently large. By Lemma 6.17, $B(s,t)$ is co-spectral with a matrix which is the direct sum of

$$\begin{bmatrix} t\lambda_j & \sqrt{1-t^2\lambda_j^2} \\ \sqrt{1-t^2\lambda_j^2} & -t\lambda_j \end{bmatrix} \quad \text{for } j = 2, \ldots, n \tag{6.6}$$

and

$$\begin{bmatrix} 0 & s \\ s & 0 \end{bmatrix}$$

Each matrix in (6.6) is orthogonal, and hence each have spectrum contained in $\{-1, 1\}$. Hence such $B(s,t)$ has at most 4 distinct eigenvalues. □

6.3. Related Graph Parameters

In this section we discuss several graph parameters related to multiplicity lists. A *partition* of the positive integer n is a non-increasing sequence $\mathbf{n} = (n_1, \ldots, n_q)$ of some number of positive integers whose sum is n. Associated with the partition $\mathbf{n}$ is its Ferrer's diagram that consists of n_i left-justified unit squares at level i for $i = 1, \ldots, q$. For example, the Ferrer's diagram for $\mathbf{n} = (3, 2, 1, 1, 1)$ is on the left in Figure 6.3. The *conjugate partition* of $\mathbf{n}$ is denoted by $\overline{\mathbf{n}}$ and is the partition $(q_1, \ldots, q_{n_1})$ where q_j is the number of boxes in column j of the Ferrer's diagram of $\mathbf{n}$. The conjugate partition of $(3, 2, 1, 1, 1)$ is $(5, 2, 1)$ and its Ferrer's diagram is on the right in Figure 6.3.

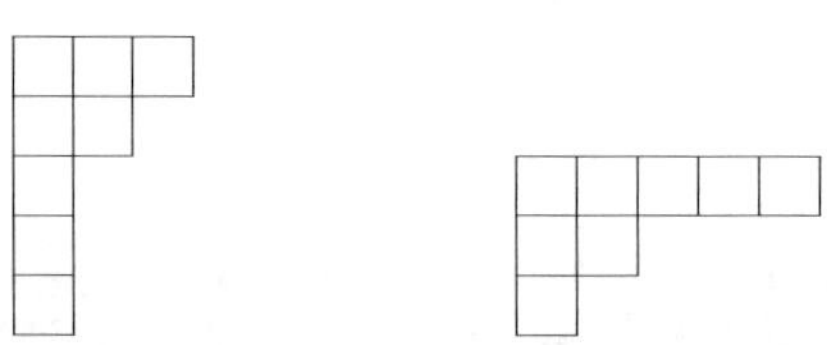

FIGURE 6.3. Ferrer's diagram of $(3, 2, 1, 1, 1)$ and its conjugate $(5, 2, 1)$

Basic facts about a partition and its conjugate partition are collected in the next lemma.

LEMMA 6.20. *Let $\mathbf{n} = (n_1, \ldots, n_q)$ be a partition of n, and $\overline{\mathbf{n}} = (q_1, \ldots, q_{n_1})$ be its conjugate partition. Then for $t = 1, \ldots, n_1$,*

(1) *the conjugate of $\overline{\mathbf{n}}$ is $\mathbf{n}$,*
(2) $q_t = |\{j : n_j \geq t\}|$,
(3) $\sum_{i=1}^{t} q_i = \sum_{j=1}^{q} \min\{n_j, t\}$ *and*
(4) $\sum_{i=t+1}^{n_1} q_i = \sum_{j=1}^{q} \max\{n_j - t, 0\}$.

The following interprets Lemma 6.20 in the case that $\mathbf{n}$ is a multiplicity list of a symmetric matrix.

THEOREM 6.21. *Let A be an $n \times n$ symmetric matrix, and let $\mathbf{n} = (n_1, \ldots, n_q)$ be its multiplicity list. Then*

(1) *n_1 is the maximum multiplicity of A;*

(2) *q is the number of distinct eigenvalues of A;*
(3) *q_t is the number of eigenvalues of A with multiplicity at least t;*
(4) $\sum_{j=1}^{t} q_j = \sum_{i=1}^{q} \min\{n_i, t\}$*; and*
(5) $\sum_{j=t+1}^{n_1} q_j = \sum_{i=1}^{q} \max\{n_j - t, 0\}$.

The Ferrer's diagram perspective suggests some natural questions: what is the maximum width (respectively, height) of a Ferrer's diagram corresponding to a matrix in $\mathcal{S}(G)$?, what is the minimum width (respectively, height) of a Ferrer's diagram corresponding to a matrix in $\mathcal{S}(G)$? or what can be said about the minimum or maximum number of simple eigenvalues? In each case either the maximum or minimum problem is known and the other is interesting and not completely resolved. In particular, we have the following basic results.

REMARK 6.22.

(1) The minimum width (maximum height, maximum number of simple eigenvalues) of a Ferrer's diagram for $\mathcal{S}(G)$ is 1 (respectively $|V(G)|$, $|V(G)|$) as there is a matrix in $\mathcal{S}(G)$ with distinct eigenvalues.
(2) The maximum width of a Ferrer's diagram for $\mathcal{S}(G)$ is the maximum multiplicity of G, which was introduced in Section 2.1.
(3) The minimum height of Ferrer's diagram for $\mathcal{S}(G)$, is $q(G)$, which is the topic of Chapter 8.
(4) For a tree T on at least 2 vertices, the minimum number $\operatorname{sim}(T)$ of simple eigenvalues satisfies $\operatorname{sim}(T) \geq 2$.
(5) For a graph G, $\operatorname{sim}(G) \leq |V(G)| - M(G) = \operatorname{mr}(G)$.

Let G be graph of order n and $i \in \{1, \ldots, n\}$. Two graph parameters related to partitions are the following. The maximum $n_1 + \cdots + n_i$ over all $\mathbf{n} = (n_1, \ldots, n_q)$ in $\mathbf{mult}(G)$, that is, the maximum sum of multiplicities of i eigenvalues of a matrix $A \in \mathcal{S}(G)$ is denoted by $M_i(G)$. The values of $M_2(G)$ have been studied in [**133**]. The minimum of $q_1 + q_2 + \cdots + q_i$ over all conjugate partitions of elements of $\mathbf{mult}(G)$ generalizes $q(G)$. Bounds on this minimum are given in Section 7.1.

6.4. Path Removal and Multiplicities for Trees

Multiplicity lists for trees have some special properties, which are interesting in their own right, but also lead to powerful generalizations and results that will be discussed in Chapters 7 and 8.

We begin with the following result that generalizes the observation that the nullity of a matrix decreases by at most one when one deletes any row and any column. The theorem was first proven for adjacency matrices of trees in [**160**]. It was extended to any symmetric matrix whose graph is a tree in [**211**] using properties of the Smith Normal Form of the matrix $xI - A$. A further extension to non-trees will be discussed in Section 7.1.

The proof presented here uses Schur complements. Recall that if $B = A[\alpha]$ is an invertible principal submatrix of A, then the *Schur complement* of A with respect to B is denoted by A/B and given by

$$A/B = A(\alpha) - A[\overline{\alpha}, \beta]B^{-1}A[\alpha, \overline{\beta}].$$

A crucial property for us is that $\operatorname{null} A = \operatorname{null}(A/B)$.

THEOREM 6.23. *Let T be a tree, α be the vertex set of a path in T, and $A \in \mathcal{S}(T)$. Then*

$$\operatorname{null} A(\alpha) \geq \operatorname{null} A - 1.$$

PROOF. Without loss of generality we may assume that $\alpha = \{1, 2, \ldots, \ell\}$,

$$1\text{–}2 - \cdots - \ell$$

is the path, the vertices of each connected component of $T(\alpha)$ occur consecutively, and that those corresponding to a component having a vertex adjacent to i proceed those corresponding to a component having a vertex adjacent to j whenever $i < j$. Thus A has the form

$$\left[\begin{array}{c|c} A[\alpha] & Y \\ \hline Y^\top & \oplus_{t=1}^{c} A[\beta_t] \end{array}\right],$$

where $A[\alpha]$ is irreducible tridiagonal, the β_j are the vertex sets of the connected components of $T(\alpha)$ and the nonzero entries of row i of Y occur in columns to the left of the nonzero entries of row $i+1$ of Y for $i = 1, \ldots, \ell - 1$.

Let $\widehat{A} = A(\ell, 1)$ and $B = A[\alpha](\ell, 1)$. Then

$$\widehat{A} = \left[\begin{array}{c|c} B & \widehat{Y} \\ \hline \widehat{X} & \oplus_{t=1}^{c} A[\beta_t] \end{array}\right],$$

where $\widehat{X}$ is obtained from $Y^\top$ by deleting its first column, and $\widehat{Y}$ is obtained from Y by deleting its last row. Note that B is lower triangular with nonzeros on its main diagonal. Thus B is invertible and its inverse is lower triangular. As deleting a row and column drops the nullity by at most one,

$$\operatorname{null} A - 1 \leq \operatorname{null}\ A(\ell, 1) = \operatorname{null}\ \widehat{A}/B. \tag{6.7}$$

The i, j-entry $\widehat{X} B^{-1} \widehat{Y}$ equals

$$\sum_{r=1}^{\ell-1} \sum_{s=1}^{\ell-1} (\widehat{X})_{i,r} (B^{-1})_{r,s} (\widehat{Y})_{s,j}.$$

This sum is nonzero only if there exist r and s with each of $(\widehat{X})_{i,r}$, $(B^{-1})_{r,s}$ and $(\widehat{Y})_{s,j}$ is nonzero. Note that $(\widehat{X})_{i,r}$ is nonzero only if there is an edge in T from $\ell + i$ to $r + 1$, $(B^{-1})_{r,s}$ is nonzero only if $r \geq s$, and $(\widehat{Y})_{s,j}$ is nonzero only if there is an edge in T from s to $\ell + j$. It follows from the ordering of the vertices that $\widehat{A}/B$ is block lower triangular with diagonal blocks $A[\beta_1], \ldots, A[\beta_c]$; that is, $\widehat{A}/B$ has the form

$$\begin{bmatrix} A[\beta_1] & O & \cdots & O \\ Z_{21} & A[\beta_2] & \ddots & \vdots \\ \vdots & \ddots & \ddots & O \\ Z_{c1} & \cdots & Z_{c,c-1} & A[\beta_c]. \end{bmatrix}$$

The algebraic multiplicity of 0 as an eigenvalue of $\widehat{A}/B$ equals the nullity of the matrix $\oplus_{t=1}^{c} A[\beta_t] = A(\alpha)$. The fact that the algebraic multiplicity of an eigenvalue is an upper bound on its geometric multiplicity and (6.7) yield $\operatorname{null} A - 1 \leq \operatorname{null} A(\alpha)$. □

Theorem 6.23 does not apply for graphs with cycles. For example, let

$$A = \begin{bmatrix} 0 & 1 & -1 & 1 & 1 \\ 1 & 5 & -3 & 1 & 1 \\ -1 & -3 & 1 & 1 & 1 \\ 1 & 1 & 1 & 6 & 0 \\ 1 & 1 & 1 & 0 & -2 \end{bmatrix}$$

Then 1–2 is a path in $\mathcal{G}(A)$, $\operatorname{null} A = 2$, but $\operatorname{null} A(\{1,2\}) = 0$.

COROLLARY 6.24. *Let F be a forest and $p_1, \ldots, p_k$ be vertex disjoint paths of F. Let $A \in \mathcal{S}(F)$ with the maximum multiplicity of an eigenvalue of A being m, and let λ be an eigenvalue of A. Then*

$$\operatorname{mult}_{A(\cup_{i=1}^k V(p_i))}(\lambda) \geq \max\{\operatorname{mult}_A(\lambda) - k, 0\}.$$

Furthermore,

$$|V(F)| - |\cup_{i=1}^k V(p_i)| \geq \sum_{i=k+1}^{m} q_i(A).$$

PROOF. The proof of the first part follows from repeated application of Theorem 6.23, and the fact that the multiplicity of an eigenvalue of a direct sum of matrices is the sum of the multiplicity of the eigenvalue for each summand.

Let $\lambda_1, \ldots, \lambda_q$ be the distinct eigenvalues of A. By the first part, if $m_A(\lambda_j) > k$ then λ_j is an eigenvalue of $A(\cup_{i=1}^k V(p_i))$ of multiplicity at least $\operatorname{mult}_A(\lambda_j) - k$. The sum of the multiplicities of the eigenvalues of $A(\cup_{i=1}^k V(p_i))$ is $|V(F)| - |\cup_{i=1}^k V(p_i)|$. Thus

$$|V(F)| - |\cup_{i=1}^k V(p_i)| \geq \sum_{i=1}^{q(A)} \max\{\operatorname{mult}_A(\lambda_i) - k, 0\} = \sum_{i=k+1}^{m} q_i(A),$$

by Theorem 6.21. □

Theorem 6.23 and Corollary 6.24 have many nice consequences. If F is a forest, we use $p_{i,j}$ to denote the unique path in F from vertex i to vertex j.

EXAMPLE 6.25. As a simple application we show that no matrix in $\mathcal{S}(T_{BF})$ has two eigenvalues of multiplicity 3 or more (or equivalently in the language of the last section, for $A \in \mathcal{S}(T_{BF})$ the conjugate $(q_1, \ldots, q_m)$ of $\mathbf{mult}(A)$ satisfies $q_3 \leq 1$).

We assume that T_{BF} has vertices labelled as in Figure 6.1. Suppose to the contrary that there is a $B \in \mathcal{S}(T_{BF})$ with eigenvalues λ and μ of B of multiplicity at least 3. Removing the vertices of the disjoint paths $p_{3,4}$ and $p_{7,10}$ results in the isolated vertices 6 and 9. Hence, by Corollary 6.24, and the symmetry of T_{BF}, $\{A[i], A[j]\} = \{\lambda, \mu\}$ whenever i and j are pendant vertices distance more than two apart. But this is not possible since T_{BF} has three pendant vertices that are mutually at distance 4 apart.

EXAMPLE 6.26. The list $\mathbf{m} = (1, 2, 4, 2, 1)$ is an example of a non-spectrally arbitrary multiplicity list for the Barioli-Fallat Tree T_{BF}. To see this we first show that the list is realizable. The matrix

$$A = \left[\begin{array}{c|ccc|ccc|ccc} 0 & 5 & 0 & 0 & 5 & 0 & 0 & 5 & 0 & 0 \\ \hline 5 & 0 & 3 & 4 & 0 & 0 & 0 & 0 & 0 & 0 \\ 0 & 3 & 0 & 0 & 0 & 0 & 0 & 0 & 0 & 0 \\ 0 & 4 & 0 & 0 & 0 & 0 & 0 & 0 & 0 & 0 \\ \hline 5 & 0 & 0 & 0 & 0 & 3 & 4 & 0 & 0 & 0 \\ 0 & 0 & 0 & 0 & 3 & 0 & 0 & 0 & 0 & 0 \\ 0 & 0 & 0 & 0 & 4 & 0 & 0 & 0 & 0 & 0 \\ \hline 5 & 0 & 0 & 0 & 0 & 0 & 0 & 0 & 3 & 4 \\ 0 & 0 & 0 & 0 & 0 & 0 & 0 & 3 & 0 & 0 \\ 0 & 0 & 0 & 0 & 0 & 0 & 0 & 4 & 0 & 0 \end{array}\right]$$

has distinct eigenvalues -10, -5, 0, 5 and 10, and ordered multiplicity list $\mathbf{m}(A) = (1, 2, 4, 2, 1)$.

Next we show that the distinct eigenvalues $\lambda_1 < \cdots < \lambda_5$ of each $B \in \mathcal{S}(T_{BF})$ with $\mathbf{m}(B) = (1, 2, 4, 2, 1)$ satisfy

$$\lambda_1 + \lambda_5 = \lambda_2 + \lambda_4. \tag{6.8}$$

Applying Corollary 6.24 to the vertex disjoint paths $p_{3,4}$, $p_{6,7}$, $p_{9,10}$, we conclude that $B[1] = [\lambda_3]$. Applying Corollary 6.24 to each diametrical path, we conclude that $B[v] = \lambda_3$ for each pendant vertex v of T_{BF}.

Using this, and again applying Corollary 6.24 to each diametrical path, we conclude that both λ_2 and λ_4 are eigenvalues of $B[\{3j-1, 3j, 3j+1\}]$ for $j = 1, 2, 3$. Thus, $\operatorname{spec}(B[\{3j-1, 3j, 3j+1\}]) = \{\lambda_2, \lambda_3, \lambda_4\}]$, which in turn implies that $\operatorname{tr} B[\{3j-1, 3j, 3j+1\}] = \lambda_2 + \lambda_3 + \lambda_4$ for $j = 1, 2, 3$.

Therefore,

$$\begin{aligned} \lambda_1 + 2\lambda_2 + 4\lambda_3 + 2\lambda_4 + \lambda_5 &= \sum_{j=1}^{5} \operatorname{mult}_B(\lambda_j)\lambda_j \\ &= \operatorname{tr} B \\ &= \operatorname{tr} B[1] + \sum_{j=1}^{3} \operatorname{tr} B[\{3j-1, 3j, 3j+1\}] \\ &= \lambda_3 + 3(\lambda_2 + \lambda_3 + \lambda_4). \end{aligned} \tag{6.9}$$

Equation (6.9) implies that (6.8) holds.

As previously noted the above was originally established by applying the Parter-Wiener Theorem (Theorem 2.19) repeatedly.

Theorem 6.23 leads to simple proofs, and additional insights, of several known results. Another simple proof of the bound $q(T) \geq \operatorname{diam}(T) + 1$ is given in Theorem 7.1.

COROLLARY 6.27. *Let T be a tree on n vertices. Then $q(T) \geq \operatorname{diam}(T) + 1$. Moreover, if $A \in \mathcal{S}(T)$, $q(T) = \operatorname{diam}(T) + 1$, and α is the vertex set of a diametrical path of A, then each eigenvalue μ of $A(\alpha)$ is an eigenvalue of A and* $\operatorname{mult}_A(\mu) = \operatorname{mult}_{A(\alpha)}(\mu) + 1$.

PROOF. Let $A \in \mathcal{S}(T)$. Let $\lambda_1, \ldots, \lambda_q$ be the distinct eigenvalues of A with multiplicities $m_1, \ldots, m_q$ respectively.

By Theorem 6.23, λ_j occurs with multiplicity at least $m_j - 1$ in spec($A(\alpha)$), and spec($A(\alpha)$) has $n - (\text{diam}(T) + 1)$ elements (counting multiplicity). Hence

$$n - \text{diam}(T) + 1 \geq \sum_{j=1}^{q} (m_j - 1) = \left(\sum_{j=1}^{q} m_j \right) - q = n - q. \tag{6.10}$$

It follows that $q \geq \text{diam}(T) + 1$. Hence $q(T) \geq \text{diam}(T) + 1$.

Now assume that $q = \text{diam}(T) + 1$. Then equality holds throughout (6.10). This implies that each non-simple eigenvalue λ_j of A is an eigenvalue of multiplicity $m_j - 1$ of $A(\alpha)$, and that $A(\alpha)$ has no eigenvalues other than these. The result now follows. □

COROLLARY 6.28. *Let T be a tree of order n. Then*

$$\text{M}(T) + \text{diam}(T) \leq n.$$

PROOF. There is a matrix $A \in \mathcal{S}(T)$ with $\text{null}\, A = M(T)$. Such a matrix has at most $n - M(T)$ distinct nonzero eigenvalues. Hence

$$\text{diam}(T) + 1 \leq q(T) \leq n - \text{M}(T) + 1,$$

and the result follows. □

REMARK 6.29. Let $p_1, \ldots, p_k$ be a path cover of the forest F obtained by recursively removing removing Nylen paths, and $A \in \mathcal{S}(F)$; that is, p_j is a Nylen path of $V(p_1) \cup \cdots \cup V(p_j - 1)$ for $j = 2, \ldots, k$. Then by Corollary 6.24, for each eigenvalue λ of A,

$$\text{mult}_{A(\cup_{i=1}^k V(p_i))}(\lambda) \geq \text{mult}_\lambda A - k.$$

As $A(\cup_{i=1}^k V(p_i))$ is the vacuous matrix, $0 = \text{mult}_{A(\cup_{i=1}^k V(p_i)}(\lambda)$, which implies that $k \geq \text{mult}_A(\lambda)$. for each eigenvalue λ of A. Hence, $\text{P}(F) \geq \text{M}(F)$ and if $\text{mult}_\lambda A = \text{M}(F)$, then λ is a (simple) eigenvalue of each $A[V(p_i)]$.

This latter observation suggests the following a way for constructing a matrix $A \in \mathcal{S}(F)$ with $\text{null}\, A = \text{M}(F)$. Let W be the set of vertices of F such that w is the unique high-degree vertex of $V(p_j)$ in the subgraph of G induced on $V(p_1) \cup \cdots \cup V(p_{j-1})$ for some $j = 1, \ldots, k - 1$. Now let A be any matrix in $\mathcal{S}(F)$ such that $A(W)$ is the Laplacian matrix of $F \setminus W$. Note that $F \setminus W$ has $k + |W|$ connected components. Hence $\text{null}\, A(W) = k + |W|$. This implies that $\text{null}\, A \geq \text{null}\, A(W) - |W| = k$. However, $k = \text{P}(F) \geq \text{M}(F) \geq \text{null}\, A$. Hence $\text{null}\, A = \text{M}(F)$, as desired.

CHAPTER 7

Rigid Linkages

7.1. Rigid Linkages

As we have seen Theorem 6.23 does not hold for arbitrary paths in arbitrary graphs. The following result suggests that it is possible to extend Theorem 6.23 to arbitrary graphs under an additional constraint on the type of path used.

THEOREM 7.1. *Let G be a graph, α be a path in G from i to j, and $A \in \mathcal{S}(G)$. If α is the unique path from u to v, then*

$$\operatorname{mult}_{A(V(\alpha))}(\lambda) \geq \operatorname{mult}_A(\lambda) - 1.$$

PROOF. Without loss of generality, we may assume that $i = 1$, $j = \ell$ and α is the path 1–2– $\cdots$ – ℓ. Let B be a connected component of $G - \alpha$. As α is the unique path in G from i to j there is at most one vertex of α adjacent to at least one vertex in B. Thus, one can now follow the proof Theorem 6.23 to get the desired result. □

It is now natural to ask if Theorem 7.1 can be extended to vertex disjoint paths. The next example shows that this is not the case.

EXAMPLE 7.2. Let G be the graph obtained from the complete graph with vertices $\{1, \ldots, 4\}$ by inserting vertex 5 and the edge joining 4 and 5. Then 1–2 and 3–4 are the unique vertex disjoint paths in G with one going from 1 to 2 and the other going from 3 to 4. The matrix

$$\begin{bmatrix} 1 & 1 & 1 & 1 & 0 \\ 1 & 1 & 1 & 1 & 0 \\ 1 & 1 & 1 & 1 & 0 \\ 1 & 1 & 1 & 2 & 1 \\ 0 & 0 & 0 & 1 & 1 \end{bmatrix}$$

has nullity 3. Deleting the vertices of the two paths results in the invertible matrix [1]. Thus, deleting the vertices on two vertex disjoint paths with such a uniqueness property can drop the nullity by more than 2.

We now describe an additional condition on sets of vertex disjoint paths that affords a generalization of Theorem 7.1. Let G be a graph. A subgraph $\mathcal{P}$ of G each of whose connected components is a path is called a *linkage* of G. We write $\mathcal{P} = \{p_1, \ldots, p_t\}$ to denote the linkage of G whose connected components are the paths $p_1, \ldots, p_t$, and define the *order* of $\mathcal{P}$ to be t. We say that $\mathcal{P}$ is *spanning* provided each vertex of G is a vertex of one of the paths of $\mathcal{P}$. The *end set* of a path is the set of its endpoints. As customary we allow a single vertex as a path, and its end set is a singleton set. The *pattern* of a linkage is the set consisting of its end sets. The linkage $\mathcal{P}$ is *unique* if it is the only linkage with its pattern. If $\mathcal{P}$ also spans G, then $\mathcal{P}$ is a *vital* linkage.

Let α and β be subsets of $V(G)$. The linkage $\mathcal{P}$ is an (α, β)-*linkage* provided α consists of one endpoint of each p_i in $\mathcal{P}$ and β consists of the other endpoints of the p_i. When p_i is a single vertex, that vertex belongs to both α and β. Thus, if $\mathcal{P}$ is an (α, β)-linkage of order t then $|\alpha| = t = |\beta|$. Note that $\mathcal{P}$ can be an (α, β)-linkage for many different (α, β); indeed for exactly 2^c where c is the number of p_j that are not isolated vertices.

The linkage $\mathcal{P}$ is (α, β)-*rigid* if $\mathcal{P}$ is the unique (α, β)-linkage in G. In the 4-cycle 1–2–3–4–1, $\mathcal{P} = \{1–4, 2–3\}$ is a $(\{1, 2\}, \{3, 4\})$-rigid linkage, and is a non-rigid $(\{1, 3\}, \{2, 4\})$-linkage. We say $\mathcal{P}$ is *rigid* when it is (α, β)-rigid for some α and β. Note that every rigid linkage is unique, and every linkage in a forest is rigid. The *rigid linkage number*, denoted by $\mathrm{rl}_G(t)$ (or $\mathrm{rl}(t)$ if G is clear), is the maximum number of vertices in an order t rigid linkage of G.

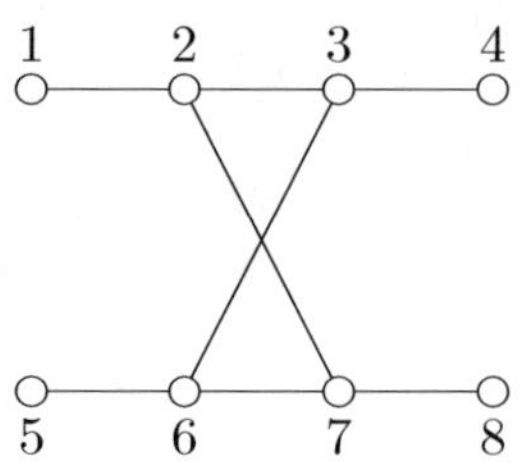

FIGURE 7.1. A graph for Example 7.3

EXAMPLE 7.3. Let G be the graph in Figure 7.1.

Let $\mathcal{P} = \{1–2–3–4, 5–6–7–8\}$. Then $\mathcal{P}$ is a linkage of G of order 2/ $\mathcal{P}$ is both a $(\{1, 5\}, \{4, 8\})$- and a $(\{1, 8\}, \{4, 5\})$-linkage. As $\mathcal{P}$ is the only linkage with $\{\{1, 4\}, \{5, 8\}\}$, $\mathcal{P}$ is a unique linkage. As $\mathcal{P}$ spans G, it is a vital linkage.

Let $\mathcal{Q} = \{1–2–7–8, 5–6–3–4\}$. Then $\mathcal{Q}$ and $\mathcal{P}$ are both $(\{1, 5\}, \{4, 8\})$-linkages. Hence neither $\mathcal{P}$ nor $\mathcal{Q}$ is a $(\{1, 5\}, \{4, 8\})$-rigid linkage. However, $\mathcal{P}$ is still rigid since it is a $(\{1, 8\}, \{4, 5\})$-rigid linkage.

The linkage $\{1–2–7, 5–6\}$ is a rigid linkage.

The linkage $\{7–6–3–4, 5\}$ is a $(\{7, 5\}, \{4, 5\})$-linkage, which is not unique.

The next theorem shows that for $A \in \mathcal{S}(G)$ and an (α, β)-rigid linkage $\mathcal{P}$ of G, $\det A(\alpha, \beta)$ is a nonzero multiple of $\det A(V(\mathcal{P}))$. The proof of this theorem requires a short foray into digraphs.

A *digraph* $\mathcal{D}$ consists of a finite set V whose elements are called vertices, and a collection E of ordered pairs (v, w) in $V \times V$. The elements of E are called arcs of $\mathcal{D}$; an arc of the form (v, v) is a loop at v. We use $v \to w$ to indicate that $(v, w) \in E$, and say that $\mathcal{D}$ has an arc from v to w. Note E need not be symmetric, that is, if $\mathcal{D}$ contains the arc from v to w, then it may or may not contain the arc from w to v. A path of length ℓ in $\mathcal{D}$ consists of distinct vertices $v_0, v_1 \ldots, v_\ell$ where $v_i \to v_{i+1}$ for $i = 0, \ldots, \ell$. A cycle of length ℓ consists of distinct vertices $v_1 \ldots, v_\ell$ such that $v_i \to v_{i+1}$ for $i = 1, \ldots, \ell - 1$ and $v_\ell \to v_1$. Note that here a cycle of length 1 or 2 is allowed. A *disjoint cycle union* γ of $\mathcal{D}$ is a collection of vertex disjoint cycles of $\mathcal{D}$. We denoted the vertex set of γ by $V(\gamma)$. When $V(\gamma) = V(\mathcal{D})$, we say γ is a *spanning disjoint cycle union*. The set of all disjoint cycle unions with vertex set S is denoted by $\mathrm{DCU}(S)$; when $S = V(\mathcal{D})$ we abbreviate this to $\mathrm{DCU}(\mathcal{D})$. The sign of a disjoint cycle union γ is defined as $(-1)^{|V(\gamma)|-c(\gamma)}$ where $c(\gamma)$ is the number of directed cycles in γ.

A weighting of $\mathcal{D}$ is a labelling $\mathrm{wt}(i \to j)$ of the arcs of $\mathcal{D}$ by nonzero real numbers, (or more generally, nonzero polynomials). The weight, $\mathrm{wt}(\alpha)$, of a subset α of arcs (or a subgraph) is the product of the weights of the arcs. An $n \times n$ matrix $M = [m_{ij}]$ determines the weighted digraph $\mathcal{D}(M)$ whose vertex set is $\{1, \dots, n\}$, $i \to j$ if and only if $m_{ij} \neq 0$, and the weight of $i \to j$ is m_{ij}. Note there is a loop at vertex i if and only if $m_{ii} \neq 0$.

LEMMA 7.4. *Let M be an $n \times n$ matrix (over $\mathbb{R}$ or $\mathbb{R}[x]$) with weighted digraph $\mathcal{D}(M)$. Then*

$$\det(M) = \sum_{\gamma \in \mathrm{DCU}(\mathcal{D})} \mathrm{sgn}(\gamma) \cdot \mathrm{wt}(\gamma).$$

PROOF. Let $M = \left[m_{ij}\right]$. From the standard definition of the determinant,

$$\det(M) = \sum_{\sigma \in S_n} \mathrm{sgn}(\sigma) \prod_{i=1}^{n} m_{i,\sigma(i)},$$

where S_n denotes the set of all permutations of $\{1, 2, \dots, n\}$. If a permutation $\sigma \in S_n$ gives a nonzero term in $\det(M)$, then σ corresponds to a subdigraph γ_σ in which each vertex has in-degree and out-degree 1. Such a subgraph corresponds to a spanning disjoint cycle union and has the properties $\mathrm{sgn}(\sigma) = \mathrm{sgn}(\gamma_\sigma)$ and $\prod_{i=1}^{n} m_{i,\sigma(i)} = \mathrm{wt}(\gamma_\sigma)$. Since there is a sign-preserving one-to-one correspondence between $\mathrm{DCU}(\mathcal{D})$ and the set of permutations giving a nonzero term in $\det(M)$, the result now follows. □

THEOREM 7.5. *Let G be a graph and $\mathcal{P} = \{p_1, \dots, p_t\}$ an (α, β)-rigid linkage of G. Let $A \in \mathcal{S}(G)$ and $\lambda \in \mathrm{spec}(A)$. Then*

(1) $\det A(\alpha, \beta)$ *is a nonzero multiple of* $\det A(V(\mathcal{P}))$, *and*
(2) $\mathrm{mult}_{A(V(\mathcal{P}))}(\lambda) \geq \mathrm{mult}_A(\lambda) - t$.

PROOF. Let $\alpha = \{i_1, \dots, i_t\}$ and $\beta = \{j_1, \dots, j_t\}$. We may assume that the endpoints of p_k are i_k and j_k. Define $\widehat{A}$ to be the matrix obtained from A by replacing column j_k with the standard basis vector $\mathbf{e}_{i_k}$ for $k = 1, \dots, t$. Then $\mathrm{null}\, \widehat{A} = \mathrm{null}\, A(\{i_1, \dots, i_t\}, \{j_1, \dots, j_t\})$ and

$$\det \widehat{A} = \pm \det A(\{i_1, \dots, i_t\}, \{j_1, \dots, j_t\}).$$

Let $\mathcal{D}$ be the weighted digraph corresponding to $\widehat{A}$. Note that $\mathcal{D}$ is obtained from $\mathcal{G}(A)$ by replacing each edge ij in $\mathcal{G}(A)$ by the arcs $i \to j$ and $j \to i$ each of weight a_{ij}, inserting a loop at i of weight a_{ii} whenever $a_{ii} \neq 0$, and for $k = 1, \dots, t$ replacing all arcs terminating at j_k by the single arc $i_k \to j_k$ of weight 1.

By Lemma 7.4, the nonzero terms in the determinant of $\widehat{A}$ are in one-to-one correspondence with the set of spanning disjoint cycle unions of $\mathcal{D}$. Consider $\gamma \in \mathrm{DCU}(\mathcal{D})$. Since $i_k \to j_k$ is the unique arc in $\mathcal{D}$ with terminal vertex j_k and γ is spanning, each of the arcs $i_k \to j_k$ is an arc of γ for $k = 1, \dots, t$. Since $\mathcal{P}$ is a rigid linkage, each cycle in γ contains at most one of the arcs $i_k \to j_k$ for $k \in \{1, 2, \dots, t\}$, and the cycle τ_k having the arc $i_k \to j_k$ consists of $i_k \to j_k$ and the path obtained from p_k by directing its edges accordingly. Thus γ consists of these k cycles $\tau_1, \ldots, \tau_t$, and a disjoint cycle union γ' of $A(V)$. Hence $\operatorname{sgn}(\gamma) = (-1)^{|V(\mathcal{P})|-t} \operatorname{sgn}(\gamma')$ and $\operatorname{wt}(\gamma) = \operatorname{wt}(\tau_t) \cdots \operatorname{wt}(\tau_t) \operatorname{wt}(\gamma')$. So

$$\det \widehat{A} = \pm \sum_{\gamma' \in \mathrm{DCU}(V(G) \setminus V(\mathcal{P}))} \operatorname{sgn}(\gamma') \cdot \operatorname{wt}(\gamma') = \det A(V(\mathcal{P})).$$

This establishes (1).

Now let $W = V(G) \setminus V(\mathcal{P})$ and τ a subset of W such that $A[\tau]$ is invertible and $|\tau| = \operatorname{rank} A(V(\mathcal{P}))$. Then $\mathcal{P}$ is a rigid linkage of the subgraph H of G induced on $V(\mathcal{P}) \cup \tau$. Let $B = A[V(H)]$. By statement (1), $\det B(\alpha, \beta)$ is a nonzero multiple of $\det B[\tau]$ But $B[\tau] = A[\tau]$ is invertible. Hence $B(\alpha, \beta)$ is invertible. In particular, $A(\alpha, \beta)$ has an invertible submatrix of size $|V(\mathcal{P})| - t + |\tau|$. Thus $\operatorname{rank} A(\alpha, \beta) \geq |V(\mathcal{P})| - t + |\tau|$, and

$$\operatorname{null} A - t \leq \operatorname{null} A(\alpha, \beta) \leq n - |V(\mathcal{P})| - |\tau| = |W| - |\tau| = \operatorname{null} A(V(\mathcal{P})). \quad \square$$

By applying Theorem 7.5 to an appropriate shift of A we obtain the following result that asserts that removing the edges of a rigid linkage of order t drops the multiplicity of an eigenvalue by at most t.

COROLLARY 7.6. *Let $\mathcal{P}$ be an (α, β)-rigid linkage of order t in a graph G. Then, for each $A \in \mathcal{S}(G)$ and each $\lambda \in \operatorname{spec}(A)$*

$$\operatorname{mult}_{A(V(\mathcal{P}))}(\lambda) \geq \operatorname{mult}_A(\lambda) - t.$$

Recall that as defined in Section 6.3, $q_j(A)$ is the number of distinct eigenvalues with multiplicity at least j.

THEOREM 7.7. *Let G be a graph, $A \in \mathcal{S}(G)$, $q_j = q_j(A)$, and t be a positive integer. Then*

(1) $\sum_{j=1}^{t} q_j \geq \operatorname{rl}(t)$ *and*

(2) $q(G) \geq \left\lceil \frac{\operatorname{rl}(t)}{t} \right\rceil$. *If the equality holds, then there is $A \in \mathcal{S}(G)$ each of whose eigenvalues has multiplicity at least t.*

PROOF. Let G be a graph on n vertices and $A \in \mathcal{S}(G)$ with $\operatorname{spec}(A) = \{\lambda_1^{(m_1)}, \ldots, \lambda_q^{(m_q)}\}$. Let $\mathcal{P}$ be a rigid linkage of order t of G with $|V(\mathcal{P})| = \operatorname{rl}(t)$. By Corollary 7.6,

$$\operatorname{mult}_{A(V(\mathcal{P}))}(\lambda_j) \geq m_j - t.$$

Since $A(V(\mathcal{P}))$ as order $n - \operatorname{rl}(t)$,

$$n - \operatorname{rl}(t) \geq \sum_{j=1}^{q} \max\{m_j - t, 0\}. \tag{7.1}$$

By Theorem 6.21(5)

$$\sum_{j=1}^{q} \max\{m_j - t, 0\} = q_{t+1} + \cdots + q_{m_1}. \tag{7.2}$$

As $\sum_{j=1}^{m_1} q_j = n$, inequality (1) follows from (7.1) and (7.2).

Now choose $A \in \mathcal{S}(G)$ with $q(A) = q(G)$. As $(q_1, q_2, \ldots, q_{m_1}))$ is in non-increasing order, $q(G)t = q_1 t \geq q_1 + q_2 + \cdots + q_t \geq \operatorname{rl}(t)$. Thus (2) holds. If equality holds, then $q(G) = q_1(A) = \cdots = q_t(A)$. By considering conjugate partitions, we see that each m_i is at least t. $\square$

In the next example we show that the inequality in Theorem 7.7(1) is tight.

EXAMPLE 7.8. [**135**] Consider the tree T shown in Figure 6.1. Computing the rigid linkage numbers is simplified by using the fact that in a tree every linkage is rigid. The rigid linkage numbers $\mathrm{rl}(t)$ are as follows.

- The diameter of T is 4. Thus, $\mathrm{rl}(1) = 5$.
- The $(\{3,7\},\{4,9\})$-linkage spans 8 vertices. Deleting any vertex of T results in a forest with path cover number at least 3. Hence $\mathrm{rl}(2) = 8$.
- The $(\{3,7,10\},\{4,9,10\})$-linkage spans 9 vertices and the path cover number of T is 4. Hence $\mathrm{rl}(3) = 9$.
- As the path cover number of T is 4, $\mathrm{rl}(4) = 10$.

The matrix adjacency matrix of A has multiplicity list $\mathbf{mult} = (4,2,2,1,1)$, which has conjugate partition $(5,3,1,1)$. Note that

$$\begin{aligned} &q_1 = 5 = \mathrm{rl}(1)\\ &q_1 + q_2 = 8 = \mathrm{rl}(2)\\ &q_1 + q_2 + q_3 = 9 = \mathrm{rl}(3) \text{ and}\\ &q_1 + q_2 + q_3 + q_4 = 10 = \mathrm{rl}(4). \end{aligned}$$

COROLLARY 7.9. *Let G be a graph and $A \in \mathcal{S}(G)$. Let $q_j = q_j(A)$. If $\mathrm{rl}(t) = \sum_{j=1}^{t} q_j$ and $\mathcal{P}$ is a rigid linkage of order t with $|V(\mathcal{P})| = \mathrm{rl}(t)$, then*

$$\mathrm{spec}(A(V(\mathcal{P})) = \{\lambda^{(\mathrm{mult}_A(\lambda)-t)} : \lambda \in \mathrm{spec}(A) \textit{ with } \mathrm{mult}_A(\lambda) > t\}.$$

PROOF. From Corollary 7.6, λ is an eigenvalue of $A(V(\mathcal{P}))$ with multiplicity at least $\mathrm{mult}_A(\lambda) - t$. Thus,

$$\mathrm{spec}(A(V(\mathcal{P})) \supseteq \{\lambda^{(\mathrm{mult}_A(\lambda)-t)} : \lambda \in \mathrm{spec}(A) \text{ with } \mathrm{mult}_A(\lambda) > t\}.$$

The order of $A(V(\mathcal{P}))$ is

$$|V(G)| - \mathrm{rl}(t) = |V(G)| - \sum_{i=1}^{t} q_i = \sum_{i=t+1}^{m_1} q_i = \sum_{\substack{\lambda \in \mathrm{spec}(A)\\ \mathrm{mult}_A(\lambda) > t}} (\mathrm{mult}_A(\lambda) - t),$$

where m_1 is the maximum multiplicity of an eigenvalue of A. Therefore $\mathrm{spec}(A)$ is exactly the described spectrum. □

7.2. Rigid Shortest Linkages

The notion of a rigid linkage is quite restrictive. For example, a $(\{u\},\{v\})$-linkage is a rigid linkage of order 1 in a graph G if and only if G can be obtained from the disjoint union of a path p from u to v and connected graphs $H_1, \ldots, H_m$ by joining some of the vertices of H_i to exactly one vertex of p for $i = 1, \ldots, m$. In particular if G has a $(\{u\},\{v\})$-rigid linkage with $u \neq v$, then each vertex a path between u and v is a cut-vertex.

The next result shows that imposing an additional constraint on the linkage one can obtain multiplicity results for non-rigid linkages. The proof of this result, and its extensions, makes use of the following concepts and results. This method was developed and used in [**211**].

Let $A = [a_{ij}]$ be an $n \times n$ real matrix, which not necessarily symmetric. The *characteristic matrix of* A is denoted by A_x and is the matrix $xI - A$ where x is an indeterminate. A $k \times k$ *minor* of A is the determinant of a $k \times k$ submatrix of

A. For $k = 1, \ldots, n$, the *kth determinantal factor* of A_x is defined as the greatest common divisor (over the ring $\mathbb{R}[x]$) of all $k \times k$ minors of A_x and is denoted by $\Delta_k(A_x)$.

By the Cauchy–Binet theorem, each $k \times k$ minor of a product of two matrices is a weighted sum of $k \times k$ minors of the first matrix (as well as the second). It follows that if P is an invertible matrix with entries in $\mathbb{R}$, or more generally a matrix with entries in $\mathbb{R}[x]$ whose determinant is ± 1, then $P^{-1}A_xP$ and A_x have the same determinantal divisors.

LEMMA 7.10. *Let A be an $n \times n$ real symmetric matrix with multiplicity list* $\mathbf{mult}(A) = (m_1, \ldots, m_{q(A)})$, *and A_x its characteristic matrix. Then*

$$\deg \Delta_{n-k+1}(A_x) = \sum_{i=1}^{q(A)} \max\{m_i - k + 1, 0\} = \sum_{j=k}^{m_1} q_j(A)$$

for $k = 1, \ldots, n$.

PROOF. As A is real symmetric there exists an invertible matrix P such that $P^{-1}AP$ is a diagonal matrix whose diagonal entries are the eigenvalues of $\lambda_1, \ldots, \lambda_{q(A)}$ of A with the appropriate multiplicities. By the comments preceding this lemma, in studying the determinantal divisors of A, we may without loss of generality assume that A is a diagonal matrix.t

The only nonzero minors of A_x are the principal ones. If $m_i \geq k$, then each $(n-k+1)\times(n-k+1)$ principal submatrix of A has at least m_i-k+1 diagonal entries equal to $x - \lambda_i$, and there is such a principal submatrix with exactly $m_i - k + 1$ diagonal entries equal to $x - \lambda_i$. If $m_i < k$ then there is a principal $(n - k + 1) \times (n - k + 1)$ submatrix having no diagonal entries equal to $x - \lambda_i$. Hence $\max\{m_i-k+1, 0\}$ is the largest power of $(x-\lambda_i)$ that divides $\Delta_{n-k+1}(A_x)$. The first equality now follows. The second equality is a consequence of $(q_1(A), \ldots, q_{m_1}(A))$ and $(m_1, \ldots, m_{q(A)})$ being conjugate partitions. □

Theorem 7.5 asserts that removing the vertices in a rigid linkage of order t decreases the nullity by at most t. The next example shows that this is not necessarily the case for a shortest path.

EXAMPLE 7.11. Consider the matrix

$$A = \begin{bmatrix} 2 & 0 & 1 & 1 \\ 0 & 2 & 1 & -1 \\ 1 & 1 & 1 & 0 \\ 1 & -1 & 0 & 1 \end{bmatrix}.$$

The $\mathcal{G}(A)$ is the 4-cycle 1–3–2–4–1, $\operatorname{null} A = 2$, 1–3 is the unique shortest path between 1 and 3, and $\operatorname{null} A(\{1,3\}) = 0$.

Next we present a different proof of Theorem 4.8.

PROOF. Let $\widehat{A_x}$ be the matrix obtained from A_x by replacing its uth column by $\mathbf{e}_v$. Consider a nonzero term τ in the determinantal expansion of $\widehat{A_x}$. Such τ corresponds to a disjoint union of cycles of $\mathcal{D}(\widehat{A_x})$, one of which contains the arc $v \to u$. Let p' be the path from u to v on the cycle containing $v \to u$ viewed as a path in G. If p' has length greater than ℓ, then $\deg(\tau) < n - \ell - 1$. Otherwise, $p' = p$ and $\deg(\tau) = n - \ell - 1$.

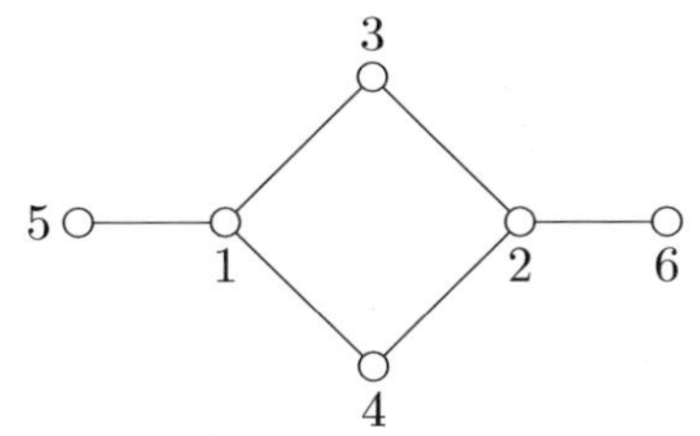

FIGURE 7.2. The double kite

Hence $\det(\widehat{A_x})$ is the sum of one term of degree $n-\ell-1$ and terms of smaller degree. Thus $\deg\det(\widehat{A_x}) = n-\ell-1$. This implies that A_x has an $(n-1)\times(n-1)$ minor whose degree is $n-\ell-1$, which in turn implies $\deg\Delta_{n-1}(xI-A) \le n-\ell-1$. Now apply Lemma 7.10 to obtain

$$n-\ell-1 \ge \deg\Delta_{n-1}(A_x) = \sum_{i=1}^{q(A)} m_i(A) - 1 = n - q(A).$$

Therefore, $q(A) \ge \ell + 1$. □

EXAMPLE 7.12. Let G be a graph obtained from a path p_1 on k vertices and a path p_2 on ℓ vertices by joining a pendant vertex of p_2 to a single non-pendant or to two consecutive vertices of p_1. Thus G has $k+\ell$ vertices. The path p_1 is the unique shortest path between its endpoints. Hence by Proposition 4.8, $q(G) \ge k$. In particular, if $\ell = 1$, then $q(G) \ge |V(G)| - 1$, and equality holds since G is not a path.

Now we move to the more general setting. Let G be a graph having an (α,β) linkage, and let s be the minimum number of vertices in an (α,β)-linkage $\mathcal{P}$ in G. An (α,β)-linkage of the graph G is an (α,β)-*rigid shortest linkage* if it is the unique (α,β)-linkage on s vertices. A linkage $\mathcal{P}$ is a *rigid shortest linkage* if it is an (α,β)-rigid shortest linkage for some (α,β).

The *rigid shortest linkage number*, denoted by $\mathrm{rsl}_G(t)$ (or $\mathrm{rsl}(t)$ if G is clear), is the maximum number of vertices in an order t rigid shortest linkage of G. Since each rigid linkage is a rigid shortest linkage, $\mathrm{rsl}(t) \ge \mathrm{rl}(t)$. A cycle of order $n \ge 3$ shows the equality needs not hold since $\mathrm{rsl}(1) = \lceil \frac{n}{2} \rceil$ and $\mathrm{rl}(1) = 1$.

It is shown in [**135**] that if $\mathcal{P}$ is a spanning (α,β)-rigid linkage for G of order t, then α is a zero-forcing set for G, and hence $t \ge \mathrm{Z}(G)$. T If (α,β)-linkage is a rigid shortest linkage that spans the graph, then every (α,β)-rigid linkage spans, and hence is necessarily a rigid shortest linkage. Hence a rigid shortest linkage that spans the graph is necessarily a rigid linkage. Thus, if $\mathcal{P}$ is a spanning shortest rigid linkage of order t, then $t \ge \mathrm{Z}(G)$. The connections between rigid linkages and zero-forcing are described more in Section 9.7.

By Corollary 9.125, any spanning rigid linkage has order at least $Z(G)$. Hence $\mathrm{rsl}(t) = |V(G)|$ implies $t \ge Z(G)$.

The next result is the RSL-analog of Theorem 7.7.

THEOREM 7.13. [**135**] *Let G be a graph, $A \in \mathcal{S}(G)$, $q_j = q_j(A)$, and t be a positive integer. Then*

(1) $\sum_{j=1}^{t} q_j \geq \operatorname{rsl}(t)$ *and*

(2) $q(G) \geq \left\lceil \frac{\operatorname{rsl}(t)}{t} \right\rceil$. *If the equality holds, then there is $A \in \mathcal{S}(G)$ each of whose eigenvalues has multiplicity at least t.*

PROOF. Let $\alpha = \{i_1, \ldots, i_t\}$ and $\beta = \{j_1, \ldots, j_t\}$ be subsets of $V(G)$ such that $\mathcal{P}$ is an (α, β)-rigid shortest linkage with $|V(\mathcal{P})| = s$. Let $A \in \mathcal{S}(G)$. Define $\widehat{A_x}$ to be the matrix obtained from A_x by replacing column j_k by $\mathbf{e}_k$ for $k = 1, \ldots, t$. By Lemma 7.4, a nonzero term in the $\det \widehat{A_x}$ corresponds to a disjoint cycle union of the weighted digraph $\mathcal{D}(\widehat{A_x})$ that contains each of the arcs $i_1 \to j_1, \ldots, i_t \to j_t$. The cycles that contain at least one of these determine an (α, β)-linkage $\mathcal{L}$ in G; namely delete these arcs from the cycles to obtain the linkage. If $\mathcal{L}$ spans more than s vertices, then the degree of this term, as a polynomial of x, is less than $n - s$. Otherwise $\mathcal{L}$ spans exactly s vertices, and thus is unique, consists of cycles formed by the k-th paths of $\mathcal{P}$ and the arc (i_k, j_k), and a disjoint union of cycles in on the vertices outside of $V(\mathcal{P})$. It follows that this term has degree $n - s$. Hence $\deg \det(\widehat{A_x}) = n - s$. Note that $\det \widehat{A_x} = \pm \det A_x(\alpha, \beta)$, and hence $\det \widehat{A_x}$ is an $(n-t) \times (n-t)$ minor of A. This implies that $\Delta_{n-t} \leq n - s$. The result now follows from Lemma 7.10. □

Theorem 7.13 was used in [**135**] to determine rigid shortest linkage numbers for the following families of graphs.

THEOREM 7.14. [**135**]

(1) *For $n \geq 2$*

$$\operatorname{rsl}_{K_n}(t) = t + 1 \text{ for } 1 \leq t \leq n - 1,$$

and there exists a matrix in $\mathcal{S}(K_n)$ that simultaneously achieves equality in each bound in Theorem 7.13(1).

(2) *For $n \geq 3$*

$$\operatorname{rsl}_{C_n}(t) = \begin{cases} \lceil \frac{n}{2} \rceil & \text{if } t = 1, \text{ and} \\ n & \text{if } t = 2, \end{cases}$$

and there exists a matrix in $\mathcal{S}(C_n)$ that simultaneously achieves equality in each bound in Theorem 7.13(1).

(3) *For $1 \leq m \leq n$ and $n \neq 1$,*

$$\operatorname{rsl}_{K_{m,n}}(t) = \begin{cases} 2 & \text{if } t = 1 \text{ and } m \geq 2, \\ 3 & \text{if } t = 1 \text{ and } 1 = m < n, \text{ and} \\ t + 2 & \text{if } 2 \leq t \leq m + n - 2. \end{cases}$$

For $m = n$ and $n \geq 3$ equality in each bound in Theorem 7.13(1) *can be achieved but not with a single matrix.*

(4) *The hypercube Q_n satisfies $\operatorname{rsl}_{Q_n}(t) = 2t$ for $1 \leq t \leq 2^{n-1}$. Moreover, there is a single matrix in $\mathcal{S}(Q_n)$ that achieves equality simultaneously in each bound in Theorem* 7.13(1).

It is natural to ask how do the rigid shortest linkages behave with respect to various graph products. For Cartesian products the following is known.

PROPOSITION 7.15. *Let G and H be any graphs and let $t \leq |V(H)|$. Then*

$$\operatorname{rsl}_{G \square H}(t) \geq t \cdot \operatorname{rsl}_G(1).$$

It is noted in [**135**] that the bound in Proposition 7.15 need not be tight.

CHAPTER 8

Minimum Number of Distinct Eigenvalues

Previous chapters have focused on a key invariant, the maximum multiplicity of a graph. As noted in Chapter 7, from the point of view of Ferrer's diagrams, the dual invariant to $M(G)$ is $q(G)$, the minimum number of distinct eigenvalues of a matrix with graph G. The invariant $q(G)$ hasn't been studied for as long as $M(G)$, but recent work suggests that it is an interesting, useful parameter. This chapter delves more deeply into $q(G)$.

8.1. Number of Distinct Eigenvalues for Adjacency Matrices

There has been interest for quite some time in studying graphs with a small number of distinct eigenvalues. In this section, which is not needed for the other sections, we survey some of these results to give some indication on how properties of a graph are related to the number of distinct eigenvalues of its adjacency matrix. Throughout this section A_G denotes the adjacency matrix of the graph G.

We begin with some simple, yet useful observations. These follow from the basic facts from linear algebra that the degree of the minimal polynomial of a matrix M is the dimension of the space spanned by $I, M, M^2, \ldots$, the roots of the minimal polynomial of M are its eigenvalues, and these roots are simple when M is symmetric.

OBSERVATION 8.1. *Let A be a symmetric matrix. Then*
(1) *$q(A) \geq k$ if and only if $\{I, \ldots, A^{k-1}\}$ is a linearly independent set;*
(2) *$q(A)$ is the smallest k such that $\{I, \ldots, A^k\}$ is linearly dependent; and*
(3) *$q(A)$ is the degree of the minimal polynomial of A.*

Doob initiated the study of graphs with a small number of distinct eigenvalues in 1970. Some simple results are collected in the next theorem.

THEOREM 8.2. [**115**]
(1) *$q(A_G) = 1$ if and only if G is an empty graph;*
(2) *$q(A_G) = 2$ if and only if for some $t \geq 2$, G is the disjoint union of some positive number of complete graphs K_t;*
(3) *if G is connected and regular, then $q(A_G) \leq 3$ if and only if G is a strongly regular graph; and*
(4) *$q(A) \geq \operatorname{diam}(G) + 1$.*

PROOF. A symmetric matrix has a single eigenvalue if and only if it is a scalar matrix. This gives (1).

The distinct eigenvalues of the adjacency matrix of a disjoint union of some number of complete graphs K_t with $t \geq 2$ are $t-1$ and -1. Hence the if part of (2) holds. Conversely, assume that $q(A(G)) = 2$. By the Perron-Frobenius theorem, if G is connected then $A_G - \lambda I$ is symmetric and rank 1, where λ is the smallest eigenvalue of A_G. Hence if G is connected, then G is a complete graph. It follows that each connected component of G is a complete graph, and the complete graphs have the same size. As $q(G) = 2$, G is not empty. Hence, for some $t \geq 2$, G is the disjoint union of some number of complete graphs K_t.

It is not difficult to argue that the connected graph G is strongly regular if and only if $A_G J$ is a multiple of J and A^2 is a linear combination of A, I and J (here J denotes the matrix of all ones). These latter conditions are equivalent to G is regular and the degree of the minimal polynomial of A_G is at most 3. As the degree of the minimal polynomial is $q(A_G)$, (3) holds.

Without loss of generality $1 - \cdots - (d+1)$ is a diametrical path in G; that is, $d = \operatorname{diam}(G)$. This and the nonnegativity of A imply that for $k = 0, \ldots, d$ the leading principal submatrix of A^k of order $d+1$ has only zeros above its k-th superdiagonal, and only nonzeros on its k-th superdiagonal. Hence $\{I, \ldots, A^d\}$ is a linearly independent set. The result follows from Observation 8.1. □

In [**75, 98**] the following question (open at the time of this book's writing) is presented: do connected graphs with $q(A_G) = 3$ have at most three different degrees? We note that there is reason to suspect that the assumption that $q(A_G) = 3$ restricts the degree sequence of G. Indeed the assumption $q(A_G) = 3$ implies that for any vertices $v_1, \ldots, v_4$ of G the determinant of the 4×4 integer matrix whose i, j-entry is the number of walks of length j in G that start at vertex v_i is zero. Results related to the question are presented below without proof.

THEOREM 8.3.
(1) *There exist infinite families of non-regular G with $q(A_G) = 3$* [**59**].
(2) *For each real number ρ there are finitely many graphs G with $q(A_G) = 3$ and spectral radius at most ρ* [**101**].
(3) *There exist connected, non-bipartite graphs with $q(A_G) = 4$ and arbitrarily many distinct degrees* [**100**].

To date only finitely many examples of G with $q(G) = 3$ and three different degrees are known [**75, 99**].

8.2. Basic Results

We now return to the study of the minimum number of distinct eigenvalues as part of the IEP-G. This section presents several basic observations about q.

THEOREM 8.4. *Let G be a graph and $A \in \mathcal{S}(G)$. Then*
(1) *$q(G)$ is the smallest degree of a monic polynomial satisfied by some matrix with graph G.*
(2) *$q(G) \geq k$ if and only if for each $A \in \mathcal{S}(G)$, the set of matrices $\{I, \ldots, A^{k-1}\}$ is linearly independent.*
(3) *If G has a unique shortest path from a vertex u to a vertex v and this path has length ℓ, then $q(G) \geq \ell + 1$* [**3**].
(4) *$q(G) \leq |V(G)| - |M(G)| + 1 = \operatorname{mr}(G) + 1$* [**3**].

PROOF. Statements (1) and (2) follow from basic linear algebra (see e.g., observation 8.1).

Assume that G has a unique shortest path between u and v, and let ℓ be the length of this path. Without loss of generality, we may assume that the path from u

to v is $1-\cdots-(\ell+1)$. Let $A \in \mathcal{S}(G)$. For each nonnegative integer k, the i,j-entry of A^k is the sum of the weights of walks in G from i to j of length k. It follows that for $k = 0,\ldots,\ell$ the k-th superdiagonal of $A^k[\{1,\ldots,k\}]$ has only positive entries, and all entries above that superdiagonal are zero. Hence the set $\{I,\ldots,A^\ell\}$ is linearly independent. Statement (3) now follows from (2).

Finally, if $A \in \mathcal{S}(G)$ with null $A = M(G)$ then A has at most $|V(G)|-|M(G)| = \text{mr}(G)$ distinct nonzero eigenvalues, and (4) is established. □

The following relates $q(G)$ to orthogonal representations of G [**168**]. Recall that an orthogonal representation of G corresponds to a matrix X such that $X^\top X \in \mathcal{S}(G)$.

THEOREM 8.5. *For each graph G, $q(G) - 1$ is the minimum number of nonzero distinct lengths among the rows of X over all orthogonal representations X of G whose rows are mutually orthogonal.*

PROOF. First take $A \in \mathcal{S}(G)$ with $q(A) = q(G)$. Without loss of generality, we may assume A is positive semidefinite and $0 \in \text{spec}(A)$. Let $A = Q^\top DQ$ be a spectral decomposition of A. Since A is positive semidefinite, each diagonal entry of D is nonnegative. Let $\sqrt{D}$ denote the diagonal matrix whose diagonals are the nonnegative square roots of the corresponding entries of D. It follows that $\sqrt{D}Q$ is an orthogonal representation having mutually orthogonal rows. The number of distinct nonzero lengths among the rows of $\sqrt{D}Q$ equals $q(\sqrt{D}) - 1 = q(A) - 1 = q(G) - 1$. Hence

$$q(G) - 1 \geq \ell. \tag{8.1}$$

If G is the empty graph, then $q(G) = 1$ and the $1 \times |V(G)|$ zero matrix is an orthogonal matrix with no row with a nonzero length. Hence the claim holds for the empty graph. Now assume that G has at least one edge, and let ℓ be the minimum number of distinct nonzero lengths among the rows of X over all orthogonal representations X of G whose rows are mutually orthogonal. Deleting the rows of X that are rows of zeros, results in an orthogonal representation of G whose rows are mutually orthogonal with $\ell > 0$ nonzero row lengths. Hence, without loss of generality, we may assume that each row of X is nonzero.

Then $X^\top X \in \mathcal{S}(G)$ and $XX^\top$ is diagonal and each of its ℓ distinct diagonal entries is positive. Recall the nonzero eigenvalues of $X^\top X$ and $XX^\top$ are equal. Hence $q(G) \leq q(X^\top X) \leq q(XX^\top) + 1 = \ell + 1$, which implies

$$q(G) - 1 \leq \ell. \tag{8.2}$$

By (8.1) and (8.2), $q(G) - 1 = \ell$, as desired. □

COROLLARY 8.6. *Let G be a graph. Then $q(G) \leq 2$ if and only if there exists an orthogonal matrix with graph G.*

PROOF. If $q(G) = 1$, then G is the empty graph, and the identity matrix is an orthogonal matrix with graph G.

Now assume $q(G) = 2$. By Theorem 8.5, G has an orthogonal representation X whose rows are mutually orthogonal and whose nonzero rows have the same length. Deleting the rows of zeros of such an X and results in a matrix Y that that $YY^\top$ is a nonzero scalar matrix. Thus there is an orthogonal matrix with graph G.

The converse follows by noting that a symmetric orthogonal matrix has at most two distinct eigenvalues, namely, each its eigenvalues is ± 1. $\square$

The parameter q has been determined for various families of graphs. Recall the (k,p)-lollipop, $L_{k,p}$, is the graph constructed by connecting a K_k to a P_p by a bridge from an endpoint of the P_p to a vertex of the K_k. The d-th hypercube graph is denoted by Q_d. Thus, Q_1 is K_2, Q_2 is a 4-cycle, and Q_3 is the graph of the cube (that is, its vertices correspond to the corners of a unit cube and two vertices are adjacent in Q_3 if and only if they are adjacent vertices of the unit cube). Each of the results below other than (6) are found in [**3**]; 6 is found in ([**17**]).

THEOREM 8.7.

(1) *$q(K_n) = 2$ for $n \geq 2$.*
(2) *$q(C_n) = \lceil \frac{n}{2} \rceil$.*
(3) *$q(L_{k,p}) = p + 2$ for $k \geq 2$ and $p \geq 1$.*
(4) *$q(K_{m,n}) = 2$ if $m = n$, and $q(K_{m,n}) = 3$ otherwise.*
(5) *$q(Q_n) = 2$ for $n \geq 1$.*
(6) *The complete m-partite graph with each part of size n satisfies $q(K_{n,n,\dots,n}) = 2$ when m is even and $m \neq 4$.*

PROOF. The $n \times n$ matrix of all ones has exactly two distinct eigenvalues, and graph K_n for $n \geq 2$. This gives (1).

Deleting a vertex of C_n results in P_n, and hence $\mathrm{M}(C_n) \leq 2$. This implies that $q(C_n) \geq \lceil \frac{n}{2} \rceil$. The matrix $C + C^\top$, where C is the flipped-cycle matrix of order n (see Theorem 4.5), achieves equality. Thus (2) holds.

The graph $L_{k,p}$ has two vertices having a unique shortest path whose length is $p+1$, and has clique cover number $p+1$. Thus (3) follows from Theorem 2.27 (4) and Theorem 4.8.

The adjacency matrix of $K_{m,n}$ has distinct eigenvalues 0, $\pm\sqrt{mn}$. Hence $q(K_{m,n}) \leq 3$. As $K_{m,n}$ has a least one edge, $q(K_{m,n}) \geq 2$. If $m = n \geq 3$,

$$\begin{bmatrix} 0 & 1 \\ 1 & 0 \end{bmatrix} \otimes \left(I - \frac{2}{n} J \right)$$

is a symmetric orthogonal matrix whose graph is $K_{m,m}$. Since $K_{2,2} = C_4$, $q(K_{2,2}) = 2$. Hence $q(K_{m,m}) = 2$ for all m.

Assume that $A \in \mathcal{S}(K_{m,n})$ with $q(A) = 2$. By shifting and scaling, we assume the distinct eigenvalues of A are -1 and 1. Hence A is a symmetric orthogonal matrix. Up to permutation similarity A has the form

$$\left[\begin{array}{c|c} D & B \\ \hline B^\top & E \end{array} \right],$$

where B is $m \times n$ and D and E are diagonal matrices. As the rows of A are mutually orthogonal, and the rows of the diagonal matrix D are mutually orthogonal, the rows of B are nonzero and mutually orthogonal. This requires $m \leq n$. Similarly, the columns of B are nonzero and mutually orthogonal. This requires $n \leq m$. Hence (4) holds.

Statement (5) is proven by induction on n. The matrix

$$\begin{bmatrix} 0 & 1 \\ 1 & 0 \end{bmatrix}$$

is an orthogonal matrix with graph Q_1. Assume $n \geq 2$ and that U is a symmetric orthogonal matrix whose graph is Q_{n-1}. Then

$$\frac{1}{\sqrt{2}}\left[\begin{array}{c|c} U & I \\ \hline I & -U \end{array}\right]$$

is a symmetric orthogonal matrix whose graph is Q_n, and hence its distinct eigenvalues are ± 1.

Statement (6) comes from a construction in [**17**] using the symmetric conferences matrices M that gives hollow symmetric matrices of each even order other than $n = 4$, and then uses tensor products to construct an orthogonal matrix in $K_{n,\ldots,n}$. □

The values of q for all graphs of order at most 6 have been determined in [**44**] (see also [**4**]).

The behavior of the parameter q for various graph products has been studied. We begin with Cartesian products.

THEOREM 8.8. *Let G be a connected graph with at least two vertices. Then*

(1) $q(G \,\square\, K_2) \leq 2q(G) - 2$ [**3**].
(2) *If* $q(G) = 2$, *then* $q(G \,\square\, K_2) = 2$ [**3**].
(3) *If G and H have spectrally arbitrary ordered multiplicity lists with q_1 and q_2 elements respectively, then* $q(G \,\square\, H) \leq q(G) + q(H) - 1$ [**44**].

PROOF. Let $A \in \mathcal{S}(G)$ with $q(A) = q(G)$. As $q(A) \geq 2$, we may shift A and without loss of generality assume that A has two nonzero eigenvalues that are opposites of each other. With suitable labeling of vertices,

$$B = \left[\begin{array}{c|c} A & I \\ \hline I & -A \end{array}\right]$$

has graph $G \square K_2$. Let Q be a orthogonal matrix such that QAQ^T is diagonal. Then $(Q \oplus Q)B(Q^T \oplus Q^\top)$ is (after permutational similarity) a direct sum of matrices of the form

$$\begin{bmatrix} \lambda & 1 \\ 1 & -\lambda \end{bmatrix},$$

where λ is an eigenvalue of A. The eigenvalues of such a matrix are $\pm\sqrt{\lambda^2 + 1}$. This gives at most $2q(A)$ possible eigenvalues for B. The two eigenvalues of A that are nonzero and opposites, give the same pairs of eigenvalues. Hence $q(G) \leq q(B) \leq 2q(A) - 2$.

Now assume $q(G) = 2$. Then by the first part of the theorem, $q(G \,\square\, K_2) \leq 2$. As G is not empty $q(G \,\square\, K_2) \geq 2$, and the result follows.

Under the assumptions of (3), choose $A \in \mathcal{S}(G)$ and $B \in \mathcal{S}(H)$ so that the distinct eigenvalues of A are $0, \ldots, q_1 - 1$, and the distinct eigenvalues of B are $0, \ldots, q_2 - 1$. The matrix $A \otimes I + I \otimes B$ has graph $G \,\square\, H$, and its eigenvalues are all sums of an eigenvalue of A and an eigenvalue of B. Hence the distinct eigenvalues of $A \otimes I + I \otimes B$ are k where $k \in \{0, \ldots, q_1 + q_2 - 2\}$, and (3) follows. □

Some families of Cartesian products whose q parameter have been determined are the following.

THEOREM 8.9. [**44**]

(1) *For* $s \geq 2$, $q(P_s \,\square\, P_2) = s$.
(2) *For* $k \geq 1$ *and* $s \geq 2$ *with* $s \not\equiv 2 \pmod 4$, $q(C_4 \,\square\, P_{2k}) = 2k$ *and* $q(C_4 \,\square\, C_s) = \lceil \frac{s}{2} \rceil$.

For tensor products of graphs we have the following.

THEOREM 8.10. [**44**]

(1) *For* $s \geq 2$, $q(C_4 \times P_s) = s$.
(2) $q(C_4 \times C_4) = 2$.
(3) $q(C_4 \times C_{2k}) \leq k$.
(4)

$$\min\{s,t\} \leq q(P_s \times P_t) \leq \begin{cases} \frac{st}{2} & \textit{for } s \textit{ and } t \textit{ even,} \\ \frac{s(t-1)}{2} + 1 & \textit{for } s \textit{ even and } t \textit{ odd, and} \\ \frac{(s-1)(t-1)}{2} + 1 & \textit{for } s \textit{ and } t \textit{ odd.} \end{cases}$$

Next we describe some special results for bipartite graphs. Let G be a bipartite graph. Then, up to permutation similarity, the hollow matrices $A \in \mathcal{S}(G)$, have the form

$$A = \left[\begin{array}{c|c} O & B \\ \hline B^\top & O \end{array}\right].$$

For such A,

$$A^2 = BB^\top \oplus B^\top B. \tag{8.3}$$

The matrix B is called the *reduced part* of A.

THEOREM 8.11. [**3**] *Let* G *be a bipartite graph and* A *be a hollow matrix in* $\mathcal{S}(G)$ *with reduced part* B. *Assume that* B *has* m *rows and* n *columns with* $m \leq n$. *Then* $q(G) \leq 2q(BB^T)+1$. *Moreover, if the parts in the bipartition of* G *have equal size, then* $q(G) \leq 2q(BB^T)$.

PROOF. By (8.3) the eigenvalues of A^2 are those of $BB^\top$ and $B^\top B$. It is known that $BB^\top$ and $B^\top B$ have the same nonzero eigenvalues, and have the same eigenvalues if B is square.

Each nonzero eigenvalue of A is the square root of an nonzero eigenvalue of $BB^\top$. As each nonzero eigenvalue of $BB^\top$ gives at most two eigenvalues of A, $q(A) \leq 2q(BB^\top) + 1$ if $m < n$, and $q(A) \leq 2q(BB^\top)$ if $m = n$. □

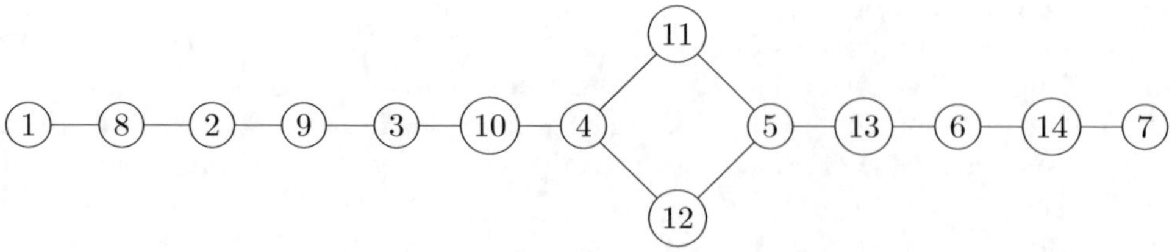

FIGURE 8.1. The graph $S_{6,4}$.

EXAMPLE 8.12. We show how Theorem 8.11 can be used to determine $q(G)$ for the graph in Figure 8.1. Using the labeling in Figure 8.1, for any nonzero $a, \ldots, \ell$, we can construct a matrix $A \in \mathcal{S}(G)$ with reduced part

$$B = \left[\begin{array}{cccc|ccc} a & 0 & 0 & 0 & 0 & 0 & 0 \\ b & c & 0 & 0 & 0 & 0 & 0 \\ 0 & d & e & 0 & 0 & 0 & 0 \\ 0 & 0 & f & g & g & 0 & 0 \\ \hline 0 & 0 & 0 & -h & h & i & 0 \\ 0 & 0 & 0 & 0 & 0 & j & k \\ 0 & 0 & 0 & 0 & 0 & 0 & \ell \end{array}\right].$$

Note that $BB^\top$ is the direct sum of a 4×4 matrix C whose graph is P_4, and a 3×3 matrix D whose graph is P_3. By the results in [**119**], it is possible to choose $a, \ldots, \ell$ such that $\operatorname{spec}(D)$ is contained in $\operatorname{spec}(C)$. By Theorem 8.11, $q(A) \leq 2q(BB^\top) = 8$. Since G has a unique shortest path from 1 to 12 of length 7, $8 \leq q(A)$ by Theorem 4.8. Hence $q(A) = 8$.

The technique in the previous example was developed in [**3**] to determine the value of q for a family of graphs $S_{m,n}$. For positive integers m and n, we denote by $S_{m,n}$ the graph obtained from $C_4 \sqcup P_m \sqcup P_n$ by inserting an edge from a pendant vertex of P_m to a vertex v of C_4, and an edge from a pendant vertex of P_n to the vertex of C_4 that is distance 2 from v. For example, the graph in Figure 8.1 is $S_{6,4}$.

THEOREM 8.13. [**3**] *For all positive integers m and n with $m \equiv n \bmod 2$,*

$$q(S(m,n)) = \max\{m,n\} + 2.$$

Recall that $G \circ K_1$ is the graph obtained from G by inserting for each vertex of G a pendant edge incident to that vertex.

THEOREM 8.14. [**3**] *For each graph G, $q(C \circ K_1) \leq 2q(A)$.*

PROOF. Let $A \in \mathcal{S}(G)$ with $q(A) = q(G)$. Then

$$B = \left[\begin{array}{c|c} A & I \\ \hline I & O \end{array}\right]$$

has graph $G \circ K_1$. Let Q be a orthogonal matrix such that $QAQ^\top$ is diagonal. Then $(Q \oplus Q)B(Q^T \oplus Q^\top)$ is (up to permutation similarity) a direct sum of matrices of the form

$$\begin{bmatrix} \lambda & 1 \\ 1 & 0 \end{bmatrix},$$

where λ is an eigenvalue of A. This matrix has eigenvalues $\frac{\lambda \pm \sqrt{\lambda^2+4}}{2}$, and hence B has at most $2q(A)$ distinct eigenvalues. □

The maximum multiplicity of a disjoint union is well-behaved: it is the sum of the maximum multiplicities of its components (see Theorem 2.27 (3)). As the number of distinct eigenvalues of a direct sum is at least the number of distinct eigenvalues in one of its summands,

$$q(G \sqcup H) \geq \max\{q(G), q(H)\}. \tag{8.4}$$

The next results give instances where equality holds. This is followed by an example that shows that equality need not hold.

THEOREM 8.15. *Let G be a graph having an ordered multiplicity list $\mathbf{m} = (m_1, \ldots, m_{q(G)})\}$ which is spectrally arbitrary, and let H be an arbitrary graph. Then $q(G \sqcup H) = \max\{q(G), q(H)\}$.*

PROOF. Choose $B \in \mathcal{S}(H)$ with $q(B) = q(H)$. If $q(H) < q(G)$, then since $\mathbf{m}$ is spectrally arbitrary for G, there exists $A \in \mathcal{S}(G)$ whose smallest $q(H)$ distinct eigenvalues are those of B. If $q(H) \geq q(G))$ then, since $\mathbf{m}$ is spectrally arbitrary for G, there exists a matrix $A \in \mathcal{S}(G)$ each of whose eigenvalues is an eigenvalue of B. In either case $q(A \oplus B) \leq \max\{q(A), q(B)\} = \max\{q(G), q(H)\}$. By (8.4), equality holds. □

We note that if $q(G) \leq 2$, then by scaling and shifting, each ordered multiplicity list of G with at most two entries is spectrally arbitrary. Hence we have the following.

THEOREM 8.16. [**44**] *If $q(G) \leq 2$ and H is an arbitrary graph, then $q(G \sqcup H) = \max\{q(G), q(H)\}$.*

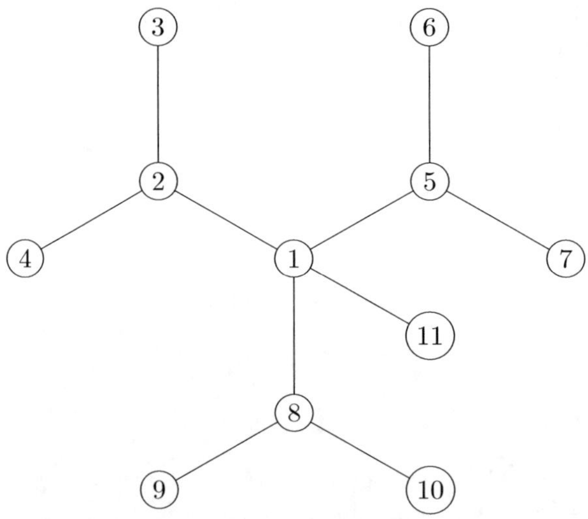

FIGURE 8.2. The extended Barioli-Fallat Tree

The next result shows that equality in (8.4) need not hold. The first example of this was found by H. Tracy Hall [**169**], and involved graphs of larger order.

COROLLARY 8.17. *Let G be the Barioli-Fallat tree labelled as in Figure* 6.1, *and H be the extended Barioli-Fallat tree in Figure* 8.2. *Then*

(1) $q(G) = 5$, *for each matrix $A \in \mathcal{S}(G)$ with $q(A) = 5$ we have $\mathbf{m}(A) = (1, 2, 4, 2, 1)$ and the distinct eigenvalues $\lambda_1 < \cdots < \lambda_5$ of A satisfy $\lambda_1 + \lambda_5 = \lambda_2 + \lambda_4$.*
(2) $q(H) = 5$, *for each matrix $B = [b_{ij}] \in \mathcal{S}(H)$ with $q(B) = 5$ we have $\mathbf{m}(B) = (1, 3, 4, 2, 1)$ or $\mathbf{m}(B) = (1, 2, 4, 3, 1)$ and the distinct eigenvalues $\mu_1 < \cdots < \mu_5$ of B satisfy $\mu_1 + \mu_3 + \mu_5 = \mu_4 + \mu_2 + b_{1,1}$, and $\mu_3 \in \operatorname{spec} B[\{1, 11\}]$.*
(3) $q(G \sqcup H) > \max\{q(G), q(H)\}$.

PROOF. Let $A \in \mathcal{S}(T_{BF})$ with distinct eigenvalues $\lambda_1 < \cdots < \lambda_5$. We claim that $\mathbf{m}(A) = (1, 2, 4, 2, 1)$. First note that each branch β of T_{BF} not containing vertex 1 is a connected component of some graph obtained from removing a diametrical path in T_{BF}. By Corollary 6.27, the spectrum $A[V(\beta)]$ is a subset of $\operatorname{spec}(A)$. By Lemma 6.6, the largest and smallest eigenvalues of a matrix whose graph is a tree strictly interlace the eigenvalues of subtree, $\operatorname{spec}(A[\beta])$ is a subset of $\{\lambda_2, \lambda_3, \lambda_4\}$. Thus $\{\lambda_2, \lambda_3, \lambda_4\} = \operatorname{spec}(A[\{3j-1, 3j, 3j+1\}])$ for $j = 1, 2, 3$, and $A[v] = [\lambda_3]$ for each pendant vertex of T_{BF}. Now we can see that for each diametrical path γ of A we have $\operatorname{spec}(A(\gamma)) = \{\lambda_2, \lambda_3^{(3)}, \lambda_4\}$. Corollary 6.27 implies that $\operatorname{spec}(A) = \{\lambda_1, \lambda_2^{(2)}, \lambda_3^{(4)}, \lambda_4^{(2)}, \lambda_5\}$, and $\mathbf{m}(A)$ is as claimed. By Example 6.26, $\lambda_1 + \lambda_5 = \lambda_2 + \lambda_4$. This establishes statement (1).

The matrix

$$\left[\begin{array}{c|ccc|ccc|ccc|c}
\sqrt{2} & 1 & 0 & 0 & 1 & 0 & 0 & 1 & 0 & 0 & \sqrt{2} \\ \hline
1 & 0 & 1 & 1 & 0 & 0 & 0 & 0 & 0 & 0 & 0 \\
0 & 1 & 0 & 0 & 0 & 0 & 0 & 0 & 0 & 0 & 0 \\
0 & 1 & 0 & 0 & 0 & 0 & 0 & 0 & 0 & 0 & 0 \\ \hline
1 & 0 & 0 & 0 & 0 & 1 & 1 & 0 & 0 & 0 & 0 \\
0 & 0 & 0 & 0 & 1 & 0 & 0 & 0 & 0 & 0 & 0 \\
0 & 0 & 0 & 0 & 1 & 0 & 0 & 0 & 0 & 0 & 0 \\ \hline
1 & 0 & 0 & 0 & 0 & 0 & 0 & 0 & 1 & 1 & 0 \\
0 & 0 & 0 & 0 & 0 & 0 & 0 & 1 & 0 & 0 & 0 \\
0 & 0 & 0 & 0 & 0 & 0 & 0 & 1 & 0 & 0 & 0 \\ \hline
\sqrt{2} & 0 & 0 & 0 & 0 & 0 & 0 & 0 & 0 & 0 & \sqrt{2}
\end{array}\right]$$

has graph H, and spectrum $\frac{\sqrt{2}-\sqrt{30}}{2}$, $-\sqrt{2}^{(2)}$, $0^{(4)}$, $\sqrt{2}^{(3)}$, $\frac{\sqrt{2}+\sqrt{30}}{2}$. Hence this matrix has 5 distinct eigenvalues and ordered multiplicity list $(1, 2, 4, 3, 1)$. So $q(H) = 5$.

We claim that the only ordered multiplicity lists of $B = [b_{ij}] \in \mathcal{S}(H)$ with $q(B) = 5$ are $(1, 2, 4, 3, 1)$ and $(1, 3, 4, 2, 1)$. As for G, we can argue that $\{\mu_2, \mu_3, \mu_4\} = \operatorname{spec} B[\{3j-1, 3j, 3j+1\}]$ for $j = 1, 2, 3$, and $B[v] = [\mu_3]$ for each pendant vertex other than vertex 11 of H. By Corollary 6.27 and the fact that $\mathrm{P}(H) = 4$, $\operatorname{mult}_B(\mu_2) \geq 2$, $\operatorname{mult}_B(\mu_4) \geq 2$, and $\operatorname{mult}_B(\mu_3) = 4$. As μ_1 and μ_5 are simple, we conclude that $\mathbf{m}(B)$ is one of the desired lists. This and Corollary 6.27 applied to a diametrical path imply that $B[11] = [\mu_2]$, or $B[11] = [\mu_4]$. Also, by applying the corollary to the three Nylen paths of H, we conclude that $\mu_3 \in \operatorname{spec} B[\{1, 11\}]$.

First assume $B[11] = [\mu_2]$, and hence $\mathbf{m}(B) = (1, 3, 4, 2, 1)$. Taking traces gives

$$\begin{aligned}
\mu_1 + 3\mu_2 + 4\mu_3 + 2\mu_4 + \mu_5 &= \operatorname{tr} B \\
&= \operatorname{tr} B[\{2,3,4\}] + \operatorname{tr} B[\{5,6,7\}] + \operatorname{tr} B[\{8,9,10\}] \\
&\quad + \operatorname{tr} B[\{1,11\}] \\
&= 3(\mu_2 + \mu_3 + \mu_4) + \mu_2 + b_{1,1}.
\end{aligned}$$

Hence $\mu_1 + \mu_3 + \mu_5 = \mu_4 + \mu_2 + m_{1,1}$.

Similarly if $B[11] = [\mu_4]$, then $\mathbf{m}(B) = (1, 2, 4, 3, 1)$ and $\mu_1 + \mu_3 + \mu_5 = \mu_4 + \mu_2 + b_{1,1}$. This establishes (2).

Now suppose that $M \in \mathcal{G} \sqcup \mathcal{H}$ with $q(M) = 5$. Then $M = A \oplus M$ for some $A \in \mathcal{S}(G)$ and $B = [b_{i,j}] \in \mathcal{S}(H)$ with $q(M) = 5$ and the distinct eigenvalues $\lambda_1 < \lambda_2 < \cdots < \lambda_5$ of A and B are the same.

By (2) and (3),

$$\lambda_1 + \lambda_5 = \lambda_2 + \lambda_4, \text{ and } \lambda_1 + \lambda_3 + \lambda_5 = \lambda_4 + b_{1,1} + \lambda_2.$$

Hence $\lambda_3 = b_{1,1}$. But this is impossible as λ_3 is an eigenvalue of $B[\{1,11\}]$. Therefore, $q(G \sqcup H) > \max\{q(G), q(H)\}$. □

8.3. Strong Properties and q

The parameter q can be poorly behaved with respect to subgraphs. Fortunately, the strong multiplicity property provides results that mitigate this bad behavior. We first give an example showing how q can increase or decrease when edges are inserted.

EXAMPLE 8.18. [**3**] First consider that graph $P_n \vee K_1$, and let v be the vertex corresponding to the K_1. As deleting v results in a path and $P_n \vee K_1$ is not a path, $M(P_n \vee K_1) = 2$. Hence $q(P_n \vee K_1) \geq \lceil \frac{n+1}{2} \rceil$. By Theorem 4.5, there is a matrix A whose graph is the cycle on $n+1$ vertices that has the SMP, and its multiplicity list $(2, 2, \ldots, 2)$ if n is even, and $(2, 2 \ldots, 2, 1)$ if n is odd. Hence by Corollary 5.33, there is a matrix in $\mathcal{S}(P_n \vee K_1)$ with one of the two multiplicity lists given above. This implies that $q(P_n \vee K_1) \leq \lceil \frac{n+1}{2} \rceil$. It follows that $q(P_n \vee K_1) = \lceil \frac{n+1}{2} \rceil$.

By Theorem 6.2, $q(K_{1,n}) = 3$ for $n \geq 2$.

As $q(P_n) = n$, $q(K_1 \sqcup P_n) \geq n$. Equality is achieved with the Laplacian matrix of $K_1 \sqcup P_n$. Hence $q(K_1 \sqcup P_n) = n$.

Thus by deleting certain edges of $q(P_n \vee K_1)$ we can attain $K_{1,n}$ which decreases the value q, or we can attain $P_n \sqcup K_1$ which increases the value of q.

The strong multiplicity property can be used to define a related, better behaved parameter. For a graph G we denote by $q_M(G)$ the smallest value of $q(A)$ over matrices in $\mathcal{S}(G)$ with the SMP. Corollary 5.33 implies the following.

COROLLARY 8.19. [**33**] *Let H be a supergraph of G with the same number of vertices as G. Then $q_M(H) \leq q_M(G)$.*

As $q_M(C_n) = \lceil \frac{n}{2} \rceil$, Corollary 8.19 shows that if G is hamiltonian then $q(G) \leq q_M(G) \leq \left\lceil \frac{|V(G)|}{2} \right\rceil$ [**44**].

Unlike the maximum multiplicity, deleting or inserting a vertex can significantly alter the value of the minimum number of distinct eigenvalues.

THEOREM 8.20. *For each graph G and each vertex v of G,*

$$\frac{q(G-v)+1}{2} \leq q(G) \leq 2q(G-v)+1.$$

Moreover, each inequality is tight.

PROOF. Let A by an $n \times n$ symmetric matrix. Removing the eigenvalues of A from the real line results in $q(A)-1$ bounded open intervals. By Cauchy interlacing (Lemma 6.5), each of these open intervals contains at most one eigenvalue of $A(v)$. Hence $A(v)$ has at most $q(A) + (q(A)-1) = 2q(A)-1$ distinct eigenvalues.

Similarly, removing the $q(A(v))$ distinct eigenvalues of $A(v)$ from the real line results in $q(A(v))+1$ open intervals each of which contains at most one eigenvalue of A. Hence $q(A) \leq [q(A(v) + q(A(v)) + 1] = 2q(A(v)) + 1$.

As $q(P_n \vee K_1) = \lceil \frac{n+1}{2} \rceil$ and $q(P_n) = n$, the first inequality can be tight. As $q(P_{2n+1}) = 2n+1$ and $q(P_{2n+1} - v) = n$ where v is the center vertex of P_{2n+1}, the second inequality can tight. □

The parameter q_M is better behaved under vertex deletion (and indeed minors), as the following corollary shows.

COROLLARY 8.21. [**239, 33**] *Let H be a graph having G as a minor. Then $q(H) \leq q_M(H) \leq q_M(G) + |V(H)| - |V(G)|$.*

PROOF. By definition there exists $A \in \mathcal{S}(G)$ having the SMP with $q(A) = q_M(G)$. By the Decontraction Lemma (Lemma 4.27), there exist $B \in \mathcal{S}(H)$ with the same ordered multiplicity list as A followed by $|V(H)| - |V(G)|$ ones. Hence

$$q(H) \leq q_M(H) \leq q(B) = q(A) + |V(H)| - |V(G)| = q_M(G) + |V(H)| - |V(G)|.$$

□

One interesting, easily proven, property of the $M(G)$ is that for each ν with $0 \leq \nu \leq M(G)$ there is a matrix $A \in \mathcal{S}(G)$ with $\operatorname{null} A = \nu$. It is not known if the analogous result holds for q. However, we do have the following theorem for q_M, which follows directly from Corollary 5.33.

COROLLARY 8.22. [**126**] *Let G be a graph. For each r with $q_M(G) \leq r \leq |V(G)|$, there exists $B \in \mathcal{S}(G)$ with $q(B) = r$.*

Another consequence of Corollary 5.32 is the following bound on $q(G)$ in terms of the chromatic number of the complement $\overline{G}$ of G. A *vertex partition* of the graph G is a collection of subgraphs of G whose vertex sets partition the vertex set of G.

THEOREM 8.23. [**33**] *Let $G_1, \ldots, G_t$ be a vertex partition of the graph G. Then*

$$q(G) \leq q_M(G) \leq q_M(G_1) + \cdots + q_M(G_t).$$

PROOF. Let $A_i \in \mathcal{S}(G_i)$ such that A has the SMP and $q_M(G_i) = q(A_i)$ ($i = 1, \ldots, t$). For some real numbers $c_1, \ldots, c_t$, the eigenvalues of the matrices $c_j A_j$ are mutually disjoint. Hence $\oplus_{i=1}^{t} c_j A_j$ has the SMP, and (up to re-labeling of its vertices) is a subgraph of G. The result now follows from Corollary 5.33, and the fact that a direct sum of matrices has the SMP if no pair of summands have a common eigenvalue. □

A bound on $q(G)$ in terms of the chromatic number of its complement is now attainable.

COROLLARY 8.24. [**33**] *Let G be a graph. Then $q(G) \leq 2\chi(\overline{G})$.*

PROOF. Take a coloring of $\overline{G}$ with the minimum number of colors. The sets of vertices of a given color determine a vertex partition of G into $\chi(\overline{G})$ cliques. As $q_M(K_n) \leq 2$ for all n, the result follows from Theorem 8.23. □

The graph G is the *partial join of H and K* if it is obtained from $H \sqcup K$ by inserting edges having one vertex in H and the other in K.

COROLLARY 8.25. [**3**] *Let G be the partial join of two cliques. Then $q(G) \leq 4$.*

PROOF. This follows from Corollary 8.24 and the observation that $\overline{G}$ is bipartite if and only if G is the partial join of two cliques. □

Another consequence of Corollary 8.24 is an upper bound on the expected value of $q(G)$ for random graphs on n vertices. Here, random means that each edge is chosen with probability 1/2. It is known [**228**] that with high probability the chromatic number of a random graph on n vertices is

$$\frac{n}{2\lg n - \lg(\lg n)} + O(1),$$

where lg denotes the base-2 logarithm. This implies the following.

COROLLARY 8.26. *With high probability,*

$$q(G) \le q_M(G) \le \frac{2n}{2\lg n - \lg(\lg n)} + O(1).$$

The strong parameter q_M gives the following forbidden graphs theorem.

THEOREM 8.27. [**33**] *Let k be a positive integer. Then there exists a finite list of graphs such that a graph H satisfies $q_M(H) \ge |V(H)| - k$ if and only if no graph in the list is a minor of H.*

PROOF. By Corollary 8.21, the graphs G for which $q_M(G) \ge |V(G)| - k$ is closed under minors. The result now follows from the Robertson-Seymour theorem [**247**]. □

Example applications of Theorem 8.27 are the characterizations of the graphs G on n vertices with $q(G)$ large as described in Section 4.1. Namely for $k = 0$: $q(G) = |V(G)|$ if and only if G is a path (see Theorem 4.1); and for $k = 1$: $q(G) \ge V(G) - 1$ if and only if G is a path, the disjoint union of a path and an isolated vertex, a path with one leaf attached to an interior vertex, or a path with an extra edge joining two vertices at distance 2 (see Theorem 4.9).

8.4. Joins and Graphs with Small q

As previously discussed, graphs G with $\mathrm{M}(G)$ small are interesting and have been characterized for $\mathrm{M}(G) = 1$ and $\mathrm{M}(G) = 2$ (see Section 2.4). In this section we discuss properties of graphs G with $q(G)$ small, in particular with $q(G) \le 2$.

The first result shows that such graphs are related to symmetric orthogonal matrices, and to orthogonal projectors. An *orthogonal projector* is a symmetric matrix P such that $P^2 = P$. The $n \times n$ orthogonal projectors are precisely the matrices P for which the action sending x to Px is the orthogonal projection of $\mathbb{R}^n$ onto some subspace of $\mathbb{R}^n$.

LEMMA 8.28. [**3**] *Let A be a symmetric $n \times n$ matrix. Then the following are equivalent:*

(1) *$q(A) \le 2$;*
(2) *There exists real numbers s and t with $s \ne 0$ such that $sA + tI$ is an orthogonal matrix; and*
(3) *There exist real numbers s and t with $s \ne 0$ such that $sA + tI$ is an orthogonal projector.*

PROOF. First assume $q(A) \le 2$. Then the eigenvalues of A can be mapped to be contained in $\{-1, 1\}$ by shifting and scaling; that is, there is a nonzero s and a real number t with such that each eigenvalue of $sA + tI$ belongs to $\{-1, 1\}$. Hence $(sA+tI)^\top(sA+tI) = (sA+tI)^2 = I$. Therefore, $sA + tI$ is an orthogonal matrix. Similarly, there exist such s and t such that each eigenvalue of $sA + tI$ belongs to $\{0, 1\}$. Hence the minimal polynomial of $sA + tI$ divides $x(x-1)$ and it follows that $sA + tI$ is an orthogonal projector. Therefore (1) implies both (2) and (3).

As symmetric orthogonal matrices have real eigenvalues of modulus 1, they have at most two eigenvalues. As an orthogonal projector has a minimal polynomial that divides $x^2 - x$, it has at most 2 distinct eigenvalues. Hence (2) implies (1), and (3) implies (1). □

Lemma 8.28 shows that graphs G with $q(G) = 2$ are inextricably related to orthogonal matrices. Understanding the combinatorial properties of orthogonal matrices is a long-standing open problem in combinatorics. Indeed at one of the early combinatorics conferences [**137**], Miroslav Fielder posed the following problem:

Characterize the sign-patterns of $n \times n$ orthogonal matrices.

There has been little headway on this problem (see [**37, 84, 137, 207, 262**]). Sign-patterns of orthogonal matrices have been fully characterized only up to order 5. A strong property for orthogonality is developed in [**93**] (see Section 3.3).

In the remainder of this section we survey some of the known results and recent work on graphs G with $q(G) = 2$.

THEOREM 8.29. [**3**] *Let G be a connected graph. Then $q(G \vee G) = 2$.*

PROOF. The result is clear if G has just one vertex. Otherwise, let A be the adjacency matrix of G, ρ the spectral radius of A. and $M = \rho I + A$. Then M is a positive definite matrix. Let $M = \sum_{j=1}^n \lambda_j \mathbf{q}_j \mathbf{q}_j^T$ be a spectral decomposition of M.

From basic results about Taylor Series,

$$\sqrt{1-y} = 1 - \frac{1}{2}\sum_{j=1}^{\infty}\left(\prod_{k=1}^{j}\frac{2k-1}{2}\right) y^j$$

for $-1 < y < 1$. Hence for ϵ sufficiently small,

$$I - \frac{1}{2}\sum_{j=1}^{\infty}\left(\prod_{k=1}^{j}\frac{2k-1}{2}\right)\epsilon^{2j}M^{2j}.$$

converges to a matrix N_ϵ having spectral decomposition $\sum_{j=1}^n \sqrt{1-\epsilon^2\lambda_j^2 \mathbf{q}_j\mathbf{q}_j^T}$ (see, for example, Theorem 4.7 [**175**] for more details about function of matrices).

As $\epsilon \to 0$, $N_\epsilon \to I$. As M is nonnegative with positive main diagonal, for $k \ne \ell$ the k,ℓ-entry of M^{2j} is positive if and only if there is a walk in $\mathcal{G}(A)$ from k to ℓ of length $2j$ or less. Hence, the connectivity of $\mathcal{G}(A)$ implies that each off-diagonal of N_ϵ is negative. Hence for ϵ sufficiently small, N_ϵ is nowhere zero and the matrix

$$\left[\begin{array}{c|c} \epsilon M & N_\epsilon \\ \hline N_\epsilon & -\epsilon M \end{array}\right]$$

lies in $\mathcal{S}(G \vee G)$. By Lemma 6.17 this matrix is similar to the direct sum of the 2×2 matrices

$$\begin{bmatrix} \epsilon\lambda_j & \sqrt{1-\epsilon^2\lambda_j^2} \\ \sqrt{1-\epsilon^2\lambda_j^2} & -\epsilon\lambda_j \end{bmatrix} \quad (j = 1, \ldots, n).$$

Each such matrix has trace 0 and determinant -1, and hence spectrum $\{-1, 1\}$. Hence $(n, n) \in \mathcal{L}(G \vee G)$, and $q(G \vee G) \leq 2$. The result follows from the fact that the only graphs H with $q(H) = 1$ are the empty graphs. □

Using techniques similar to the strong spectral property the following result was established in [**234**].

THEOREM 8.30. [**234**] *Let G be a connected graph on n vertices and let $\lambda_1, \ldots, \lambda_n$ be n distinct real numbers. Then there exists $A \in \mathcal{S}(G)$ with* $\operatorname{spec}(A) = \{\lambda_1, \ldots, \lambda_n\}$ *having a basis of eigenvectors each of which is nowhere zero.*

A consequence of this is the following (the case of $|V(G)| = |V(H)|$ is proven in [**234**], and the general case in [**218**]).

COROLLARY 8.31. *Let G and H be connected graphs with $|V(H)| = |V(G)| + m$ for some $m \geq 0$. Then $q(G \vee H) \leq m + 2$.*

PROOF. Let $|V(G)| = n$. By Theorem 8.30, there exists $A \in \mathcal{S}(G)$ having spectrum $\{1, 2, \ldots, n\}$ and a nowhere zero eigenbasis $\mathbf{q_1}, \ldots, \mathbf{q_n}$. Also, there exists $B \in \mathcal{S}(H)$ each having spectrum $\{1, 2, \ldots, n+m\}$ and a nowhere zero eigenbasis $\mathbf{r_1}, \ldots, \mathbf{r_{m+n}}$. Thus $A = \sum_{j=1}^{n} j\mathbf{q_j}\mathbf{q_j}^T$ and $B = \sum_{k=1}^{n+m} k\mathbf{r_k}\mathbf{r_k}^\top$.

For t in $(0, \frac{1}{n+m})$, let

$$M_t = \left[\begin{array}{c|c} tA & \sum_{j=1}^{n} \sqrt{1-t^2j^2}\mathbf{q_j}\mathbf{r_j}^\top \\ \hline \sum_{j=1}^{n} \sqrt{1-t^2j^2}\mathbf{q_j}\mathbf{r_j}^\top & -tB \end{array}\right]$$

Using the fact that the bases are nowhere zero one can show that there exists a t' such that each entry of $\sum_{j=1}^{n} \sqrt{1-t'^2j^2}\mathbf{q_j}\mathbf{r_j}^T$ is nonzero (see [**234**] for more details). Hence $M_{t'} \in \mathcal{S}(G \vee H)$. Let Q be the matrix whose columns are $\mathbf{q_1}, \ldots, \mathbf{q_n}$, and R be the matrix whose columns are $\mathbf{r_1}, \ldots, \mathbf{r_{n+m}}$. Then $Q \oplus R$ is an orthogonal matrix and $(Q \oplus R)^\top M_{t'}(Q \oplus R)$ is the direct sum of

$$\left[\begin{array}{c|c} \operatorname{diag}(t', \ldots, nt') & \operatorname{diag}(\sqrt{1-(t')^2}, \ldots, \sqrt{1-(nt')^2}) \\ \hline \operatorname{diag}(\sqrt{1-(t')^2}, \ldots, \sqrt{1-(nt')^2}) & -\operatorname{diag}(t', \ldots, nt') \end{array}\right] \tag{8.5}$$

and $-t'\operatorname{diag}(n+1, \ldots, n+m)$. Up to permutation similarity the matrix in (8.5) is the direct sum of the matrices

$$\begin{bmatrix} jt' & (\sqrt{1-(t'j)^2} \\ (\sqrt{1-(t'j)^2} & -jt' \end{bmatrix} \quad j = 1, \ldots, n. \tag{8.6}$$

and $-t'\operatorname{diag}(n+1, \ldots, n+m)$. Each of the matrix in (8.6) has spectra $\{-1, 1\}$. Hence $q(M_t) \leq m + 2$. □

For $n \geq 2$, $q(K_{n,n+1}) = 3$ and $K_{n,n+1}$ is join of an empty graph on n vertices with an empty graph $n+1$ vertices. Hence, for $m = 1$, equality holds in Corollary 8.31 infinitely often. It is not known if there exists an $m \geq 2$ for which equality holds infinitely often.

The threshold graphs are recursively defined as follows: K_1 is a threshold graph, and if G is a threshhold graph then so is $G \sqcup K_1$ and the $G \vee K_1$. Corresponding to a threshhold graph is a sequence $(0, a_2, \ldots, a_n)$ of 0s and 1s where a_i is 0 if the vertex added as stage i is isolated and is 1 if it is adjacent to all previous vertices.

Let G be a threshold graph. If one orders the vertices of G so that those with $a_i = 0$ precede those with $a_i = 1$, and the ordering within each set remains as initially ordered, then the adjacency matrix of G has the form

$$\left[\begin{array}{c|c} O & B \\ \hline B^\top & J - I \end{array}\right],$$

where J is the matrix of all 1s, and $B = [b_{i,j}]$ is a *staircased* $(0,1)$-*matrix*; that is if $b_{i,j} = 1$, then $b_{k,\ell} = 1$ for all k and ℓ with $k \leq i$ and $\ell \geq j$.

Thus each matrix in $\mathcal{S}(G)$ has the form

$$A = \left[\begin{array}{c|c} D & C \\ \hline C^\top & E \end{array}\right], \tag{8.7}$$

where D is diagonal, C has the same pattern as B, and the graph of E is complete. We call C the *reduced part of* A.

The parameter q for threshhold graphs is studied in [**130**]. In particular they give the following partial characterization of threshold graphs with $q(G) = 2$.

THEOREM 8.32. [**130**] *Let G be a non-empty threshold graph with reduced part B and $A \in \mathcal{S}(G)$. Then $q(A) = 2$ if and only if there is matrix C with the same zero-nonzero pattern as B whose rows are mutually orthogonal.*

PROOF. Let B have m rows and n columns.

First suppose $\mathcal{A} \in \mathcal{S}(G)$ with $q(A) = 2$. By shifting and scaling we may assume that the distinct eigenvalues of A are ± 1, and hence that A is an orthogonal matrix. We also may take A to have the form in (8.7). Since A is orthogonal and the rows of the diagonal matrix D are mutually orthogonal, C is a matrix with zero-nonzero pattern B whose rows are mutually orthogonal.

Conversely, suppose there is a matrix C with zero-nonzero pattern B whose rows are mutually orthogonal. Since G is nonempty, the first row of C is nowhere zero. For sufficiently small $\epsilon > 0$, the matrix $\widehat{C}$ obtained from C by scaling its first row by $(1 - \epsilon)$ has the property that the inner product of any pair of columns of $\widehat{C}$ is nonzero. Moreover the rows of $\widehat{C}$ remain mutually orthogonal. Let ℓ be the largest Euclidean length of a row of $\widehat{C}$. Then there is a diagonal matrix $Y = \operatorname{diag}(y_1, \ldots, y_m)$ such the matrix M defined by

$$M = \begin{bmatrix} Y & \widehat{C} \end{bmatrix}$$

has mutually orthogonal rows each of length ℓ; that is $MM^\top = \ell^2 I_m$. Thus ℓ^2 is the only eigenvalue of $MM^\top$. Since $M^\top M$ and $MM^\top$ have the same nonzero eigenvalues, $q(M^\top M) \leq 2$. Since

$$M^\top M = \left[\begin{array}{c|c} Y^\top Y & Y^\top \widehat{C} \\ \hline \widehat{C}^\top Y & \widehat{C}^\top \widehat{C} \end{array}\right],$$

the fact that Y is a diagonal matrix and assumptions on the columns of $\hat{C}$ imply that $M^\top M \in \mathcal{S}(G)$. As G is not the empty graph, $q(G) = 2$. $\square$

Further results regarding q and threshhold graphs, as well as several open problems can be found in [**130**].

8.5. A Nordhaus-Gaddum Conjecture for q

The Graph Complement Conjecture (Conjecture 2.66) would give a tight Nordhaus-Gaddum type sum upper bound for minimum rank. It is natural, as was done in [**218**], to ask for a tight Nordhaus-Gaddum type sum upper bound result for q. Here we briefly survey the results of this work.

As $q(G) \leq 1 + \mathrm{mr}(G)$ for every graph, if Conjecture 2.66 is true, then

$$q(G) + q(\overline{G}) \leq |V(G)| + 4.$$

There are four known graphs G for which $q(G) + q(\overline{G}) > |V(G)| + 2$; namely, P_4, P_5 and its complement, and the graph $T_{2,2}$ illustrated in Figure 8.3. the following is conjectured [**218**] and is based on results for small graphs, and the results given below.

CONJECTURE 8.33. **Nordhaus-Gaddum Conjecture for q**
$q(G) + q(\overline{G}) \leq |V(G)| + 2$ *unless G is P_4, P_5 or its complement, or $T_{2,2}$.*

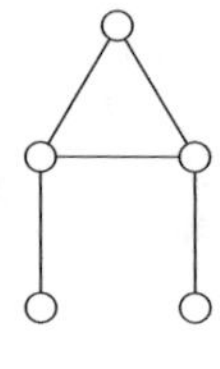

FIGURE 8.3. The graph $T_{2,2}$.

It is known (see Theorem 8.33) that $q(\overline{P_n}) = 2$ for $n \geq 6$. Hence $q(P_n) + q(\overline{P_n}) = n + 2$ for $n \geq 6$; that is the bound in Conjecture 8.33 is achieved for all $n \geq 6$.

Let G be a graph which is the partial join of K_m and K_n. Note that the complement of each bipartite graph is such a partial join. Then G determines a bipartite graph B_G with one part being the vertices of K_m, the other part being vertices of K_n, and uv and edge if and only if $u \in V(K_m)$, $v \in V(K_n)$ and uv an edge of G. We identify the vertices of K_m in B_G by $1, 2, \ldots, m$ and the vertices of K_n in B_G by $1', 2', \ldots, n'$. The *bi-adjacency matrix* of B_G is the $m \times n$ matrix $B = [b_{ij}]$ with $b_{ij} = 1$ if ij' is an edge of B_G and $b_{ij} = 0$ otherwise.

The following result shows that most complements of bipartite graphs have minimum number of distinct eigenvalues equal to 2.

THEOREM 8.34. [**218**] *Let G be the partial join of K_m and K_n with corresponding bi-adjacency matrix B. If both the graph of $BB^\top$ and of $B^\top B$ are connected, then $q(G) = 2$.*

PROOF. Without loss of generality assume $m \leq n$. By the singular value decomposition, there exist orthonormal bases $\mathbf{u}_1, \ldots, \mathbf{u}_m$ of $\mathbb{R}^m$ and $\mathbf{v}_1, \ldots, \mathbf{v}_n$ of $\mathbb{R}^n$, and nonnegative real numbers such that $B = \sum_{i=1}^m \sigma_i \mathbf{u}_i \mathbf{v}_i^\top$. Here we allow some of σ_i may be 0. As in Theorem 8.29, there is a sufficiently small, nonzero ϵ

such that

$$
\begin{aligned}
M = \sqrt{I_m - \epsilon^2 BB^\top} &= \sum_{i=1}^{m} \sqrt{1-\epsilon^2{\sigma_i}^2}\mathbf{u}_i\mathbf{u}_i^\top \in \mathcal{S}(K_m) \text{ and} \\
N = \sqrt{I_n - \epsilon^2 B^\top B} &= \sum_{i=1}^{m} \sqrt{1-\epsilon^2{\sigma_i}^2}\mathbf{v}_i\mathbf{v}_i^\top + \sum_{i=m+1}^{n} \mathbf{v}_i\mathbf{v}_i^\top \in \mathcal{S}(K_n).
\end{aligned}
$$

Thus the matrix

$$
A = \left[\begin{array}{c|c} M & \epsilon B \\ \hline \epsilon B^\top & -N \end{array}\right].
$$

has graph G. Let U be the $m \times m$ matrix whose columns are $\mathbf{u}_1, \ldots, \mathbf{u_n}$, and V be the $n \times n$ matrix whose columns are $\mathbf{v}_1, \ldots, \mathbf{v}_n$. Then $Q = U \oplus V$ is an orthogonal matrix and $QMQ^\top$ is permutationally similar to the direct sum of 2×2 matrices of the form

$$
\begin{bmatrix} \sqrt{1-\epsilon^2\sigma_i^2} & \epsilon\sigma_i \\ \epsilon\sigma_i & -\sqrt{1-\epsilon^2\sigma_i^2} \end{bmatrix} \quad (i = 1, \ldots, m)
$$

and $-I_{n-m}$. It follows that $QMQ^\top$ is an orthogonal matrix and $q(M) \leq 2$. □

Using Theorem 8.34, the q parameter for the complement of any tree is determined in [**218**].

For positive integers k, m and n, $S^k_{m,n}$ denotes the tree obtained from the path P_k by adding m leaves to one end of P_k and n leaves to the other end. The tree obtained from the disjoint union of P_2, $K_{1,\ell}$ and d copies of K_1 by inserting a new vertex that is adjacent to a center vertex of each of the components is denoted by $W(2, \ell, (1, d))$.

THEOREM 8.35. [**218**] *Let T be a tree. Then*

$$
q(\overline{T}) = \begin{cases} 4 & \text{if } T \text{ is } P_4; \\ 3 & \text{if } T \neq P_4 \text{ and either } T = S^k_{m,1} \text{ for some } k \in \{2,3\} \text{ and } m \geq 1, \\ & \text{or } T = W(2,\ell,(1,d)) \text{ with } \ell \geq 0 \text{ and } d \geq 1; \text{ and} \\ 2 & \text{otherwise.} \end{cases}
$$

Conjecture 8.33 has been verified for many families of graphs.

THEOREM 8.36. [**218**]
Conjecture 8.33 *holds for*

(1) *Graphs satisfying the graph complement conjecture for minimum rank (Conjecture*
(2) *Bipartite graphs and their complements (in particular trees);*
(3) *graphs of minimum rank at most* 3 *and their complements;*
(4) *joins $G \vee H$ and disjoint unions $G \sqcup H$ where the number of vertices of G and of H differ by at most* 2*;*
(5) *unicyclic graphs; and*
(6) *graphs with at most* 7 *vertices.*

8.6. Minimum Number of Distinct Eigenvalues for Trees

As we've seen in Corollary 6.27, $q(T) \geq \operatorname{diam}(T) + 1$ when T is a tree. This bound does not hold for all graphs as $q(K_{2,2}) = 2$ and $\operatorname{diam}(K_{2,2}) = 2$. Thus, restricting to trees can yield different results than for graphs. In this section we discuss some particular problems and results related to $q(T)$ where T is a tree.

We begin with several examples where $q(T)$ is larger than $\operatorname{diam} T + 1$.

EXAMPLE 8.37. Let T be the tree given in Figure 8.4. Note that T has diameter 6. Suppose to the contrary that there exists $A \in \mathcal{S}(T)$ with 7 distinct eigenvalues, say $\lambda_1 < \cdots < \lambda_7$. Let B, C, D be a nested sequence of branches along a path from 1 to a pendant vertex of T (e.g., $V(B) = \{2, 5, 6, 11, 12, 13, 14\}$, $V(C) = \{5, 11, 12\}$ and $V(D) = \{11\}$).

As in Example 6.26, Corollary 6.27 implies that the distinct eigenvalues of $A[V(B)]$ are $\lambda_2, \ldots, \lambda_6$; $A[V(C)]$ are $\lambda_3, \lambda_4, \lambda_5$; and $A[V(D)]$ are λ_4. Thus if P is a diametrical path of T, then each of λ_3 and λ_5 is an eigenvalue of $A(V(P))$ of multiplicity 3. By Corollary 6.27, each of λ_3 and λ_5 is an eigenvalue of A of multplicity 4. Now let P_1, P_2 and P_3 be the diametrical paths in the three branches of T at 1. By Corollary 6.24, each of λ_3 and λ_5 is an eigenvalue of

$$A(V(P_1) \cup V(P_2) \cup V(P_3)).$$

The connected components of $T - (V(P_1) \cup V(P_2) \cup V(P_3))$ are pendant vertices in T and the isolated vertex 1. We are lead to the contradiction that both λ_3 and λ_5 are eigenvalues of the 1×1 matrix $A[1]$.

Thus for the tree T in Figure 8.4, we have that $q(T) > \operatorname{diam} T + 1$.

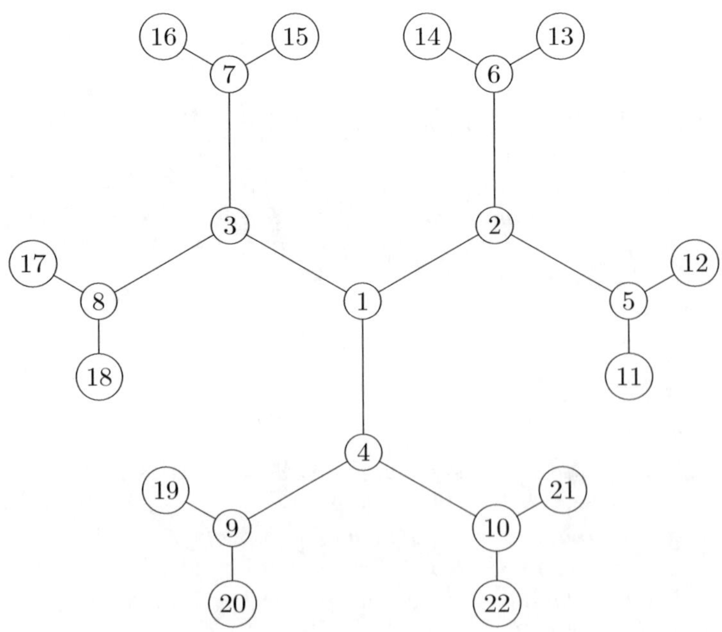

FIGURE 8.4. A tree with $q(T) > d(T) + 1$.

The following corollary implies that the graph in Figure 8.4 is not a connected component of $G - V(\gamma)$ where G is a tree of diameter 8, $q(G) = \text{diam}(G) + 1$ and γ is a diametrical path of G.

COROLLARY 8.38. *Let T be a tree, and let γ be a diametrical path of T. If there is a connected component K of $T(V(\gamma))$ with $q(K) > \operatorname{diam} T - 1$, then $q(T) > \operatorname{diam} T + 1$.*

PROOF. Suppose to the contrary such a K exists and $q(T) \leq \operatorname{diam} T + 1$. Then by Corollary 6.27, $q(T) = \operatorname{diam} T + 1$. Furthermore, the eigenvalues of $A[V(K)]$ are contained in those of A, and are distinct from the largest and smallest eigenvalue of A. Hence A has at least $q(A[V(K)]) + 2 > \operatorname{diam} T + 1$ distinct eigenvalues, a contradiction. □

We call the tree in Figure 8.4 the *tri-tree* of height 3, and denote it by Tri_3. For $k \geq 4$, we denoted by Tri_k the tree obtained 3 disjoint copies of Tri_{k-1}, by inserting a new vertex that is joined to each of the centers of the copies of Tri_{k-1}. Note that the new vertex becomes the center of Tri_k. Example 8.37, Corollary 8.38 and inductive argument yields the following.

COROLLARY 8.39. $q(\text{Tri}_k) > \text{diam}(\text{Tri}_k) + 1$, *for* $k \geq 3$.

The $(3, \ell)$-whirl $(\ell \geq 1)$ is denoted by $W_{3,\ell}$ and is shown in Figure 8.5. Note that $W_{3,\ell}$ has $6\ell + 4$ vertices and diameter $2\ell + 2$. Also $W_{3,1}$ is the Barioli-Fallat tree, and in this case the bound in Corollary 8.40 only gives the known result that $q(T_{BF}) \geq \operatorname{diam} T_{BF} + 1$.

COROLLARY 8.40. **[211]** $q(W_{3,\ell}) \geq \frac{9}{8}\text{diam}(W_{3,\ell}) + \frac{1}{2}$, *for* $\ell \geq 1$.

PROOF. Let $G = W_{3,\ell}$. Note $\text{P}(G) = 4$, and hence $\text{M}(G) = 4$. Take $A \in \mathcal{S}(G)$ with $q(A) = q(G)$. For $i = 1, \ldots, 4$, let x_i be the number of eigenvalues of A of multiplicity i. Then

$$x_1 + 2x_2 + 3x_3 + 4x_4 = |V(G)| = 6\ell + 4,$$

and

$$x_1 + x_2 + x_3 + x_4 = q(A).$$

Thus,

$$6\ell + 4 = 2q(A) - x_1 + x_3 + 2x_4 \leq 2q(A) - 2 + x_3 + 2x_4, \tag{8.8}$$

where we are using that $x_1 \geq 2$ since G is a tree.

Call the 6 paths P_ℓ in Figure 8.5 the legs of G. Using Corollary 6.24 we can show that $x_4 \leq 1$, an eigenvalue of multiplicity 4 is an eigenvalue of each of the 6 legs, and each eigenvalue of multiplicity 3 is an eigenvalue of at least 4 of the legs of G.

Hence $4x_3 + 6x_4 \leq 6\ell$. So

$$\begin{aligned} x_3 + 2x_4 &= (1/4)(4x_3 + 6x_4) + x_4/2 \\ &\leq 3\ell/2 + x_4/2 \\ &\leq 3\ell/2 + 1/2. \end{aligned}$$

The result now follows from (8.8) and $\text{diam}(W(3, \ell)) = 2\ell + 2$. □

A more general result holds for graphs that whirl-like (see Figure 8.6).

COROLLARY 8.41. [**211**] *Let G be a connected graph on n vertices that consists of a connected subgraph H on m vertices containing vertices v_1, v_2, v_3 such that for all r, s and t with $\{r,s,t\} = \{1,2,3\}$, there exists a unique shortest path from v_s to v_t that does not go through v_r; legs L_1, ...,L_6 each each ahve ℓ vertices and one of each of the end vertices of L_1, L_2 (respectively L_3, L_4 and L_5, L_6) is adjacent to vertex v_1 (respectively v_2 and v_3). Then $q(G) \geq \frac{9\ell}{4} - 2m + \frac{15}{2}$.*

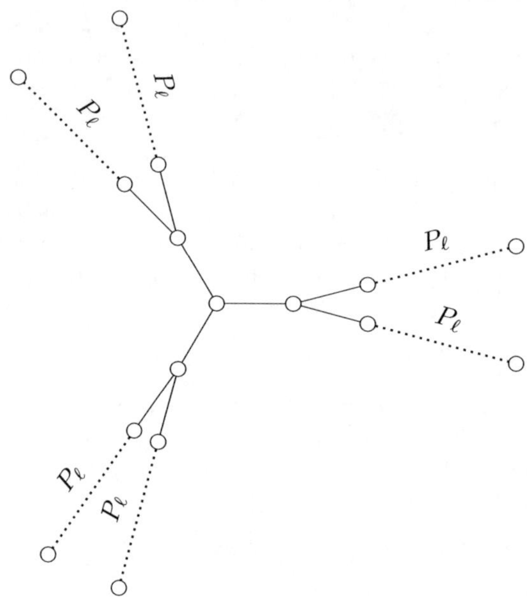

FIGURE 8.5. The $(3,\ell)$-whirl, $\ell \geq 1$.

In the remainder of this section we discuss the following intriguing result.

THEOREM 8.42. [**204**].
Let d be a fixed positive integer. Then there exists a constant c_d such that $q(T) \leq c_d$ for each tree T of diameter d.

We note that Theorem 8.42, says $q(T) \leq c_d$ regardless of how many vertices T has! Some reasons that this is perhaps surprising are the following.

(a) The analog for Theorem 8.42 does not hold for adjacency matrices of graphs, since $q(K_1 \vee P_n) = \lceil \frac{n+1}{2} \rceil$ and $\operatorname{diam}(K_1 \vee P_n) = 2$.
(b) For each d and each interval $[a, b]$ of the positive reals, there exist matrices A whose nonzero entries lie in $[a, b]$ having graph a tree of diameter at most d and $q(A)$ arbitrarily large [**120**].

To prove Theorem we need some terminology and notation. A *rooted tree* is a tree with a designated vertex called its root. The *height* of a rooted tree T is the largest distance between the root and another vertex of T. If T is rooted at r and v is a vertex of T, then a *branch* B at v is a connected component of $T - v$ not containing r. The root of the branch B is taken to be unique vertex of B that is adjacent to v. The branch B is a *duplicate branch* of T provided there is another

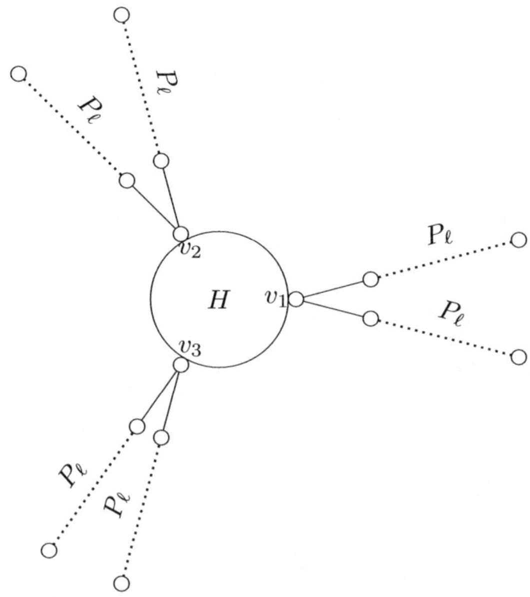

FIGURE 8.6. The $(3, \ell)$ whirl-like graph.

branch at v which is isomorphic to B as a rooted tree. The *replication* number of B is the number of duplicate branches of B at v (including B itself).

A *seed* of height h is a rooted tree T of height h such that no vertex has a branch of replication number two or more. For example, the Barioli-Fallat tree with its central vertex is a tree of height two, and is not a seed. The seeds of height two are P_3 with a pendant vertex as the root, and P_4 with a non-pendant vertex as its root.

LEMMA 8.43. **[204]** *For each positive integer h, there are, up to rooted isomorphism, only finitely many seeds of height at most h.*

PROOF. The proof is by induction on h. If $h = 1$, then the only seeds of height at most 1 are K_1 and K_2 with one of the vertices identified as a root. Assume $h \geq 2$ and there are only finitely many seeds of height $h - 1$. Note that a tree T of height at most h with root r is a seed if and only if each of the branches at r is a seed and no two of these branches are isomorphic as rooted trees. Hence there are at most a finite number of choices for the branches at r. Therefore there are finite number of seeds of height at most h. □

We now describe a general method for exploiting symmetry in a graph G to have some control on $q(G)$, or equivalently to produce a matrix in $\mathcal{S}(G)$ with multiple eigenvalues. Let G be a graph with cut-set α, and let H be one of the connected components of $G - \alpha$. For a positive integer k, the *k-duplication of G at α by H* is the graph obtained from the disjoint union of G and $k - 1$ copies of H by inserting edges between a vertex $v \in \alpha$ and a vertex w in each copy of H if and only if v and the vertex corresponding to w in the original H are adjacent in G. If G is a tree and α has cardinality 1, then H is a branch at the unique vertex in α. In this case duplication is referred to as *branch duplication* (see [**204**] for more details in this case).

Note that that with suitable labeling, the matrices whose graph is the k-duplication of G at α by H is the set of matrices of the form

$$\left[\begin{array}{c|c|c|c|c|c} D & U^\top & O & O & \cdots & O \\ \hline U & B & V_1^\top & V_2^\top & \cdots & V_k^\top \\ \hline O & V_1 & C_1 & O & \cdots & O \\ \hline O & V_2 & O & C_2 & \ddots & \vdots \\ \hline \vdots & \vdots & \vdots & \ddots & \ddots & O \\ \hline O & V_k & O & \cdots & O & C_k \end{array}\right],$$

where the rows of B correspond to the elements of α,

$$\left[\begin{array}{c|c|c} D & U^\top & O \\ \hline U & B & V_j{}^\top \\ \hline O & V_j & C_j \end{array}\right] \in \mathcal{S}(G), \text{ and } C_j \in \mathcal{S}(H) \text{ for } i = 1, \ldots, k.$$

When one takes the C_j to be equal and the V_j to be equal, the matrix can be more easily expressed and the spectrum can be readily determined. Namely, such matrices have the form

$$\left[\begin{array}{c|c|c} D & U^\top & O \\ \hline U & B & \frac{1}{\sqrt{k}}\mathbb{1}_k^\top \otimes V^\top \\ \hline O & \frac{1}{\sqrt{k}}\mathbb{1}_k \otimes V & I_k \otimes C \end{array}\right],$$

where

$$\left[\begin{array}{c|c|c} D & U^\top & O \\ \hline U & B & V^\top \\ \hline O & V & C \end{array}\right] \in \mathcal{S}(G), \text{ and } C \in \mathcal{S}(H). \tag{8.9}$$

LEMMA 8.44. *Let the graphs G and H, and the matrices C, D, and U be as in (8.9). For $k \geq 1$ define*

$$A_k = \left[\begin{array}{c|c|c} D & U^\top & O \\ \hline U & B & \frac{1}{\sqrt{k}}\mathbb{1}_k^\top \otimes V^\top \\ \hline O & \frac{1}{\sqrt{k}}\mathbb{1}_k \otimes V & I_k \otimes C \end{array}\right]$$

Then for $k \geq 2$, the multiset $\operatorname{spec} A_k$ *is the union of* $\operatorname{spec} A_1$ *and $k-1$ copies of* $\operatorname{spec} C$. *Hence for $k \geq 2$, $q(A_k) = q(A_2)$.*

PROOF. There is a $k \times k$ orthogonal matrix Q that takes $\frac{1}{\sqrt{k}}\mathbb{1}_k$ to the first column $\mathbf{e_1}$ of I_k. Let $R = I_n \oplus (Q \otimes I_\ell)$, where n is the sum of the orders of D and B, and ℓ is the order of C. Then R is an orthogonal matrix and

$$RA_kR^\top = \left[\begin{array}{c|c|c} D & U^\top & O \\ \hline U & B & \mathbf{e_1}^\top \otimes V^\top \\ \hline O & \mathbf{e_1} \otimes V & I_k \otimes C \end{array}\right] = A_1 \oplus (\oplus_{i=1}^{k-1} C).$$

Hence A_k and $A_1 \oplus (\oplus_{i=1}^{k-1} C)$ are co-spectral and the claims follow. □

Lemma 8.44 immediately implies the following result which shows that $q(G_H(k))$ can be bounded in terms of $q(G)$ and $q(H)$ in a way that does not depend on k.

COROLLARY 8.45. *Let $G_H(k)$ be the k-duplication of a graph G at the cut-set α by component H with $k \geq 2$. Then $q(G_H(k)) \leq q(G) + q(H)$.*

LEMMA 8.46. *Let T be a rooted tree of height h. Then there exists a matrix $A \in \mathcal{S}(T)$ whose spectrum is contained in the union of the spectra of all adjacency matrices of seeds of height at most h.*

PROOF. We prove this by induction on $|V(T)|$. Let T be a rooted tree with root r. We construct a special matrix $A = [a_{ij}] \in \mathcal{S}(T)$ as follows. Each diagonal entry of A is 0. For $i \neq j$ with ij not in T, set $a_{ij} = 0$. For $i \neq j$ with ij an edge of T we may assume that i is closer to the root r than j, and we set

$$a_{ij} = \frac{1}{\sqrt{k_{ij}}},$$

where k_{ij} is the number of branches of T at i that as rooted trees are isomorphic to the branch at i that contains j.

If T is a seed, then A is the adjacency matrix of T and the claim holds. Otherwise T is not a seed, and there is a vertex v of T such that some branch γ at v has $k \geq 2$ duplicate branches. Among all such v choose one whose distance from r is largest. Up to relabelling, A has the form in Lemma 8.44 where the C's correspond to the branches at v that are rooted isomorphic to γ. By the choice of v, C is the adjacency matrix of γ. By Lemma 8.44, spec(A) is the union of $k-1$ copies of spec(C), and spec$(\hat{A})$ where $\hat{A}$ is the special matrix for the rooted tree obtained from T by deleting all but one branch at v that is rooted isomorphic to γ. Hence by induction spec$(\hat{A})$ is contained in the desired union of spectra, and the result follows. $\square$

By choosing a center vertex as the root, a tree of diameter d is a rooted tree of of height at most $\lceil \frac{d}{2} \rceil$. Hence Lemma 8.46 implies Theorem 8.42. At the time of this writing there is little known about c_d other than $c_d \geq \frac{9d}{8} + 1/2$ by Corollary 8.40. In [**86**] it is shown that $q(d)$ is bounded above the largest number of vertices in a seed of height d; this is a super-exponential upper bound.

Part 4

Zero Forcing, Propagation Time, and Throttling

In this part we focus on various types of zero forcing and related parameters. We also consider the time it takes to fill all vertices (propagation time) and combinations of time and number of vertices used to fill all vertices (throttling). To make the material in this part accessible to readers whose primary interests lie in graph theory, this part is independent of Parts 2 and 3; however, familiarity with Part 1 is assumed.

Most of the zero forcing parameters are motivated by maximum nullity problems. Additional maximum nullity parameters are introduced in Chapter 9, and results for these and maximum nullity parameters already introduced are presented. In particular, the parameters $\widehat{\mathrm{Z}}(G)$ and $\mathrm{Z}_q(G)$ in Section 9.6 are specifically motivated by improving bounds on maximum nullities of classes of matrices described by a graph, and rigid linkage zero forcing (see Section 9.7) is motivated by the application of rigid linkages to the IEP-G (see Chapter 7). The relationship between maximum nullity and zero forcing is synergistic: Not only does a zero forcing number provide an upper bound for the corresponding maximum nullity, a maximum nullity is also used as a lower bound for the corresponding zero forcing number. One example of this is that the maximum (standard, positive semidefinite, or skew) nullity provides a lower bound on the (standard, positive semidefinite, or skew) zero forcing number of a Cartesian product of graphs. The connection with the IEP-G, and especially with maximum nullity, has been fundamental to the development of the entire area of zero forcing, propagation time, and throttling. Zero forcing also has other origins (which are described in Section 9.1) and has also branched out and established connections with other graph searching parameters, such as Cops and Robbers.

Chapter 9 discusses zero forcing and its variants; connections with other graph parameters, such as Cops and Robbers and power domination, are also discussed there. Zero forcing is a coloring game on a graph, where the goal is to color or fill all the vertices, starting with a given initial set of filled vertices.[1] There are numerous variations and applications, with each variant distinguished by its color change rule, which defines when an unfilled vertex can be filled. The interpretation of a filled vertex varies with the application, such as a zero in a null vector of a matrix or control of node in a quantum system (see Section 9.1), a node in an electrical network that can be monitored (see Section 9.9), a part of a graph that has been searched for an adversary (see Section 9.10), or a person who has heard a rumor in a social network. Note that while the discussions of zero forcing and its variants are relatively comprehensive, the discussions of other graph parameters, such as power domination and Cops and Robbers, are brief and intended primarily as a foundation for discussion of connections with zero forcing and for discussion of propagation time/capture time and throttling in subsequent chapters (there are entire books written on the subject of Cops and Robbers, e.g., [**48**]).

Chapter 10 discusses the 'time' or number of rounds needed to complete filling all the vertices (or to capture the robber), called propagation time or capture time. Chapter 11 discusses minimizing some combination of number of initially filled vertices and the time needed to fill all vertices, called throttling. While capture time

[1]Most early papers colored the vertices black or white, and later black was replaced by blue. Recent work replaces the terms 'blue' and 'white' by 'filled' and 'unfilled,' which has the advantage of displaying well in all media. We have adopted 'filled' and 'unfilled,' but we retain other terms associated with coloring, such as 'color change rule.'

for Cops and Robbers and propagation time for zero forcing arose independently, the study of throttling was introduced first for zero forcing and then extended to other graph search parameters by building on the results for zero forcing throttling.

CHAPTER 9

Zero Forcing, Variants, and Related Parameters

We begin this chapter with discussion of the original (standard) zero forcing as defined in [**6**] and [**70**], and discussed in Section 2.2. We then present a universal approach to definitions for many variants of zero forcing, as was done in [**19**]. The universal approach is more convenient when discussing multiple zero forcing parameters. In subsequent sections we discuss individually several variants of zero forcing and summarize what is known about them. Then we discuss related parameters including power domination and Cops and Robbers. Finally we discuss these parameters for random graphs.

For each of the zero forcing parameters discussed, it is common to determine values for specific families of graphs, to develop computation methods and bounds, to characterize graphs with extreme values, to establish useful bounds on the effects of various graph operations, and to consider Nordhaus-Gaddum sum problems; these are themes of this chapter. In each section we summarize what is known related to these themes (although not necessarily in the order just listed), in addition to other results; in some cases where there were no relevant results for a theme in the literature, we present some.

9.1. Standard Zero Forcing, $\mathrm{Z}(G)$

We remind the reader that zero forcing and the zero forcing number have been defined in Definition 2.8 and illustrated in Section 2.2. Because of possible confusion with other types of zero forcing, in this part we will call this parameter the *standard zero forcing number*, refer to *standard zero forcing* or Z*-forcing*, and adapt the remaining terminology accordingly, such as the *standard color change rule* or Z*-color change rule*, a *standard zero forcing set* or Z*-forcing set*, and u Z*-forces* w.

Zero forcing arose through the study of maximum nullity as described in Chapter 2 and was introduced independently at approximately the same time by Burgarth and Giovannetti in control of quantum systems [**70**], where it was called *graph infection*. The same process was later rediscovered and called *fast mixed graph searching* in [**267**]. The zero forcing process was introduced even earlier as part of the power domination process in the study of optimal placement for monitoring units in an electrical network (see Section 9.9).

Section 4 of [**128**] provides a succinct survey of what was known about the standard zero forcing number through 2012. In this section we reproduce many of the main results as well as most newer results. We begin with elementary observations. It is common to focus on connected graphs, because the standard zero forcing number sums across connected components.

OBSERVATION 9.1. *If the connected components of G are $G_1, \dots, G_c$, then* $\mathrm{Z}(G) = \sum_{i=1}^{c} \mathrm{Z}(G_i)$.

OBSERVATION 9.2. *For any graph G of order n, $\delta(G) \leq \mathrm{Z}(G) \leq n$ and if G contains at least one edge, $\mathrm{Z}(G) \leq n-1$.*

REMARK 9.3. Suppose G is a graph of order n and G' is an induced subgraph of G of order n'. Observe that $\mathrm{Z}(G) \leq n - n' + \mathrm{Z}(G')$ by choosing $B = B' \cup (V(G) \setminus V(G'))$ where B' is a minimum standard zero forcing set of G'.

REMARK 9.4. It is immediate that $\mathrm{Z}(K_n) = n-1$, but a stronger result is true. By Theorems 2.14 and 2.55, for any graph G, $\mathrm{Z}(G) \geq \mathrm{M}(G) \geq \eta(G) - 1$ where $\eta(G)$ is the order of the largest clique minor of G.

Families of Graphs. Values of $\mathrm{Z}(G)$ for various families of graphs were established in conjunction with determining maximum nullity $\mathrm{M}(G)$. The next theorem lists the values of $\mathrm{M}(G)$ and $\mathrm{Z}(G)$ for many families of graphs G for which $\mathrm{Z}(G) = \mathrm{M}(G)$ (some were already listed and proved in Theorem 2.15). In most cases where no reference is given, [**6**] established the value of $\mathrm{Z}(G)$ and in some cases $\mathrm{M}(G)$; in other cases $\mathrm{M}(G)$ was known previously.

THEOREM 9.5. *Let G be a graph.*

(1) $\mathrm{Z}(T) = \mathrm{M}(T) = \mathrm{P}(T)$ *for any tree T.*
(2) *For $n \geq 2$, $\mathrm{Z}(K_n) = \mathrm{M}(K_n) = n-1$.*
(3) *For $n \geq 1$, $\mathrm{Z}(\overline{K_n}) = \mathrm{M}(\overline{K_n}) = n$.*
(4) *For $p, q \geq 1$ and $p+q \geq 3$, $\mathrm{Z}(K_{p,q}) = \mathrm{M}(K_{p,q}) = p+q-2$.*
(5) *For $n \geq 3$, $\mathrm{Z}(C_n) = \mathrm{M}(C_n) = 2$.*
(6) *For $n \geq 4$, $\mathrm{Z}(W_n) = \mathrm{M}(W_n) = 3$, where W_n is the wheel on n vertices.*
(7) $\mathrm{Z}(Q_d) = \mathrm{M}(Q_d) = 2^{d-1}$, *where Q_d is the kth hypercube.*
(8) $\mathrm{Z}(T_k) = \mathrm{M}(T_k) = k$, *where T_k is the kth supertriangle.*
(9) $\mathrm{Z}(F_k) = \mathrm{M}(F_k) = k+1$, *where F_k is the kth friendship graph.*
(10) [**108**] $\mathrm{Z}(\mathrm{Half}_s) = \mathrm{M}(\mathrm{Half}_s) = s$, *where* Half_s *is the sth half-graph.*
(11) [**108**] $\mathrm{Z}(N_k) = \mathrm{M}(N_k) = k+2$, *where N_k is the kth necklace.*
(12) [**198**] $\mathrm{Z}(B_r) = \mathrm{M}(B_r) = r$, *where $B_r = K_{1,r} \,\square\, P_2$ is the rth book graph.*
(13) $\mathrm{Z}(K_s \,\square\, P_t) = \mathrm{M}(K_s \,\square\, P_t) = s$.
(14) $\mathrm{Z}(P_s \,\square\, P_t) = \mathrm{M}(P_s \,\square\, P_t) = \min\{s,t\}$.
(15) $\mathrm{Z}(C_s \,\square\, P_t) = \mathrm{M}(C_s \,\square\, P_t) = \min\{s, 2t\}$.
(16) $\mathrm{Z}(C_s \,\square\, K_t) = \mathrm{M}(C_s \,\square\, K_t) = 2t$ *for $s \geq 4$.*
(17) $\mathrm{Z}(K_s \,\square\, K_t) = \mathrm{M}(K_s \,\square\, K_t) = st - s - t + 2$.
(18) [**38, 198**] *For $t \geq 2$ and $n \geq 3$,*
$$\mathrm{Z}(P_t \times K_r) = \mathrm{M}(P_t \times K_r) = \begin{cases} (r-2)t+2 & \text{if } t \text{ is odd, and} \\ (r-2)t & \text{if } t \text{ is even.} \end{cases}$$
(19) [**38**] *For $t, n \geq 3$,*
$$\mathrm{Z}(C_t \times K_r) = \mathrm{M}(C_t \times K_r) = \begin{cases} (r-2)t+2 & \text{if } t \text{ is odd, and} \\ (r-2)t+4 & \text{if } t \text{ is even.} \end{cases}$$
(20) $\mathrm{Z}(P_s \boxtimes P_t) = \mathrm{M}(P_s \boxtimes P_t) = s+t-1$.
(21) [**25**] $\mathrm{Z}(C_s \circ K_1) = \mathrm{P}(C_s \circ K_1) = \lceil \frac{s}{2} \rceil$. $\mathrm{M}(C_s \circ K_1) = \lfloor \frac{s}{2} \rfloor$ *for $s \geq 4$.*
(22) $\mathrm{Z}(\overline{C_n}) = \mathrm{M}(\overline{C_n}) = n-3$.
(23) *If $T \neq K_{1,n-1}$ is a tree of order at least 4, then $\mathrm{Z}(\overline{T}) = \mathrm{M}(\overline{T}) = n-3$.*
(24) $\mathrm{Z}(L(G)) = \mathrm{M}(L(G)) = |L(G)| - |G| + 2$ *if G has a Hamiltonian path or contains $K_{k,|V(G)|-k}$ with $1 < k < |V(G)| - 1$ as a subgraph.*
(25) $\mathrm{Z}(L(T)) = \mathrm{M}(L(T)) = \ell - 1$ *if T is a tree and ℓ is the number of leaves (pendent vertices).*

(26) [**198**] *If G is a connected unit interval graph, then* $\mathrm{Z}(G) = \mathrm{M}(G)$.

Theorem 9.5 shows the bound in Theorem 2.14 is tight. If $n \geq 5$ is odd, then $\mathrm{M}(C_n \circ K_1) < \mathrm{Z}(C_n \circ K_1)$ [**6**] where $\circ$ denotes the corona. A more extensive collection of values of $\mathrm{M}(G)$ and $\mathrm{Z}(G)$ for various families of graphs can be found in [**7**].

From the time of its origin in the study of maximum nullity, many graphs G having $\mathrm{M}(G) = \mathrm{Z}(G)$ have been known. A main open problem suggested in the paper [**6**], which introduced zero forcing in this context, is the following.

PROBLEM 9.6. [**6**] *Characterize graphs G for which* $\mathrm{M}(G) = \mathrm{Z}(G)$.

From the study of average minimum rank and random graphs, it is known that $\mathrm{M}(G) < \mathrm{Z}(G)$ for most graphs (it is known that $\mathrm{Z}(G) - \mathrm{M}(G) \geq 0.14n$ for almost all graphs of order n sufficiently large [**122, 170**]). However, $\mathrm{M}(G) = \mathrm{Z}(G)$ for most graphs for which $\mathrm{M}(G)$ is known. This may be cause and effect: the zero forcing number is an important tool for determining maximum nullity, and it is most effective in the case of equality.

Most graphs for which it is known that $\mathrm{M}(G) = \mathrm{Z}(G)$ are sparse (notable exceptions include complete graphs and complete multipartite graphs). Most graphs for which it is known that $\mathrm{M}(G) = \mathrm{Z}(G)$ have a lot of symmetry and structure. It may be of interest to see if there is a connection between $\mathrm{M}(G) = \mathrm{Z}(G)$ and symmetry and structure.

Additional Tools for Zero Forcing. For a given standard zero forcing set B, we construct the final coloring, recording the forces. Depending on context, the symbol $\mathcal{F}$ is used to denote the *set of forces* that produces the final coloring, or an ordered list of forces (in the order they were performed), called a *chronological list of forces*. For a given standard zero forcing set B, there are often choices as to which vertex forces a particular vertex, so a set of forces or a chronological list of forces, is usually not unique. However, the final coloring is unique, as stated in the next observation. It is easy to justify this using chronological lists of forces, and very similar reasoning is presented in Remark 9.67 (for skew zero forcing, where this result is not in the literature).

OBSERVATION 9.7. [**6**] *For any graph G and set $B \subseteq V(G)$, the standard final coloring of B is unique.*

Given a standard zero forcing set B and a set $\mathcal{F}$ of forces, a *forcing path* or *forcing chain* is a maximal sequence of vertices $(v_1, v_2, \ldots, v_s)$ such that for $i = 1, \ldots, s-1$, $v_i \to v_{i+1}$ (if $v_1 \in B$ does not force, then (v_1) is a forcing chain). The *forcing chain cover* of $\mathcal{F}$ is the set of subgraphs induced by the vertices of each of the forcing chains of $\mathcal{F}$ (so every vertex is in one forcing chain in the forcing chain cover). Note that each vertex in B is the first vertex of exactly one forcing chain and every forcing chain starts with an element of B, so the number of forcing chains is the cardinality of the zero forcing set. Forcing chains are used to show the zero forcing number is at least as large as the path cover number.

REMARK 9.8. [**180**] For any set of forces for a zero forcing set of G, the forcing chain cover is a path cover, because the vertices of a forcing chain necessarily induce a path in G. This implies $\mathrm{P}(G) \leq \mathrm{Z}(G)$.

There are ways to make new (minimum) standard zero forcing sets from known ones.

REMARK 9.9. Let G be a graph, let B be a standard zero forcing set, and let $\mathcal{F}$ be a chronological list of forces. Suppose that the first force in $\mathcal{F}$ is $v \to w$ and $u \neq w$ is a neighbor of v. Then $u \in B$ and $B' = B \setminus \{u\} \cup \{w\}$ is a standard zero forcing set of G with a chronological list of forces obtained from $\mathcal{F}$ by replacing $v \to w$ by $v \to u$. If B is minimum, so is B'. Switching from B to B' is called *neighbor trading.* Neighbor trading works for any one neighbor of any vertex in B that can perform the first force, but not usually for other vertices in B.

A *reversal* of B is the set of last vertices of the zero forcing chains of a set of forces $\mathcal{F}$, i.e., the vertices that do not perform forces. Note that the cardinality of a reversal of B is the same as the cardinality of B. The *reverse chronological list of forces of a chronological list of forces* $\mathcal{F}$ is obtained from $\mathcal{F}$ by reversing each force and then writing the forces in reverse order.

THEOREM 9.10. [**18**] *Let G be a graph.*

(1) *If B is a standard zero forcing set, then any reversal of B is a standard zero forcing set.*
(2) *If G connected and of order greater than one, then no vertex is in every minimum standard zero forcing set of G.*

Let G be a graph of order n. The parameters minimum rank and maximum nullity can be thought of as complementary since they sum to n. For the standard zero forcing number, define the complementary parameter $\mathrm{mz}(G) = n - \mathrm{Z}(G)$. Notice that $\mathrm{mz}(G) \leq \mathrm{mr}(G)$ for every graph G, and $\mathrm{mz}(G) = \mathrm{mr}(G)$ is equivalent to $\mathrm{Z}(G) = \mathrm{M}(G)$. The proof of the next result is analogous to that of Proposition 9.40. See also the discussion of connected OS number following Theorem 9.41.

PROPOSITION 9.11. [**6**] *If H is an induced subgraph of a graph G, then* $\mathrm{mz}(H) \leq \mathrm{mz}(G)$.

Extreme Values. The next theorem collects several results about the graphs having extreme values of the zero forcing number. A graph G is a *graph of two parallel paths* if $\mathrm{P}(G) = 2$ and the graph can be drawn in the plane in such a way that the paths are parallel line segments, the edges between the two paths do not cross, and each edge is drawn as a straight line segment. A graph that consists of two connected components, each of which is a path, is a graph of two parallel paths, but a single path is not. Example ZF.1-2 in [**183**] is a graph of two parallel paths. A *polygonal path* is a "path" of cycles constructed one cycle at a time by identifying an edge of a new cycle with an edge of the most recently added cycle that is incident to a vertex of degree 2. A graph of two parallel paths is a polygonal path if and only if it is 2-connected.

THEOREM 9.12. *Let G be a graph of order n.*

(1) [**6**] $\mathrm{Z}(G) = 1$ *if and only if G is a path.*
(2) [**251**] $\mathrm{Z}(G) = 2$ *if and only if G is a graph on two parallel paths.*
(3) [**185, 251**] *Suppose G is a 2-connected graph. Then the following are equivalent:*
    (*a*) $\mathrm{Z}(G) = 2$.
    (*b*) $\mathrm{M}(G) = 2$.
    (*c*) $\xi(G) = 2$.

(d) *G is a polygonal path.*

(4) $\mathrm{Z}(G) = n$ *if and only if* $G = nK_1$.

(5) [**251**] *If G is a connected graph of order at least two, then* $\mathrm{Z}(G) = n - 1$ *if and only if G is a complete graph. More generally,* $\mathrm{Z}(G) = n - 1$ *if and only if* $G = K_r \sqcup (n-r)K_1$ *for* $r \geq 2$.

(6) [**6**] $\mathrm{Z}(G) = n - 2$ *if and only if* $G \neq K_r \sqcup (n-r)K_1$ $(r \geq 1)$ *and G does not contain* P_4, $P_3 \sqcup K_2$, dart, $\ltimes$, *or* $3K_2$ *as an induced subgraph.*

Graph Operations. Next we discuss the behavior of the zero forcing number under various graph operations.

PROPOSITION 9.13. [**121, 239**] *Let G be a graph.*

(1) *For any vertex v of G,* $-1 \leq \mathrm{Z}(G) - \mathrm{Z}(G-v) \leq 1$.

(2) *For any edge e of G,* $-1 \leq \mathrm{Z}(G) - \mathrm{Z}(G-e) \leq 1$.

All bounds are tight.

The proofs of the bounds can of course be found in the references cited, but we also point out the same techniques are used to prove related bounds for throttling, and the proof of Proposition 11.16 illustrates these techniques. Many of the results below are also known to researchers (they might be described as folklore); we provide proofs when we are unaware of a reference.

REMARK 9.14. If G is a graph and $\deg_G(v) = 1$, then $0 \leq \mathrm{Z}(G) - \mathrm{Z}(G-v) \leq 1$ and both bounds are tight. The lower bound is not difficult to see and also follows immediately from Theorem 9.18(1) below. The path and the star show the bounds are tight.

Twins play a role in analysis of maximum nullity and standard zero forcing number (see, for example, [**8, 111**]). For maximum nullity, the cases of adjacent twins and independent twins are analyzed separately in [**111**] based on whether a diagonal entry is nonzero or zero.

PROPOSITION 9.15. *If v and w are twins in G, then at least one of v or w must be in every standard zero forcing set and*

$$0 \leq \mathrm{Z}(G) - \mathrm{Z}(G-v) \leq 1.$$

Both bounds are tight for both independent and adjacent twins.

PROOF. From Proposition 9.13 we know that $\mathrm{Z}(G) \leq \mathrm{Z}(G-v)+1$. Let B be a minimum standard zero forcing set of G. Since no neighbor of v or w can perform a force until at least one of v or w is filled, at least one of v or w must be in B. Then $B \setminus \{v\} \cup \{w\}$ is a standard zero forcing set of $G - v$ and $|B \setminus \{v\} \cup \{w\}| \leq |B|$.

The star $K_{1,n-1}$ with $n \geq 4$ shows that the upper bound is tight for independent twins, and it is verified in [**183**, Example ZF.1-3] that $\mathrm{Z}(G-v) = \mathrm{Z}(G) = 2$ for the graph G in Figure 9.1; this shows the lower bound is tight for independent twins.

Let F_k be the friendship graph with $k \geq 4$. Suppose v, w, x, y, z are distinct vertices of degree two such that $v \sim w$ and $y \sim z$. Then v is an adjacent twin of w, $\mathrm{Z}(F_k) = k+1$ and $\mathrm{Z}(F_k - v) = k$, so the upper bound is tight for adjacent twins. In $F_k - \{v, x\}$, y is an adjacent twin of z, and $\mathrm{Z}(F_k - \{v,x\}) = k-1 = \mathrm{Z}(F_k - \{v,x,y\})$, so the lower bound is tight for adjacent twins. Examples are presented in [**183**, Examples ZF.1-4, ZF.1-5]. □

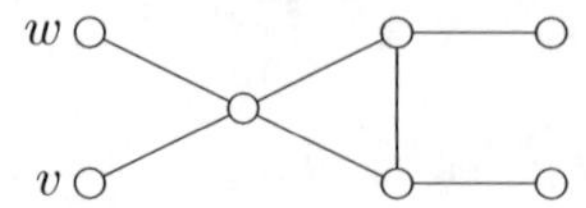

FIGURE 9.1. A graph G with independent twins w and v such that $\mathrm{Z}(G-v) = \mathrm{Z}(G)$.

When there are no isolated vertices in H, the effect of adding a universal vertex to H is determined.

PROPOSITION 9.16. *Let H be a graph with no isolated vertices and let $G = H \vee K_1$, where $V(K_1) = \{u\}$. Then $\mathrm{Z}(G) = \mathrm{Z}(H) + 1$ and there is a minimum standard zero forcing set $B \subset V(G)$ such that $u \in B$, and a set of forces for B such that u does not perform a force.*

PROOF. Since $H = G - u$, $\mathrm{Z}(G) - 1 \leq \mathrm{Z}(H)$ by Proposition 9.13(1). To prove $\mathrm{Z}(H) \leq \mathrm{Z}(G) - 1$, let $B \subset V(G)$ be a minimum standard zero forcing set of G and let $\mathcal{F}$ be a chronological of forces for B. Since H has no isolated vertices, $\deg_H(v) \geq 1$ for all $v \in V(H)$.

Suppose first that $u \notin B$. Let $x \neq u$ be the vertex that performs the first force, so $x \in V(H)$. Since $u \in N_G(x)$ and $u \notin B$, the first force must be $x \to u$. Since H has no isolated vertices, there is a vertex $y \neq x$ such that $y \in N_H[x]$. Let $B' = B \setminus \{y\} \cup \{u\}$, so $|B'| = |B|$ and construct $\mathcal{F}'$ from $\mathcal{F}$ by replacing the force $x \to u$ by $x \to y$. Then $\mathcal{F}'$ is a chronological list of forces for the minimum zero forcing set B', and B' contains u.

So assume $u \in B$. Let $\widehat{B} = B \setminus \{u\}$. If u does not force, then $\widehat{B}$ is a standard zero forcing set of H with $\mathcal{F}$. Suppose $u \to w \in \mathcal{F}$. This must be the last force performed, because every other vertex except w must be filled in order for u to force. Since $\deg_H w \geq 1$, we can replace $u \to w$ by $v \to w$ where $v \in N_H(w)$. Thus $\widehat{B}$ is a minimum standard zero forcing set of H. □

Proposition 9.16 is not true without the assumption that H is has no isolated vertices (recall that $K_{1,n} = K_1 \vee \overline{K_n}$) and $\mathrm{Z}(K_{1,n}) = n - 1$. Proposition 9.16 allows the extension of small examples that show bounds are tight to infinite families of examples.

A cut-vertex reduction theorem for the standard zero forcing number was established by Row in [**251**]. A definition is needed: Let H be a graph and $v \in V(H)$. The graph $H - v$ is called *optimal chain set extendible to* v if there exists a minimum standard zero forcing set and set of forces for H such that one chain in the forcing chain set ends in v, and if v is deleted from this chain, a minimum forcing chain set for $H - v$ is obtained.

THEOREM 9.17. *Let v be a cut-vertex of G and let $G = \bigcup_{i=1}^{h} G_i$ be the cut-vertex decomposition at v. Let $s = \min_{1 \leq j \leq h} \mathrm{Z}(G_i) - \mathrm{Z}(G_i - v)$, and let t denote the number of connected components of $G - v$ that are optimal chain set extendible to v. Then*

$$\mathrm{Z}(G) - \mathrm{Z}(G - v) = \begin{cases} 1 & \text{if } s = 1, \\ 0 & \text{if } s = 0 \text{ and } t \leq 1, \\ -1 & \text{if } s = -1 \text{ or } (s = 0 \text{ and } t \geq 2). \end{cases}$$

The next result summarizes possible effects of one edge subdivision and also of a *complete edge subdivision* of a graph G, obtained from G by subdividing each edge once, and denoted by $\mathring{\overline{G}}$.

THEOREM 9.18. *Let G be a graph.*

(1) [**239**] *For any edge e of G, $0 \leq \mathrm{Z}(G_e) - \mathrm{Z}(G) \leq 1$ and both bounds are tight.*
(2) [**30**] $\mathrm{M}(\mathring{\overline{G}}) = \mathrm{Z}(\mathring{\overline{G}})$ *and there is a method to compute this common value.*

The next result summarizes possible effects of one edge contraction.

PROPOSITION 9.19. [**239**] *For any edge e of a graph G, $-1 \leq \mathrm{Z}(G) - \mathrm{Z}(G/e) \leq 1$ and both bounds are tight.*

The proofs of the bounds for edge subdivision and contraction in Theorem 9.18(1) and Proposition 9.19 can of course be found in the references cited, but we also point out the same techniques are used to prove related bounds for throttling, and the proofs of Propositions 11.8 and 11.9 illustrate these techniques.

The next result gives an upper bound on the standard zero forcing number of the Cartesian product of two graphs, which has been used to determine maximum nullity and zero forcing number for various families of Cartesian products of graphs. Examples for which equality occurs include the graphs in Theorem 9.5(13)–(16).

PROPOSITION 9.20. [**6**] *If G and G' are graphs of order n and n', then*

$$\mathrm{Z}(G \,\square\, G') \leq \min\{n' \, \mathrm{Z}(G), \; n \, \mathrm{Z}(G')\}.$$

and this bound is tight.

The proof of Proposition 9.55 below is adapted from the proof of the upper bound in Proposition 9.20 in [**6**] (and thus serves to illustrate the proof of Proposition 9.20).

Examination of $\mathrm{Z}(K_s \,\square\, K_s) = s^2 - 2s + 2 = \mathrm{Z}(K_s)\,\mathrm{Z}(K_s) + 1 < s^2 - s = s\,\mathrm{Z}(K_s)$ for $n \geq 3$ (see Theorem 9.5(17)) shows that the inequality in Proposition 9.20 may be strict. The next proposition describes another family of graphs $G \square G'$ such that $\mathrm{Z}(G \,\square\, G') = \mathrm{Z}(G)\,\mathrm{Z}(G') + 1$. For $s \geq 2$ and $n \geq s + 2$, the *broom graph* $B_n(s)$ is the graph on n vertices constructed from a $K_{1,s}$ and a P_{n-s} by identifying an end point of the path with the center of the star. The value of $\mathrm{Z}(B_9(4) \,\square\, K_3)$ is computed in [**183**, Example ZF.1-6]

PROPOSITION 9.21. $\mathrm{Z}(B_n(s) \,\square\, K_r) \leq s(r-1) + 1 = \mathrm{Z}(B_n(s))\,\mathrm{Z}(K_r) + 1.$

PROOF. For the broom, number the leaves $1, 2, \ldots, s$, let the star center be $s+1$, and number the remaining vertices of the path $s+2, \ldots, n$. Label the vertices of the complete graph as $a_1, \ldots, a_r$. Then $V(B_n(s) \,\square\, K_r) = \{(i, a_j) : i = 1, \ldots, n, j = 1, \ldots, r\}$. We show that $B = \{(n, a_j) : j = 1, \ldots, r\} \cup \{(i, a_j) : i = 2, \ldots, s, j = 1, \ldots, r-1\}$ is a zero forcing set of $B_n(s) \,\square\, K_r$; observe that $|B| = r + (s-1)(r-1) = s(r-1) + 1$. First for $j = 1, \ldots, r$, we simultaneously force down the path,

$$(n, a_j) \to (n-1, a_j) \to \cdots \to (s+2, a_j) \to (s+1, a_j)$$

(i.e., first $(n, a_j) \to (n-1, a_j)$ for all j, etc.). Then $(i, a_1) \to (i, a_r)$ for $i = 2, \ldots, s$. Finally $(s+1, a_j) \to (1, a_j)$ for $j = 1, \ldots, r$. □

Note that the upper bounds for $\mathrm{Z}(B_n(s) \square K_r)$ in Theorem 9.20 are rs and $(r-1)n$, both of which can be substantially larger than $(r-1)s+1$. The value $\mathrm{Z}(B_n(s) \square K_r) = (r-1)s+1$ is established in Corollary 9.23 by using linear algebra to provide a lower bound for $\mathrm{Z}(G \square G')$ in the next result, which generalizes [**6**, Theorem 3.10]. Let $m_2(G)$ denote the maximum possible multiplicity of a nonzero eigenvalue of a matrix $A \in \mathcal{S}(G)$ such that $\operatorname{null} A = \mathrm{M}(G)$.

THEOREM 9.22. *Let G and G' be graphs each of which has an edge. Then*

$$\mathrm{M}(G \square G') \geq \mathrm{M}(G)\,\mathrm{M}(G') + m_2(G)m_2(G').$$

In particular,

$$\mathrm{M}(G \square G') \geq \mathrm{M}(G)\,\mathrm{M}(G') + 1.$$

PROOF. Let n and n' denote the orders of G and G', respectively. Choose $A \in \mathcal{S}(G)$ and $A' \in \mathcal{S}(G')$ such that $\operatorname{null} A = \mathrm{M}(G)$, $\operatorname{null} A' = \mathrm{M}(G')$ and there exist nonzero real numbers λ and λ' with $\operatorname{mult}_A(\lambda) = m_2(G)$ and $\operatorname{mult}_{A'}(\lambda') = m_2(G')$. Define $B = \frac{1}{\lambda}A$ and $B' = -\frac{1}{\lambda'}A'$. and $L = B \otimes I_n + I_{n'} \otimes B'$. Then $L \in \mathcal{S}(G \square G')$ and

$$\operatorname{null} L \geq \operatorname{null} A \cdot \operatorname{null} A' + \operatorname{mult}_A(\lambda) \cdot \operatorname{mult}_{A'}(\lambda') = \mathrm{M}(G)\,\mathrm{M}(G') + m_2(G)m_2(G'). \qquad \square$$

COROLLARY 9.23. *Let G and G' be graphs such that* $\mathrm{M}(G) = \mathrm{Z}(G)$ *and* $\mathrm{M}(G') = \mathrm{Z}(G')$. *Then*

$$\mathrm{Z}(G \square G') \geq \mathrm{Z}(G)\,\mathrm{Z}(G') + 1.$$

In particular, $\mathrm{Z}(B_n(s) \square K_r) = \mathrm{Z}(B_n(s))\,\mathrm{Z}(K_r) + 1$.

A tight upper bound on the standard zero forcing number of a tensor product of a graph with a complete graph is presented in Theorem 9.99; because it involves the skew zero forcing number, it is stated in Section 9.4.

Computation and Additional Bounds. It is known that computing the standard zero forcing number is NP-complete [**1, 2**].

There is code available [**183**] for finding a minimum standard zero forcing set (and thus computing the zero forcing number) of a specific graph that is not too large (e.g., of order up to 15 or 20 on a typical laptop, depending on which zero forcing function is called and how long one is willing to have the program run) using the free open-source software *Sage*. This code has two different functions for this purpose: The function `zero_forcing_set_bruteforce(G)` returns a minimum standard zero forcing set and operates by brute force, starting with sets of order one and working its way up until it finds a zero forcing set. Since it starts each cardinality with a set of lowest numbered vertices in the graph, it will return the minimum standard zero forcing set that is first lexicographically among all minimum standard zero forcing sets. The function `zero_forcing_set_wavefront(G)` uses the *wavefront* algorithm introduced by H. Tracy Hall and documented in [**73**]; further information about computation can be found in that paper and [**62**]. The function `zero_forcing_set_wavefront(G)` returns a list of three parameters items: $\mathrm{Z}(G)$, a minimum standard zero forcing set, and a computational time measure. The use of these functions are illustrated on the half-graph in Example ZF.1-1 in [**183**].

The most widely used method (so far) for determining the standard zero forcing number is to produce a matrix $A \in \mathcal{S}(G)$ and a standard zero forcing set B such that $\operatorname{null} A = |B|$, but this is doomed to failure when $\mathrm{M}(G) < \mathrm{Z}(G)$. We now turn our attention to readily computable bounds on the standard zero forcing number, in addition to $\delta(G) \le \mathrm{Z}(G) \le |V(G)| - 1$ for a graph with at least one edge.

We begin with upper bounds.

PROPOSITION 9.24. *Let G be a graph of order n.*

(1) [**251**] *If G has a connected component of order at least three that is not complete, then* $\mathrm{Z}(G) \le n - 2$.
(2) [**13**] *If* $\delta(G) \ge 1$, *then* $\mathrm{Z}(G) \le \frac{\Delta(G)}{\Delta(G)+1}n$. *If* $\delta(G) \ge 2$, *then* $\mathrm{Z}(G) \le \frac{(\Delta(G)-2)n+2}{\Delta(G)-1}$.

Next we review lower bounds. The concept of *transmission* of zero forcing across a boundary can be used to bound the zero forcing number from below. The idea is that one edge can transmit at most one force. For a graph G and subset $W \subset V(G)$, the *boundary* $\partial(W)$ equals the number of edges with one endpoint in W and one endpoint in the complement $\overline{W}$. The next result also follows from Proposition 9.13(2), because deleting $\partial(W)$ edges results in a disconnected graph $G[W] \sqcup G[\overline{W}]$.

PROPOSITION 9.25. [**121**] *For any graph G and subset $W \subseteq V(G)$,*

$$\mathrm{Z}(G) \ge \mathrm{Z}(G[W]) + \mathrm{Z}(G[\overline{W}]) - \partial(W).$$

The lower bound on the zero forcing number in the next result, in terms of the girth and minimum degree, was conjectured by Davila and Kenter in 2015 [**107**] and proved by Davila, Kalinowski, and Stephen in [**106**], building on results in [**107, 155, 156**].

THEOREM 9.26. [**106**] *If G is a graph with girth $g(G) \ge 3$ and minimum degree $\delta(G) \ge 2$, then*

$$\mathrm{Z}(G) \ge \delta(G) + (\delta(G) - 2)(g(G) - 3).$$

The bound in Theorem 9.26 is trivially tight for cycles, and less trivially $\mathrm{Z}(C_4 \,\square\, P_s) = 4 = 3 + (3-2)(4-3) = \delta(C_4 \,\square\, P_s) + (\delta(C_4 \,\square\, P_s) - 2)(g((C_4 \,\square\, P_s)) - 3)$ for $s \ge 2$.

THEOREM 9.27. [**74**] *If G is a graph of order n and* $\mathrm{Z}(G) = k$, *then*

$$|E(G)| \le nk - \frac{k(k+1)}{2}.$$

The bound in Theorem 9.27 is tight because a linear k-tree realizes this bound (see Section 9.8 for the definition of linear k-tree).

COROLLARY 9.28. [**74**] *If G is a graph of order n, then*

$$\mathrm{Z}(G) \ge \frac{|E(G)|}{n}.$$

Let $b(G)$ denote the number of 2-connected blocks of G, i.e., $b(G)$ is the number of blocks of order at least three in G. The next result is a consequence of Theorem 9.39 and was established earlier by a somewhat different method for connected zero forcing (see Section 9.5). The friendship graph F_k establishes the tightness of the bound in all cases; here we note that $\mathrm{Z}(F_k) = \mathrm{M}(F_k) = k + 1$ and $b(F_k) = k$.

THEOREM 9.29. *For any graph G,* $\mathrm{Z}(G) \ge \mathrm{M}(G) \ge b(G) + 1$, *and this bound is tight.*

Nordhaus-Gaddum Problems. The Graph Complement Conjecture $\mathrm{M}(G)+\mathrm{M}(\overline{G}) \geq |V(G)| - 2$ (see Section 2.7) is a Nordhaus-Gaddum sum lower bound for maximum nullity. In contrast, the analogous tight Nordhaus-Gaddum sum lower bound for standard zero forcing is known, and is stated in the next theorem.

THEOREM 9.30. [**122**] *Let G be a graph of order n.*

(1) $\mathrm{Z}(G) + \mathrm{Z}(\overline{G}) \geq n - 2$, *and* $\mathrm{Z}(P_n) + \mathrm{Z}(\overline{P_n}) = n - 2$ *for* $n \geq 4$.
(2) $\mathrm{Z}(G) + \mathrm{Z}(\overline{G}) \leq 2n - 1$ *for* $n \geq 2$, *and* $\mathrm{Z}(K_n) + \mathrm{Z}(\overline{K_n}) = 2n - 1$ *for* $n \geq 2$.

See Section 9.11 for discussion of the average value of $\mathrm{Z}(G) + \mathrm{Z}(\overline{G})$.

9.2. Universal Definitions for Zero Forcing

Each variant of zero forcing is defined by means of a color change rule that defines when a vertex may fill another vertex, i.e., perform a force. Given a color change rule X, we can define the sets that can fill every vertex in the graph (X-zero forcing sets) and the associated graph parameter, the X-zero forcing number $X(G)$, for all variants with one set of definitions [**19**]: For a graph G, a subset $B \subseteq V(G)$ defines an initial set of filled vertices (with all vertices not in B unfilled); this is called an *initial coloring of* G. Given an initial coloring B of G, an *X-final coloring* of B is the set of filled vertices obtained by applying the X-color change rule until no more changes are possible. An *X-zero forcing set* for G is a subset of vertices B such that $V(G)$ is an X-final coloring of B. The *X-zero forcing number* $X(G)$ is the minimum of $|B|$ over all X-zero forcing sets $B \subseteq V(G)$. That is,

$$X(G) = \min\{|B| : B \subseteq V(G) \text{ is a } X\text{-zero forcing set of } G\}.$$

Once the X-color change rule is defined, the terminology and notation *u X-forces w* and $u \xrightarrow{X} w$ (or u forces w and $u \to w$ if the type of force is clear) are also applied to the X-color change rule.

The final coloring is unique for standard, PSD, skew, and connected zero forcing (see Sections 9.1, 9.3, 9.4, and 9.5) but not for monotone floors (see Section 9.8). Other terms have been used for the final coloring, including the original term *derived set* and the more recent term *closure*.

The code [**183**] computes the X-zero forcing number for some variant color change rules in addition to the standard color change rule.

As with standard zero forcing, we can define a set of X-forces or chronological list of X-forces: For a given X-zero forcing set B, we construct the final coloring, recording the X-forces. Depending on context, $\mathcal{F}$ is used to denoted the *set of X-forces* that produces the final coloring, or an ordered list of forces (in the order they were performed), called a *chronological list of X-forces.*

In the literature when the term *zero forcing* or *color change rule* is used without mention of variants, the standard color change rule is intended. As noted earlier, in this part we use Z-*forcing* or *standard zero forcing*, Z-*forcing set* or *standard zero forcing set*, etc. when we mean zero forcing terms associated with the standard color change rule, since we discuss other types of zero forcing.

9.3. Positive Semidefinite Zero Forcing, $\mathrm{Z}_+(G)$

Recall that positive semidefinite zero forcing was designed as the zero forcing upper bound for maximum positive semidefinite nullity (see Theorem 2.43) and is defined by the Z_+-*color change rule* or *PSD color change rule* (see also Definition

2.41): Let B be the set of filled vertices and let $W_1, \dots, W_k$ be the sets of vertices of the $k \geq 1$ components of $G - B$. If $u \in B$, $w \in W_i$, and w is the only unfilled neighbor of u in $G[W_i \cup B]$, then fill w.

Note that it is possible that there is only one component of $G - B$, and in that case the first iteration of Z_+-forcing with exactly the vertices in set B filled is a first iteration of standard zero forcing when only the vertices in B are filled. Next we list some elementary observations.

OBSERVATION 9.31. *Let G be a graph. If the connected components of G are $G_1, \dots, G_t$, then*

$$\mathrm{Z}_+(G) = \sum_{i=1}^{t} \mathrm{Z}_+(G_i) \text{ and } \mathrm{M}_+(G) = \sum_{i=1}^{t} \mathrm{M}_+(G_i).$$

REMARK 9.32. Let G be a graph of order n. Any valid standard force is a valid PSD force, so $\mathrm{Z}_+(G) \leq \mathrm{Z}(G)$. Thus, $1 \leq \mathrm{Z}_+(G) \leq n$, and $\mathrm{Z}_+(G) \leq n - 1$ if G contains at least one edge.

REMARK 9.33. Suppose G is a graph of order n and G' is an induced subgraph of G of order n'. Observe that $\mathrm{Z}_+(G) \leq n - n' + \mathrm{Z}_+(G')$ by choosing $B = B' \cup (V(G) \setminus V(G'))$ where B' is a minimum PSD zero forcing set of G'.

REMARK 9.34. Since the positive semidefinite Colin de Verdière type parameter ν is minor monotone and $\nu(K_n) = n - 1$, $\nu(G) \geq \eta(G) - 1$ for any graph G. Thus $\mathrm{Z}_+(G) \geq \mathrm{M}_+(G) \geq \nu(G) \geq \eta(G) - 1$.

REMARK 9.35. For any graph G and set $B \subseteq V(G)$, the PSD final coloring of B is unique. This can be seen by examining two PSD zero forcing chronological list of forces $\mathcal{F}_1$ and $\mathcal{F}_2$ with final colorings $D_1, D_2 \subseteq V(G)$: Suppose $D_1 \not\subseteq D_2$. Among vertices in $D_1 \setminus D_2$, let w be the first vertex filled by $\mathcal{F}_1$, with PSD force $u \to w$. Then with the vertices of D_2 filled, u is filled and w is in component $(G - D_2)[W]$ of $(G - D_2)$ such that all the other neighbors of u in W are filled, so we can perform the PSD force $u \to w$. Thus D_2 was not a final coloring.

As with standard zero forcing, every vertex is excluded from some minimum PSD zero forcing set. But for PSD we have more, namely that every vertex is included in some minimum PSD zero forcing set.

THEOREM 9.36. [**122, 231**] *If G is a connected graph of order greater than one and $v \in V(G)$, then there exist minimum positive semidefinite zero forcing sets B_1 and B_2 such that $v \in B_1$ and $v \notin B_2$.*

Let G be a graph, let B be a PSD zero forcing set, and let $\mathcal{F}$ be a chronological list of forces. If the first force is a standard force, then neighbor trading applies (see Remark 9.9). In addition, another kind of trading applies to PSD forcing, as in the next result.

PROPOSITION 9.37. [**243**, Lemma 2.1.1] *Let G be a graph, let B be a PSD zero forcing set. Suppose that it is possible to have $v \to w$ be the first force. Then $B \setminus \{v\} \cup \{w\}$ is a PSD zero forcing set of G.*

All the inequalities in the next theorem are well known except $\mathrm{tw}(G) \leq \mathrm{Z}_+(G)$, which is established in [**19**].

THEOREM 9.38. *Let G be a graph of order n. Then*

$$\kappa(G) \le \delta(G) \le \operatorname{tw}(G) \le \mathrm{Z}_+(G) \le \mathrm{Z}(G).$$

Recall $b(G)$ denotes the number of blocks of order at least three in G.

THEOREM 9.39. *For any graph G, $\mathrm{Z}_+(G) \ge \mathrm{M}_+(G) \ge b(G)+1$, and this bound is tight.*

PROOF. Assume first that G is connected. The proof is by induction on $b(G)$. The result is immediate for $b(G) = 0$ or $b(G) = 1$, so let $k \ge 2$, assume $\mathrm{M}_+(H) \ge b(H) + 1$ for all H such that $b(H) < k$, and let $b(G) = k$. Let G' be the graph obtained from G by successively removing leaves until no leaves remain, so $\delta(G') \ge 2$; observe that $b(G') = b(G)$. Deleting a leaf from a graph does not change the maximum PSD nullity (see Corollary 2.46), so $\mathrm{M}_+(G) = \mathrm{M}_+(G')$. There exists a block B of G' that contains exactly one cut-vertex of G', denote this cut-vertex by v, let $U = V(B) \setminus \{v\}$, and define $G_1 = G - U$. Then $G' = G_1 \cup B$ and $V(G_1) \cap V(B) = \{v\}$. Then $\mathrm{M}_+(G') = \mathrm{M}_+(G_1) + \mathrm{M}_+(B) - 1$ by Theorem 2.45. Also,

$$\mathrm{M}_+(G) = \mathrm{M}_+(G') = \mathrm{M}_+(G_1) + \mathrm{M}_+(B) - 1 \ge b(G_1) + 1 + b(B) + 1 - 1 = b(G) + 1$$

since $\mathrm{M}_+(B) \ge 2$, $b(G_1) = b(G) - 1 < k$, and $b(B) = 1$.

If G is not connected, use the fact that both $b(G)$ and $\mathrm{M}_+(G)$ sum over connected components. To see the bound is tight, note that $\mathrm{Z}_+(F_k) = \mathrm{M}_+(F_k) = k+1$ and $b(F_k) = k$. □

Let G be a graph of order n. For the PSD zero forcing number, define the complementary parameter $\mathrm{mz}_+(G) = n - \mathrm{Z}_+(G)$. Notice that $\mathrm{mz}_+(G) \le \mathrm{mr}_+(G)$ for every graph G, and $\mathrm{mz}_+(G) = \mathrm{mr}_+(G)$ is equivalent to $\mathrm{Z}_+(G) = \mathrm{M}_+(G)$.

PROPOSITION 9.40. *If H is an induced subgraph of a graph G, then $\mathrm{mz}_+(H) \le \mathrm{mz}_+(G)$.*

PROOF. Let B be a minimum PSD zero forcing set of H. Then $B \cup (V(G) \setminus V(H))$ is a PSD zero forcing set of G, so $\mathrm{Z}_+(G) \le \mathrm{Z}_+(H) + |V(G)| - |V(H)|$. Thus $|V(H)| - \mathrm{Z}_+(H) \le |V(G)| - \mathrm{Z}_+(G)$, i.e., $\mathrm{mz}_+(H) \le \mathrm{mz}_+(G)$. □

We have used the notation $\mathrm{mz}_+(G)$ for the complementary parameter $|V(G)| - \mathrm{Z}_+(G)$ in parallel with the notation $\mathrm{mz}(G)$ for standard zero forcing introduced in [**6**] and discussed in Section 9.1. But a parameter with this value was introduced earlier: Suppose $S = (v_1, v_2, \dots, v_m)$ is an ordered subset of vertices from a given graph G. For each k with $1 \le k \le m$, let G_k be the subgraph of G induced by $\{v_1, v_2, \dots, v_k\}$, and let H_k be the connected component of G_k that contains v_k. If for each k, there exists a vertex w_k that satisfies: $w_k \ne v_l$ for $l \le k$, $\{w_k, v_k\} \in E$, and $\{w_k, v_s\} \notin E$, for all v_s in H_k with $s \ne k$, then S is called an *ordered set of vertices* in G, or an OS-set. As defined in [**166**], the *OS number* of a graph G, denoted by $\mathrm{OS}(G)$, is the maximum of $|S|$ over all OS-sets S of G.

THEOREM 9.41. [**18**] *For any graph G and any ordered set S, $V(G) \setminus S$ is a PSD zero forcing set of G, and for any PSD zero forcing set B for G, there is an order that makes $V \setminus B$ an ordered set for G. Thus $\mathrm{Z}_+(G) + \mathrm{OS}(G) = |G|$ and $\mathrm{OS}(G) = \mathrm{mz}_+(G)$.*

A related parameter, the *connected OS number* of a graph G, denoted by $\mathrm{COS}(G)$ was defined by Mitchell, Narayan, and Zimmer in [**231**], where it is shown that $\mathrm{COS}(G) = \mathrm{mz}(G)$.

Computation. It is known that computing the PSD zero forcing number is NP-complete [**268**]. The function `Zplus(G)` $= Z_+(G)$ is available for computing the PSD zero forcing number of a specific graph (that is not too large). Its use is illustrated in the *Sage* program [**183**]. When the keyword `return_track` is set to be `True`, the function `Zplus(G, return_track=True)` will return a minimum PSD zero forcing set along with the forcing process.

Tree Covers. Note that any one vertex is a PSD zero forcing set of a tree. The *tree cover number* of G, denoted by $T(G)$, is the minimum number of vertex disjoint trees occurring as induced subgraphs of G that cover all the vertices of G. Tree covers and the tree cover number are natural graph theoretical objects of study that have arisen from the interplay between linear algebra and graph theory, especially the study of maximum PSD nullity. For a vertex b in a Z_+-forcing set B and a Z_+-set of forces $\mathcal{F}$ of B, define V_b to be the set of all vertices w such that there is a sequence of forces $b = v_1 \to v_2 \to \cdots \to v_k = w$ in $\mathcal{F}$ (the empty sequence of forces is permitted, i.e., $b \in V_b$). The *forcing tree* T_b is the induced subgraph $T_b = G[V_b]$ (observe that T_b must be a tree, because if T_b contains a cycle then an illegal force would have been performed). The *forcing tree cover* (for a set of forces $\mathcal{F}$) is $\{T_b : b \in B\}$. Note that each vertex in a Z_+-forcing set is the first vertex of a forcing tree, so the number of forcing trees is the cardinality of the Z_+-forcing set. For any set of forces for a Z_+-forcing set, the forcing tree cover is a tree cover, which implies $T(G) \leq Z_+(G)$ [**122**]. Recall that $P(G) \leq Z(G)$ but $P(G)$ and $M(G)$ are noncomparable. This naturally raises the next question, which has been open for more than ten years.

PROBLEM 9.42. [**28**] *Is $T(G) \leq M_+(G)$ for every graph G?*

THEOREM 9.43. [**28, 123**] *If G is a partial 2-tree (this includes every outerplanar graph), then $Z_+(G) = M_+(G) = T(G)$.*

Families of Graphs. Values of $Z_+(G)$ for various families of graphs have been established. Some are listed in the next theorem.

THEOREM 9.44.

(1) $Z_+(T) = M_+(T) = 1$ *for any tree T.*
(2) [**6**] *For $n \geq 2$, $Z_+(K_n) = M_+(K_n) = n - 1$.*
(3) *For $p, q \geq 1$, $Z_+(K_{p,q}) = M_+(K_{p,q}) = \min\{p, q\}$.*
(4) *For $n \geq 3$, $Z_+(C_n) = M_+(C_n) = 2$.*
(5) *For $n \geq 4$, $Z_+(W_n) = M_+(W_n) = 3$.*
(6) [**244**] $Z_+(Q_d) = M_+(Q_d) = 2^{d-1}$.
(7) [**6**] $Z_+(T_k) = M_+(T_k) = k$.
(8) [**108, 244**] $Z_+(\mathrm{Half}_s) = M_+(\mathrm{Half}_s) = s$.
(9) [**108, 244**] $Z_+(N_k) = M_+(N_k) = k + 2$.
(10) $Z_+(B_k) = M_+(B_k) = 2$.
(11) $Z_+(F_k) = M_+(F_k) = k + 1$.
(12) [**260**] $Z_+(P_s \,\square\, P_t) = \min\{s, t\}$.
(13) [**260**] *For $s \leq 2t$, $Z_+(C_s \,\square\, P_t) = s$.*

(14) [**6**] *If* $T \neq K_{1,n-1}$ *is a tree of order* $n \geq 4$, *then* $\mathrm{Z}_+(\overline{T}) = \mathrm{M}_+(\overline{T}) = n-3$.
(15) [**6, 51**] *For* $n \geq 5$, $\mathrm{Z}_+(\overline{C_n}) = \mathrm{M}_+(\overline{C_n}) = n-3$.
(16) [**6, 244**] $\mathrm{Z}_+(L(G)) = \mathrm{M}(L(G)) = |L(G)| - |V(G)| + 2$ *if* G *has a Hamiltonian path.*
(17) [**6**] $\mathrm{Z}_+(L(T)) = \mathrm{M}_+(L(T)) = \ell - 1$ *where* T *is a tree and* ℓ *is the number of leaves (pendent vertices) of* T.
(18) $\mathrm{Z}_+(G) = \mathrm{M}_+(G) = |V(G)| - b$ *if* G *is a block-clique graph and* b *is the number of blocks in* G.
(19) [**166**] *For any chordal graph* H *of order* n, $\mathrm{Z}_+(H) = \mathrm{M}_+(H) = n - \mathrm{cc}(H)$ *where* $\mathrm{cc}(G)$ *is the (edge) clique cover number of* G.
(20) *For a consecutive circulant with* $k \leq \frac{n-1}{2}$, $\kappa(\mathrm{Circ}(n,[k])) = \delta(\mathrm{Circ}(n,[k])) = \mathrm{M}_+(\mathrm{Circ}(n,[k])) = \mathrm{M}(\mathrm{Circ}(n,[k])) = \mathrm{Z}_+(\mathrm{Circ}(n,[k])) = \mathrm{Z}(\mathrm{Circ}(n,[k])) = 2k$.

PROOF. (1) is immediate (and well-known). (3), (4), (5), and (10) follow from Theorem 2.47 and the known value of $\mathrm{Z}(G)$ or by exhibiting a PSD zero forcing set. For (20), this reasoning is coupled with the fact that $\kappa(\mathrm{Circ}(n,[k])) = \delta(\mathrm{Circ}(n,[k]))$ [**263**, Theorem 4.1.5 (Harary)]. (18) and (11) follow from (19) ((18) also appears in [**244**]). □

Extreme Values. The next theorem collects several results about the graphs having extreme values of the PSD zero forcing number.

THEOREM 9.45. *Let* G *be a graph of order* n.
(1) $\mathrm{Z}_+(G) = 1$ *if and only if* G *is a tree.*
(2) [**122**] *The following are equivalent.*
  (a) $\mathrm{Z}_+(G) = 2$,
  (b) $\mathrm{M}_+(G) = 2$,
  (c) *Either*
    (i) G *is the disjoint union of two trees, or*
    (ii) G *is connected, exactly one block of* G *has a cycle, and* G *does not have a* K_4 *or* T_3 *minor* (T_3 *is the third supertriangle).*
(3) [**122**] *If* $\mathrm{Z}_+(G) \leq 3$, *then* $\mathrm{Z}_+(G) = \mathrm{M}_+(G)$.
(4) $\mathrm{Z}_+(G) = n-1$ *if and only if* $G = K_r \sqcup (n-r)K_1$ *with* $r \geq 2$.
(5) [**122**] *The following are equivalent.*
  (a) $\mathrm{Z}_+(G) = n-2$,
  (b) $\mathrm{M}_+(G) = n-2$,
  (c) $G \neq K_r \sqcup (n-r)K_1$ $(r \geq 1)$ *and* G *has no induced* $P_4, K_{1,3}, P_3 \sqcup K_2, 3K_2$.

Graph Operations. Next we discuss the behavior of the PSD zero forcing number under graph operations. Examples ZF.3-1 to ZF.3-5 in [**183**] illustrate that the bounds listed are tight.

PROPOSITION 9.46. [**122**] *Let* G *be a graph.*
(1) *For any vertex* v *of* G, $\mathrm{Z}_+(G) - \mathrm{Z}_+(G-v) \leq 1$. *The star* $K_{1,n-1}$ *shows there is no useful lower bound for* $\mathrm{Z}_+(G) - \mathrm{Z}_+(G-v)$.
(2) *For any edge* e *of* G, $-1 \leq \mathrm{Z}_+(G) - \mathrm{Z}_+(G-e) \leq 1$.

All bounds are tight.

PROPOSITION 9.47. *Let* G *be a graph.*

(1) [**122**] *If v and w are adjacent twins in a connected graph G of order at least three, then* $Z_+(G) = Z_+(G-v)+1$.
(2) *If v and w are independent twins in a graph G, then* $Z_+(G-v) \le Z_+(G) \le Z_+(G-v)+1$ *and both bounds are tight.*

PROOF. For (2), let v and w be independent twins. Let B be a minimum PSD zero forcing set of G. If $v \notin B$, then define $B' = B$. If $v \in B$, then define $B' = B \setminus \{v\} \cup \{w\}$. Then B' is a PSD forcing set of $G - v$ and $|B'| \le |B|$ (with strict inequality if $v, w \in B$). Consider $K_{2,n-2}$ with $n \ge 5$. Then choosing v so that $\deg v = 2$ shows that the lower bound is tight and choosing v so that $\deg v = n-2$ shows that the upper bound is tight. □

The proof of the next result is essentially the same as the proof of Proposition 9.16, noting that $u \notin B$ implies $G - B$ is connected and if u performs a force $u \to w$, this is the last force among vertices in the connected component of H that contains w, rather than the last in the entire graph.

PROPOSITION 9.48. *Let H be a graph with no isolated vertices and let $G = H \vee K_1$, where $V(K_1) = \{u\}$. Then $Z(G) = Z(H)+1$ and there is a minimum PSD forcing set $B \subset V(G)$ such that $u \in B$, and a set of forces for B such that u does not perform a force.*

Proposition 9.48 is not true without the assumption that H is has no isolated vertices (recall that $K_{1,n} = K_1 \vee \overline{K_n}$) and $Z(K_{1,n}) = 1$). Proposition 9.48 allows the extension of small examples that show bounds are tight to infinite families of examples.

PROPOSITION 9.49. [**122**] *Let v be a cut-vertex of G and let $G = \bigcup_{i=1}^h G_i$ be the cut-vertex decomposition at v. Then* $Z_+(G) = \left(\sum_{i=1}^h Z_+(G_i)\right) - h + 1$

COROLLARY 9.50. *If G is a graph and $\deg_G(v) = 1$, then*

$$Z_+(G-v) = Z_+(G).$$

PROPOSITION 9.51. [**122**] *Let G be a graph and $e = \{u,v\} \in E(G)$. Then $Z_+(G_e) = Z_+(G)$ and any positive semidefinite zero forcing set of G is a positive semidefinite zero forcing set of G_e.*

PROPOSITION 9.52. [**122**] *For any edge e of a graph G, $Z_+(G) - Z_+(G/e) \le 1$, and this bound is tight.*

The next example shows there is no useful lower bound for $Z_+(G) - Z_+(G/e)$.

EXAMPLE 9.53. Recall the book graph is $B_k = K_{1,k} \,\square\, P_2$; B_3 is shown on the left in Figure 9.2. The contraction B_3/e is shown on the right in Figure 9.2. Observe that $B_k/e = F_k$ (the kth friendship graph). In each diagram, the filled vertices are minimum positive semidefinite zero forcing sets for each of the graphs B_k and B_k/e. Then $Z_+(B_k) = 2$ and $Z_+(B_k/e) = k+1$, so $Z_+(B_k/e) = Z_+(B_k) + (k-1)$.

Transmission across a boundary also applies to PSD zero forcing. The next result follows from Proposition 9.46(2), because deleting $\partial(W)$ edges results in a disconnected graph $G[W] \sqcup G[\overline{W}]$.

PROPOSITION 9.54. *For any graph G and subset $W \subseteq V(G)$,*

$$Z_+(G) \ge Z_+(G[W]) + Z_+(G[\overline{W}]) - \partial(W).$$

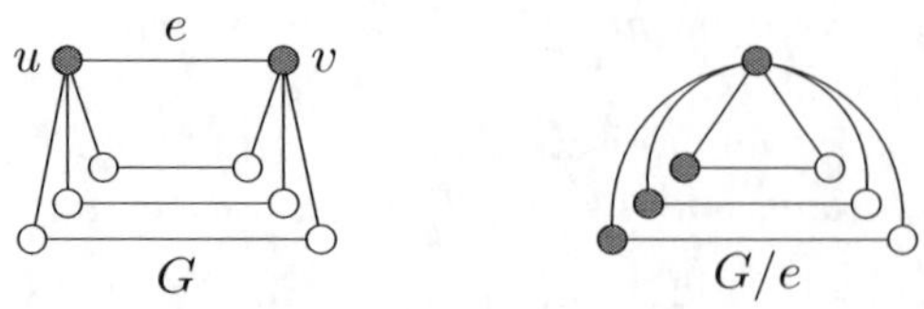

FIGURE 9.2. A book graph $G = B_3$ with $\mathrm{Z}_+(G/e) = \mathrm{Z}_+(G) + 2$.

The proof of the next result is analogous to that in [**6**] for the standard zero forcing number.

PROPOSITION 9.55. *Let G and G' be graphs of orders n and n'. Then*

$$\mathrm{Z}_+(G \square G') \leq \min\{n' \mathrm{Z}_+(G),\ n \mathrm{Z}_+(G')\}$$

and this bound is tight.

PROOF. The set of vertices associated with (the same) PSD zero forcing set in each copy of G is a PSD zero forcing set of $G \square G'$, so $\mathrm{Z}_+(G \square G') \leq n' \mathrm{Z}_+(G)$. Similarly, $\mathrm{Z}_+(G \square G') \leq n \mathrm{Z}_+(G')$. Theorem 9.44(12) shows the bound is tight. □

EXAMPLE 9.56. Let T be a tree and $r \geq 2$. Then $\mathrm{Z}_+(T \square K_r) = r$ because $r = \delta(T \square K_r) \leq \mathrm{Z}_+(T \square K_r) \leq |V(K_r)| \mathrm{Z}_+(T) = r$.

Proposition 9.55 gives an upper bound on the PSD zero forcing number of Cartesian product of the graphs. We also obtain lower bound results similar to that in Theorem 9.22, but we can no longer use negative eigenvalues. Thus we obtain a slightly different lower bound.

PROPOSITION 9.57. *Let G and G' be graphs each of which has an edge. Then*

$$\mathrm{M}_+(G \square G') \geq \mathrm{M}_+(G) \mathrm{M}_+(G').$$

PROOF. Let n and n' denote the orders of G and G', respectively. Choose $A \in \mathcal{S}_+(G)$ and $A' \in \mathcal{S}_+(G')$ such that $\operatorname{null} A = \mathrm{M}_+(G)$ and $\operatorname{null} A' = \mathrm{M}_+(G')$. Define $L = A \otimes I_n + I_{n'} \otimes A'$. Then $L \in \mathcal{S}_+(G \square G')$ and $\operatorname{null} L \geq \operatorname{null} A \cdot \operatorname{null} A'$. □

COROLLARY 9.58. *Let G and G' be graphs such that $\mathrm{M}_+(G) = \mathrm{Z}_+(G)$ and $\mathrm{M}_+(G') = \mathrm{Z}_+(G')$. Then*

$$\mathrm{Z}_+(G \square G') \geq \mathrm{Z}_+(G) \mathrm{Z}_+(G').$$

Observe that from Example 9.56, $\mathrm{Z}_+(T \square K_r) = r = \mathrm{Z}_+(T) \mathrm{Z}_+(K_r) + 1$.

Nordhaus-Gaddum Problems. The PSD Graph Complement Conjecture and ν-Graph Complement Conjecture are Nordhaus-Gaddum sum lower bounds [**20**].

CONJECTURE 9.59. (GCC$_+$) *For any graph G of order n,*

$$\mathrm{M}_+(G) + \mathrm{M}_+(\overline{G}) \geq n - 2.$$

(GCC$_\nu$) *For any graph G of order n,*

$$\nu(G) + \nu(\overline{G}) \geq n - 2.$$

In contrast, the analogous tight Nordhaus-Gaddum sum lower bound for PSD zero forcing is known as stated in the next theorem.

THEOREM 9.60. [**122**] *Let G be a graph of order n.*

(1) $\mathrm{Z}_+(G) + \mathrm{Z}_+(\overline{G}) \geq n - 2$, *and* $\mathrm{Z}_+(P_n) + \mathrm{Z}_+(\overline{P_n}) = n - 2$ *for* $n \geq 4$.
(2) $\mathrm{Z}_+(G) + \mathrm{Z}_+(\overline{G}) \leq 2n - 1$ *for* $n \geq 2$, *and* $\mathrm{Z}_+(K_n) + \mathrm{Z}_+(\overline{K_n}) = 2n - 1$ *for* $n \geq 2$.

See Section 9.11 for discussion of the average value of $\mathrm{Z}_+(G) + \mathrm{Z}_+(\overline{G})$.

9.4. Skew Forcing, $\mathrm{Z}_-(G)$

We begin this section with a discussion of two other inverse eigenvalue problems of a graph that share one zero forcing parameter as an upper bound for their maximum nullities. The first of these, the zero-diagonal or hollow inverse eigenvalue problem of a graph, was actually introduced after the second, the skew symmetric inverse eigenvalue problem of a graph, from which skew forcing takes its name. A *hollow matrix* is a square matrix all of whose diagonal entries are zero. The *hollow inverse eigenvalue problem of a graph* is to determine the possible spectra of

$$\mathcal{S}_0(G) = \{A \in \mathcal{S}(G) : a_{ii} = 0,\ 1 \leq i \leq |V(G)|\}.$$

The matrices in $\mathcal{S}_0(G)$ can be viewed as weighted adjacency matrices of G. The *maximum hollow nullity* and *minimum hollow rank* of G, defined in [**163**] as *maximum zero-diagonal nullity* and *minimum zero-diagonal rank*, are

$$\mathrm{M}_0(G) = \max\{\operatorname{null} A : A \in \mathcal{S}_0(G)\} \text{ and } \mathrm{mr}_0(G) = \min\{\operatorname{rank} A : A \in \mathcal{S}_0(G)\}.$$

As usual, $\mathrm{mr}_0(G) + \mathrm{M}_0(G) = |V(G)|$.

We can also modify the IEP-G to consider skew symmetric matrices (rather than symmetric matrices) with off-diagonal nonzero pattern described by the edges of G: The *skew symmetric inverse eigenvalue problem of a graph* is to determine the possible spectra of matrices in

$$\mathcal{S}_-(G) = \{A \in \mathbb{R}^{n\times n} : A \text{ is skew symmetric and } \mathcal{G}(A) = G\}.$$

Again, a somewhat more tractable subproblem is to determine the *maximum skew nullity* or *minimum skew rank* [**199**], i.e., to determine

$$\mathrm{M}_-(G) = \max\{\operatorname{null} A : A \in \mathcal{S}_-(G)\} \text{ or } \mathrm{mr}_-(G) = \min\{\operatorname{rank} A : A \in \mathcal{S}_-(G)\}.$$

As usual, $\mathrm{mr}_-(G) + \mathrm{M}_-(G) = |V(G)|$. The connection between these two separate maximum nullity problems is that a real skew symmetric matrix is hollow.

Note that arbitrary shifting is not allowed when the diagonal must be zero, and thus the maximum nullity need not be the maximum multiplicity of an eigenvalue. There are many graphs G for which $\mathrm{M}_0(G) = 0$ and $\mathrm{M}_-(G) = 0$ (e.g., Example 9.61), whereas the maximum multiplicity of an eigenvalue is at least one for every graph. The *maximum hollow multiplicity* of G is

$$\mathrm{MM}_0(G) = \max\{\mathrm{mult}_A(\lambda) : A \in \mathcal{S}_0(G), \lambda \in \operatorname{spec}(A)\}.$$

It should also be noted that the hollow matrix constraint means that neither the maximum hollow rank of G nor the maximum skew rank of G need be the order of G. See Example ZF.4-1 in [**183**], and [**163**] and [**199**] for more information.

Skew zero forcing, or *skew forcing*, which was designed as the zero forcing upper bound for $\mathrm{M}_-(G)$, is defined by the *skew color change rule*: If a vertex $v \in V$ has exactly one unfilled neighbor, w, fill w. Observe that it is possible for an unfilled vertex to perform a skew force, and this is called an *unfilled vertex force.*[1] The skew forcing number is denoted by $\mathrm{Z}_-(G)$. Unfilled vertex forcing means that it is possible to have $\mathrm{Z}_-(G) = 0$, as in the next example.

EXAMPLE 9.61. Let P_{2k} be the path $1\text{–}2\text{–}\cdots\text{–}(2k-1)\text{–}2k$. Then every vertex on the path can be filled with no initial filled vertices as follows. First, let $\ell = k$ and $m = k+1$ if k is even, and $\ell = k+1$ and $m = k$ if k is odd. Then

$$\begin{array}{cccc} 1 \to 2 & 3 \to 4 & \dots & \ell - 1 \to \ell \\ 2k \to 2k-1 & 2k-2 \to 2k-3 & \dots & m+1 \to m \end{array}$$

Now let $\ell' = k$ and $m' = k+1$ if k is even, and $\ell' = k-1$ and $m' = k+1$ if k is odd. Then

$$\begin{array}{cccc} \ell' \to \ell' - 1 & \ell' - 2 \to \ell' - 3 & \dots & 2 \to 1 \\ m' \to m'+1 & m'+2 \to m'+3 & \dots & 2k-1 \to 2k \end{array}$$

Thus $\emptyset$ is a skew forcing set of P_{2k} and $\mathrm{Z}_-(P_{2k}) = 0$.

THEOREM 9.62. [**163, 199**] *For every graph G,*

$$\mathrm{M}_-(G) \le \mathrm{Z}_-(G) \text{ and } \mathrm{M}_0(G) \le \mathrm{Z}_-(G).$$

The proofs that $\mathrm{M}_-(G) \le \mathrm{Z}_-(G)$ and $\mathrm{M}_0(G) \le \mathrm{Z}_-(G)$ are similar to that for standard zero forcing (see Theorem 2.14) using the fact that the diagonal entries are all zero. The bounds in Theorem 9.62 are tight; examples of graphs G for which $\mathrm{M}_-(G) = \mathrm{Z}_-(G) = \mathrm{M}_0(G)$ include any graph G for which $\mathrm{Z}_-(G) = 0$. Strict inequality is also possible: For n odd, $\mathrm{M}_0(C_n) = 0 < 1 = \mathrm{Z}_-(C_n)$ (see Theorem 9.100(5)). The Petersen graph P provides a contrasting example for maximum skew nullity, since $\mathrm{M}_-(P) = 2$ [**199**] and $\mathrm{Z}_-(P) = 4$ [**183**, Example ZF.4-2].

Next we present some basic observations about skew forcing.

REMARK 9.63. Let G be a graph of order n.

(1) Since any Z-force is a valid Z_--force, $\mathrm{Z}_-(G) \le \mathrm{Z}(G)$.
(2) Since all but one neighbor must be filled for a Z_--force to occur, $\delta(G) - 1 \le \mathrm{Z}_-(G) \le n$.
(3) If G contains at least one edge, then $\mathrm{Z}_-(G) \le n-2$ (because a set consisting of all vertices except an adjacent pair is a skew forcing set).

OBSERVATION 9.64. *Let G be a graph with connected components $G_1, \dots, G_t$. Then*

$$\mathrm{Z}_-(G) = \sum_{i=1}^{t} \mathrm{Z}_-(G_i),\ \mathrm{M}_0(G) = \sum_{i=1}^{t} \mathrm{M}_0(G_i),\ \textit{and}\ \mathrm{M}_-(G) = \sum_{i=1}^{t} \mathrm{M}_-(G_i).$$

Let G be a graph of order n. For the skew forcing number, define the complementary parameter $\mathrm{mz}_-(G) = n - \mathrm{Z}_-(G)$. Notice that $\mathrm{mz}_-(G) \le \mathrm{mr}_-(G)$ and $\mathrm{mz}_-(G) \le \mathrm{mr}_0(G)$ for every graph G. Furthermore $\mathrm{mz}_-(G) = \mathrm{mr}_-(G)$ is equivalent to $\mathrm{Z}_-(G) = \mathrm{M}_-(G)$ and $\mathrm{mz}_-(G) = \mathrm{mr}_0(G)$ is equivalent to $\mathrm{Z}_-(G) = \mathrm{M}_0(G)$. The proof of the next result is analogous to that of Proposition 9.11.

PROPOSITION 9.65. *If H is an induced subgraph of a graph G, then* $\mathrm{mz}_-(H) \le \mathrm{mz}_-(G)$.

[1]This is called a white vertex force in the literature.

REMARK 9.66. In contrast to $Z(G) \geq \eta(G) - 1$ and $Z_+(G) \geq \eta(G) - 1$, $Z_-(K_r \circ K_1) = 0$ whereas $\eta(K_r \circ K_1) - 1 = r - 1$ for $r \geq 2$.

REMARK 9.67. For any graph G and set $B \subseteq V(G)$, the skew final coloring of B is unique. This can be seen by examining two chronological lists of skew forces $\mathcal{F}_1$ and $\mathcal{F}_2$ with final colorings $D_1, D_2 \subseteq V(G)$: Suppose $D_1 \not\subseteq D_2$. Among vertices in $D_1 \setminus D_2$, let w be the first vertex filled by $\mathcal{F}_1$, with skew force $u \to w$. When all the vertices of D_2 are filled, all of the other neighbors of u are filled (since they were filled at this stage using the forces in $\mathcal{F}_1$), so we can perform the skew force $u \to w$. Thus D_2 was not a final coloring.

Any graph G with $Z_-(G) = 0$ has a unique minimum skew forcing set, namely the empty set. However, Thomas Cameron pointed out that the only graphs with unique minimum skew forcing sets are those where each connected component has skew forcing number equal to zero or is an isolated vertex [**76**]:

PROPOSITION 9.68. *If G is a connected graph of order greater than one such that $Z_-(G) \geq 1$, then no vertex is in every minimum standard zero forcing set of G. A connected graph G of order greater than one has a unique minimum skew forcing set if and only if $Z_-(G) = 0$.*

PROOF. We establish the first statement. Suppose G is a connected graph of order greater than one such that $Z_-(G) \geq 1$. Let B be a minimum skew forcing set, let $v \in B$, and let $\mathcal{F}$ be a chronological lists of skew forces. Denote the forces in $\mathcal{F}$ in order by $u_1 \to w_1, \dots, u_r \to w_r$ (the u_i are distinct, the w_i are distinct, but it is possible that $u_i = w_j$ for $i \neq j$). If no neighbor of v performs a force, then $B \setminus \{v\}$ would a skew forcing set (after all other forces are done, any neighbor of v can force v). Since B is minimum, some neighbor of v performs a force. Let k denote the least index such that $u_k \in N(v)$. Define $B' = B \setminus \{v\} \cup \{w_k\}$, so $|B'| = |B|$. Since $u_i \notin N(v)$ for $i = 1, \dots, k-1$, $u_i \to w_i$ are valid skew forces for B for $i = 1, \dots, k-1$. Then $u_k \to v$. Now every vertex in B is filled, so B' is a skew forcing set. □

REMARK 9.69. Although $Z_-(G) \leq Z(G)$ and $M_0(G) \leq M(G)$, the parameter $M_-(G)$ is noncomparable with both $M(G)$ and $M_0(G)$. It is shown in Examples 9.71 and 9.84 below that the graph G_6 shown in Figure 9.3 has $M(G_6) = 3$ and $M_0(G_6) = 1 > 0 = M_-(G_6)$. For the other direction, $M_-(K_{3,3,3}) = 7$ and $M(K_{3,3,3}) = M_0(K_{3,3,3}) = 6$ [**34, 199**].

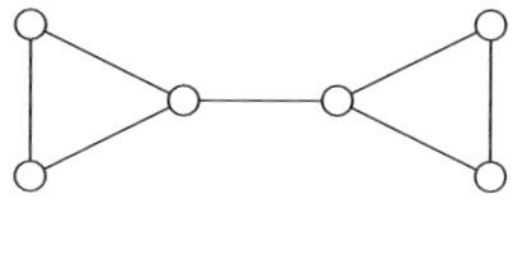

FIGURE 9.3. The graph G_6.

PROPOSITION 9.70. [**199**] *For any graph G, $\mathrm{mr}_-(G)$ is even.*

Proposition 9.70 is equivalent to $|V(G)| - M_-(G)$ is even for all graphs G, which raises the following question. Is $|V(G)| - Z_-(G)$ even for all graphs G? The next example provides a negative answer.

EXAMPLE 9.71. Consider the graph G_6 shown in Figure 9.3. Since any one vertex is a skew forcing set and $\delta(G) = 2$, $\mathrm{Z}_-(G) = 1$. Since $6 - \mathrm{M}_-(G) \geq 6 - \mathrm{Z}_-(G) = 5$ and $6 - \mathrm{M}_-(G)$ is even, $6 - \mathrm{M}_-(G) = 6$ and $\mathrm{M}_-(G) = 0$. Also $\mathrm{M}(G_6) = 3$ because $\mathrm{cc}(G_6) = 3$ and $\mathrm{Z}(G_6) = 3$.

Given a skew forcing set B and a set of forces (or chronological list of forces) $\mathcal{F}$, a *skew forcing chain* is a maximal sequence of vertices $(v_1, v_2, \dots, v_s)$ such that for $i = 1, \dots, s-1$, $v_i \to v_{i+1}$ (if $v_1 \in B$ does not force, then (v_1) is a forcing chain). The *skew forcing chain cover* of $\mathcal{F}$ is the set of all skew forcing chains of $\mathcal{F}$ (so every vertex is in one forcing chain in the forcing chain cover). It is shown in Proposition 9.73 below that a skew forcing chain must be a path or a 2-cycle (forcing chain (v_1, v_2, v_1), i.e., $v_1 \to v_2$ and $v_2 \to v_1$). A chain that is a path need not be induced, as seen in the next example.

EXAMPLE 9.72. Consider the graph G shown in Figure 9.4, which has a minimum skew forcing set $\{1\}$. Choose a chronological list of skew forces $2 \to 3, 4 \to 5, 1 \to 2, 3 \to 4$, giving the forcing chain $(1, 2, 3, 4, 5)$. Choose another chronological list of skew forces $2 \to 3, 5 \to 4, 4 \to 5, 3 \to 2$, giving the forcing chains $(2, 3, 2)$, $(5, 4, 5)$, and (1).

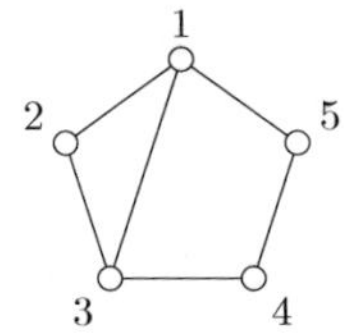

FIGURE 9.4. A graph that has a (non-induced) path on 5 vertices or two disjoint 2-cycles and a P_1 as skew forcing chains.

PROPOSITION 9.73. [**212**] *For any set of skew forces for a skew forcing set B, a forcing chain is a path or 2-cycle. The elements of the skew forcing set B are the initial vertices of the maximal skew forcing paths. If $\mathrm{Z}_-(G) = 0$, then every skew forcing chain is a 2-cycle and G has even order.*

PROOF. Think of a skew forcing chain as a directed graph, with the direction indicating a force. A vertex can perform at most one skew force, so the out-degree of every vertex in the directed skew forcing chain is at most one. Thus a directed skew forcing chain is a directed path or a directed cycle. If $k \geq 3$, then each vertex in a directed k-cycle has two neighbors in the (undirected) graph and so cannot perform a force if none of the vertices are in B (and a cycle of forces would not contain any vertex in B since every vertex is forced). $\square$

Computation. The function `Zskew(G)` $= \mathrm{Z}_-(G)$ is available for computing the skew forcing number of a specific graph (that is not too large). Its use is illustrated in the *Sage* program [**183**]. It calls the function `find_loopedZ(G,[])` $= \mathrm{Z}_-(G)$. (The function `find_loopedZ(G,S)` finds the looped zero forcing number of a loop graph G that has loops on the vertices in S; see [**56**] for more information about maximum nullity and zero forcing on loop graphs. The empty brackets [] denote that there are no loops, which means matrices associated with such a loop-graph have all diagonal entries equal to zero.) Furthermore, `find_loopedZ_sets(G,[])` will return a minimum skew forcing set.

To explain how $\mathrm{mr}_-(G)$, $\mathrm{M}_-(G)$, $\mathrm{mr}_0(G)$, and $\mathrm{M}_0(G)$ can be computed for specific graphs, we need to examine principal submatrices, the characteristic polynomial, and how the determinant of a hollow matrix can be computed by using the graph.[2] Denote the sum of all $k \times k$ principal minors of a real matrix A by $S_k(A)$. Recall that the characteristic polynomial of A is

$$p_A(x) = x^n - S_1(A)x^{n-1} + \cdots + (-1)^k S_k(A)x^{n-k} + \cdots + (-1)^n S_n(A)$$

[**196**, p. 53]. A real symmetric (or skew symmetric) matrix A has rank r if and only if $S_r(A) \neq 0$ and $S_k(A) = 0$ for $k = r+1, \ldots, n$ (because the algebraic and geometric multiplicities of eigenvalues are equal).

A *generalized cycle* of G is a subgraph of G each of whose connected components is a single edge (meaning an edge and its two endpoints) or a cycle (note that a vertex can be in at most one connected component). The *order* of a generalized cycle is the number of vertices in the generalized cycle. The set of all generalized cycles of order k of a graph G is denoted $\mathrm{cyc}_k(G)$. A generalized cycle that has the same order as the graph G is also called a *spanning generalized cycle*. Spanning generalized cycles appear in the literature under a variety of other names, including *perfect* $[1,2]$*-factors* [**177**] and *linear subgraphs* [**171**].

Let $A = [a_{ij}]$ be a hollow symmetric matrix. Given a generalized cycle $\mathcal{C}$ of $\mathcal{G}(A)$, let $\mathrm{nc}(\mathcal{C})$ denote the number of distinct cycles in $\mathcal{C}$, and $\mathrm{ne}(\mathcal{C})$ denote the number of even components of $\mathcal{C}$, i.e., the number of cycles of even order at least four plus the number of edges. With a generalized cycle $\mathcal{C}$, we can associate a permutation $\pi_{\mathcal{C}}$ of the vertices of $\mathcal{C}$ as follows: For each cycle in $\mathcal{C}$, fix an orientation and then associate a directed graph cycle $(v_{j_1}, v_{j_2}, \ldots, v_{j_\ell})$ with the cyclic permutation $(v_{j_1} v_{j_2} \cdots v_{j_\ell})$. Each edge component $\{v_{i_1}, v_{i_2}\}$ of $\mathcal{C}$ is associated with the transposition $(v_{i_1} v_{i_2})$. The permutation $\pi_{\mathcal{C}}$ is defined to be the product of these associated permutation cycles. There are $2^{\mathrm{nc}(\mathcal{C})}$ different choices for the orientation of the cycles of $\mathcal{C}$, and each choice yields a permutation that has the same sign as $\pi_{\mathcal{C}}$, namely $(-1)^{\mathrm{ne}(\mathcal{C})}$. It follows from [**171**, Theorem 3] that $S_k(A)$ can be computed using generalized cycles of $\mathcal{G}(A)$:

$$S_k(A) = \sum_{\mathcal{C} \in \mathrm{cyc}_k(\mathcal{G}(A))} (-1)^{\mathrm{ne}(\mathcal{C})} 2^{\mathrm{nc}(\mathcal{C})} a_{i_1 \pi_{\mathcal{C}}(i_1)} \cdots a_{i_k \pi_{\mathcal{C}}(i_k)}, \tag{9.1}$$

where the sum over the empty set is zero.

Equation (9.1) assists with the computation of the characteristic polynomial (and thus the nullity) of a specific matrix. More generally, it is useful for computing $\mathrm{mr}_0(G)$. Observe that $\mathrm{mr}_0(G) \leq r$ if G has no generalized cycles of order greater than r, and $\mathrm{mr}_0(G) = r$ if G has no generalized cycles of order greater than r and G has a unique generalized cycle of order r.

For skew symmetric matrices, observe that the two orientations of an odd cycle cancel each other out. Since even cycles have perfect matchings, this leads to the study of the pfaffian of a skew symmetric matrix $A = [a_{ij}]$: For a perfect matching $L = \{v_1v_2, \ldots, v_{n-1}v_n\}$ of $\mathcal{G}(A)$ ordered so that $v_1 < v_2, v_3 < v_4, \ldots, v_{n-1} < v_n$ and $v_1 < v_3 < \cdots < v_{n-1}$, π_L is the permutation of $\{1, \ldots, n\}$ that maps k

[2]These ideas can be used for symmetric matrices that are not hollow or matrices that are not symmetric, but then graphs with loops or digraphs must be used; see [**56**] and [**66**, Section 9.1].

to v_k. Let $\mathcal{M}_G$ be the set of all perfect matchings in G. The *pfaffian* of A is $\operatorname{pf}(A) = \sum_{L \in \mathcal{M}_{\mathcal{G}(A)}} \operatorname{sgn}(\pi_L) a_{i_1,i_2} \cdots a_{i_{n-1},i_n}$, where the sum over the empty set is 0.

THEOREM 9.74. [**66**, Theorem 9.5.2] *If A is a real skew symmetric of matrix even order, then* $\det A = \operatorname{pf}(A)^2$.

Extreme Values. Graphs G having the highest possible skew forcing numbers and lowest possible skew forcing number have been determined. We begin with the highest values.

REMARK 9.75. Let G be a graph of order n. By considering the ranks of the matrices associated with G and Remark 9.63, we see that the following are equivalent:

(1) G has no edges.
(2) $\mathrm{Z}_-(G) = n$.
(3) $\mathrm{M}_-(G) = n$.
(4) $\mathrm{M}_0(G) = n$.

It follows from Remark 9.63 that no graph G of order n has $\mathrm{Z}_-(G) = n - 1$, which implies $\mathrm{M}_-(G) \neq n - 1$ and $\mathrm{M}_0(G) \neq n - 1$. The next result follows from [**199**, Theorem 2.1] and Theorem 9.62.

THEOREM 9.76. *Let G be a connected graph of order $n \geq 2$. Then the following are equivalent:*

(1) $\mathrm{M}_-(G) = n - 2$.
(2) $\mathrm{Z}_-(G) = n - 2$.
(3) $G = K_{n_1,n_2,\ldots,n_t}$ *for some* $t \geq 2, n_i \geq 1, i = 1, \ldots, t$ *and* $n_1 + \cdots + n_t = n$.
(4) *G does not contain P_4 or the paw (see Figure 9.5) as an induced subgraph.*

Without the assumption that G is connected, $\mathrm{M}_-(G) = n-2$ if and only if $\mathrm{Z}_-(G) = n - 2$ if and only if G is a union of one $K_{n_1,n_2,\ldots,n_t}$ and possibly some isolated vertices.

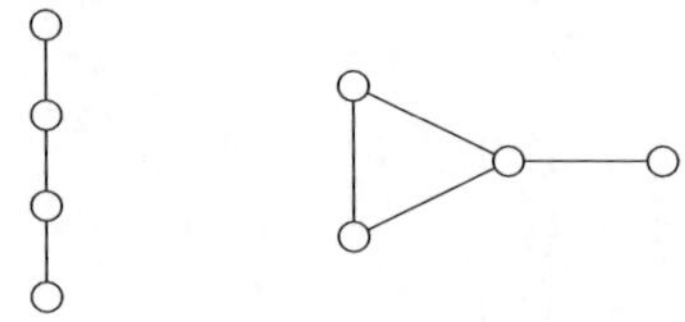

P_4 Paw

FIGURE 9.5. Forbidden induced subgraphs for $\mathrm{Z}_-(G) = \mathrm{M}_-(G) = 2$.

THEOREM 9.77. [**163**] *Let G be a connected graph of order $n \geq 2$. Then the following are equivalent:*

(1) $\mathrm{M}_0(G) = n - 2$.
(2) $G = K_{r,n-r}$ *for some r such that* $1 \leq r \leq n - 1$.
(3) *G does not contain P_4 or C_3 as an induced subgraph.*

Without the assumption that G is connected, $\mathrm{M}_0(G) = n - 2$ if and only if G is a union of one K_{n_1,n_2} and possibly some isolated vertices.

COROLLARY 9.78. *Let G be a graph of order n. If $M_0(G) = n-2$, then $M_-(G) = Z_-(G) = n-2$, but not conversely.*

Next we consider graphs for which one or more of skew forcing number, skew maximum nullity, or hollow maximum nullity is equal to zero. The next algorithm provides a method to determine whether a graph G has $Z_-(G) = 0$. Define $Z_-(\emptyset) = 0$.

ALGORITHM 9.79 (Leaf-Stripping Algorithm). [**188, 212**]
Input: *Graph G.*
Output: *Graph $\hat{G}$ with $\delta(\hat{G}) \neq 1$, or $\hat{G} = \emptyset$.*

$\hat{G} := G$.
While *$\hat{G}$ has a leaf u with neighbor v:*
  $\hat{G} := \hat{G}[V(\hat{G}) \setminus \{u, v\}]$.
Return $\hat{G}$.

THEOREM 9.80. [**188, 212**] *Let G be a graph and let $\hat{G}$ be the graph returned by Algorithm 9.79. Then:*
(1) $Z_-(G) = Z_-(\hat{G})$.
(2) $Z_-(G) = 0$ *if and only if* $\hat{G} = \emptyset$.

Since applying the Leaf-Stripping Algorithm to a forest F results in $\hat{F}$ being a set of isolated vertices (or empty), the algorithm computes the skew forcing number of a forest.

COROLLARY 9.81. *If F is a forest, then $Z_-(F) = |V(\hat{F})|$ where $\hat{F}$ is produced by Algorithm 9.79.*

It is not the case that $M_-(G) = 0$ implies $Z_-(G) = 0$, as shown by the graph G_6 in Example 9.71. The next theorem characterizes graphs G such that $M_-(G) = 0$.

THEOREM 9.82. [**199**] *Let G be a graph. Then $M_-(G) = 0$ if and only if G has a unique perfect matching.*

The next theorem characterizes graphs G such that $M_0(G) = 0$.

THEOREM 9.83. [**163**] *For every graph G, $M_0(G) = 0$ if and only if G has a unique spanning generalized cycle.*

An algorithm for determining whether a graph has a unique spanning generalized cycle is presented in [**163**].

EXAMPLE 9.84. Consider the graph G_6 shown in Figure 9.3. Then $M_0(G_6) = 1$ since G_6 does not have a unique spanning generalized cycle and $Z_-(G_6) = 1$ (see Example ZF.4-3 in [**183**]), but $M_-(G_6) = 0$ since G_6 has a unique perfect matching.

An odd cycle has a unique spanning generalized cycle, so $M_0(C_{2k+1}) = 0$, whereas $M_-(C_{2k+1}) = Z_-(C_{2k+1}) = 1$.

Graph Operations. Next we discuss the behavior of the skew forcing number under graph operations.

REMARK 9.85. For any graph H, $M_0(H \circ K_1) = M_-(H \circ K_1) = Z_-(H \circ K_1) = 0$, where $H \circ K_1$ is constructed from H by adding a leaf to each vertex of H. This implies that there cannot be a nonzero "strong" property below M_0 or M_- that is induced subgraph monotone.

PROPOSITION 9.86. [**182**] *Let G be a graph.*
(1) *For any vertex v of G, $-1 \leq \mathrm{Z}_-(G) - \mathrm{Z}_-(G-v) \leq 1$.*
(2) *For any edge e of G, $-2 \leq \mathrm{Z}_-(G) - \mathrm{Z}_-(G-e) \leq 2$.*
All bounds are tight.

Even if v is a leaf, the difference $\mathrm{Z}_-(G) - \mathrm{Z}_-(G-v)$ can realize all possible values $1, -1$, and 0, unlike standard and PSD zero forcing numbers (see Remark 9.14 and Corollary 9.50). Examples include odd paths, even paths, and the graph G constructed from G_6 in Figure 9.3 by adding a leaf v to a vertex of degree three (see Example ZF.4-4 in [**183**]). Examples ZF.4-5 and ZF.4-6 in [**183**] illustrate the edge deletions bounds.

PROPOSITION 9.87. *Suppose v and w are independent twins in G. Then one of v and w is in every skew forcing set and $\mathrm{Z}_-(G) = \mathrm{Z}_-(G-v)+1$.*

PROOF. Let B be a minimum skew forcing set of G. Since no neighbor of v and w can perform a force until at least one of v or w is filled, at least one of v or w must be in any zero forcing set B; without loss of generality, $v \in B$. Then $B \setminus \{v\}$ is a skew forcing set of $G-v$, because if $v \to u$ in G, then $w \to u$ in $G-v$ using unfilled vertex forcing. Thus $\mathrm{Z}_-(G-v) \leq \mathrm{Z}_-(G) - 1$. From Proposition 9.86 we know that $\mathrm{Z}_-(G) \leq \mathrm{Z}_-(G-v)+1$. □

It is not true that one of v and w is in every skew forcing set and $\mathrm{Z}_-(G) = \mathrm{Z}_-(G-v)+1$ when v and w are adjacent twins in G. For example, consider the friendship graph F_k with $k \geq 2$. The set B consisting of the center vertex is a minimum skew forcing set and if $\deg v = \deg w = 2$ and $w \in N(v)$, then v and w are adjacent twins and neither is in B.

The proof of the next result is similar to the proof of Proposition 9.16, but white vertex forcing requires a minor modification: If $u \notin B$ and u performs the first force, then only one vertex of H is unfilled, so we can replace B by any set of vertices that contains u and omits exactly two adjacent vertices of H without changing the cardinality of B.

PROPOSITION 9.88. *Let H be a graph with no isolated vertices and let $G = H \vee K_1$, where $V(K_1) = \{u\}$. Then $\mathrm{Z}_-(G) = \mathrm{Z}_-(H)+1$, there is a minimum skew forcing set $B \subset V(G)$ such that $u \in B$, and a set of forces for B such that u does not perform a force.*

Proposition 9.88 is not true without the assumption that H is has no isolated vertices (recall that $K_{1,n} = K_1 \vee \overline{K_n}$) and $\mathrm{Z}_-(K_{1,n}) = n-1$. Proposition 9.16 allows the extension of small examples that show bounds are tight to infinite families of examples.

It is well known that adding an universal vertex to a graph with no isolated vertices does not change the minimum rank (or PSD minimum rank): If $A \in \mathcal{S}(G)$ and $\operatorname{rank} A = \operatorname{mr}(G)$, choose $\mathbf{v}$ such that every entry of $A\mathbf{v}$ is nonzero and construct the matrix B as in the next proof. It is not so simple to find such a $\mathbf{v}$ when the diagonal must be zero, and it is not always possible, as shown by C_4 ad W_5 (see [**97**]).

LEMMA 9.89. [**97**] *Let A be an $n \times n$ hollow symmetric matrix that has a nonzero eigenvalue λ with $\operatorname{mult}_A(\lambda) \geq 2$ and an eigenvector $\mathbf{w}$ for λ that has every* *entry nonzero. Then there exists a vector* $\mathbf{v} \in \mathbb{R}^n$ *such that* $\mathbf{v}^\top A\mathbf{v} = 0$ *and every entry of* $A\mathbf{v}$ *is nonzero. Furthermore,* $B = \begin{bmatrix} A & A\mathbf{v} \\ \mathbf{v}^\top A & \mathbf{v}^\top A\mathbf{v} \end{bmatrix} \in \mathcal{S}_0(\mathcal{G}(A) \vee K_1)$ *and* $\operatorname{rank} B = \operatorname{rank} A$.

LEMMA 9.90. **[97]** *If* G *has no isolated vertices and* $(m_1, \dots, m_r) \in \mathbf{m}(G)$*, then* $\mathrm{MM}_0(G \vee K_1) \geq m_1 + 1$.

COROLLARY 9.91. **[97]** *If there exists* $A \in \mathcal{S}_0(G)$ *such that the maximum hollow multiplicity* $\mathrm{MM}_0(G)$ *is realized by the least (or greatest) eigenvalue of* A*, then* $\mathrm{MM}_0(G \vee K_1) = \mathrm{MM}_0(G) + 1$.

PROPOSITION 9.92. *Let* G *be a graph and let* e *be an edge of* G*. Then*

$$-1 \leq Z_-(G) - Z_-(G/e) \leq 1 \text{ and } -1 \leq Z_-(G_e) - Z_-(G) \leq 1.$$

All bounds are tight.

PROOF. Let $e = uw$. First we consider G/e. Let v_e be the new vertex created by contracting e.

$Z_-(G/e) \leq Z_-(G) + 1$: Let B be a minimum skew forcing set of G and let $\mathcal{F}$ be a chronological list of skew forces. If at least one of u or v performs a force in $\mathcal{F}$, assume $u \to x$ is the first such force that appears in $\mathcal{F}$. Then $B \cup \{x\}$ is a skew forcing set of G/e with chronological list of skew forces obtained from $\mathcal{F}$ by deleting $u \to x$ and replacing $w \to y$ (if such a force appears) by $v_e \to y$. If neither u nor v performs a force in $\mathcal{F}$, then B is a skew forcing set of G/e with chronological list of skew forces $\mathcal{F}$. Thus $Z_-(G/e) \leq Z_-(G) + 1$.

$Z_-(G) \leq Z_-(G/e)+1$: Now let B be a minimum skew forcing set of G/e and let $\mathcal{F}$ be a chronological list of skew forces. If $v_e \in B$, then define $B' = B \setminus \{v_e\} \cup \{u, w\}$. If $v_e \notin B$, then define B' as follows: If v_e does not perform a force, let $B' = B$. Finally suppose $v_e \to x$ in G/e. If $x \in N_G(u)$, let $B' = B \cup \{w\}$, whereas if $x \in N_G(w)$, let $B' = B \cup \{u\}$. Then B' is a skew forcing set of G with a chronological list of skew forces obtained from $\mathcal{F}$ by replacing v_e in a force $v_e \to x$ by whichever of u or w is a neighbor of x in G. Since $|B'| \leq |B| + 1$, $Z_-(G) \leq Z_-(G/e) + 1$.

The bounds for G_e follow from those for G/e, because $(G_e)/ux_e = G$ where x_e is the new vertex created in the subdivision of e. Comparison of the skew forcing numbers of the even path P_{2k} and the odd paths P_{2k-1} and P_{2k+1} shows all bounds are tight. □

Transmission across a boundary also applies to skew forcing, although now transmission may occurs in two directions (which is the reason the bound in Proposition 9.86(2) is ± 2). The next result follows from Proposition 9.86(2), because deleting $\partial(W)$ edges results in a disconnected graph $G[W] \sqcup G[\overline{W}]$.

PROPOSITION 9.93. *For any graph* G *and subset* $W \subseteq V(G)$*,*

$$Z_-(G) \geq Z_-(G[W]) + Z_+(G[\overline{W}]) - 2\partial(W).$$

It is not the case that $Z_-(G \,\square\, G') \leq \min\{n' Z_-(G), \, n Z_-(G')\}$ for all graphs G and G' of orders n and n'. Consider graphs for which $Z_-(G) = Z_-(G') = 0$, such as even paths. This implies $\delta(G) = \delta(G') = 1$. Since $\delta(G \,\square\, G') = \delta(G) + \delta(G')$, $Z_-(G \,\square\, G') \geq 1$. But $Z_-(G \,\square\, G')$ can be much larger, as the next result shows; Example ZF.4-7 in **[183]** illustrates this.

PROPOSITION 9.94. *For $n \geq 1$,*

$$\mathrm{Z}_-(P_n \,\square\, P_n) = n = \mathrm{M}_-(P_n \,\square\, P_n) = \mathrm{M}_0(P_n \,\square\, P_n).$$

PROOF. Note first that $\mathrm{Z}_-(P_n \square P_n) \leq \mathrm{Z}(P_n \square P_n) = n$. Let A be the adjacency matrix of P_n and let $L = A \otimes I_n + I_n \otimes (-A)$. Then $L \in \mathcal{S}_0(P_n \square P_n)$ and $\operatorname{null} L = n$, because for each $\lambda \in \operatorname{spec}(A)$, $0 = \lambda + (-\lambda) \in \operatorname{spec}(L)$. An analogous argument works for $\mathrm{M}_-(P_n \,\square\, P_n)$ using $A \in \mathcal{S}_-(P_n)$. $\square$

We used linear algebra to establish Proposition 9.94 and we have linear algebra lower bounds analogous to that for standard zero forcing. Let $m_2^-(G)$ denote the maximum possible multiplicity of a nonzero eigenvalue of a matrix $A \in \mathcal{S}_-(G)$ such that $\operatorname{null} A = \mathrm{M}_-(G)$. Let $m_2^0(G)$ denote the maximum possible multiplicity of a nonzero eigenvalue of a matrix $A \in \mathcal{S}_0(G)$ such that $\operatorname{null} A = \mathrm{M}_0(G)$. The proof of the next result is analogous to that of Theorem 9.22, except that a factor of two appears in the skew bound because the purely imaginary nonzero eigenvalues of a skew symmetric matrix appear in complex conjugate pairs.

PROPOSITION 9.95. *Let G and G' be graphs, each of which has an edge. Then*

$$\mathrm{M}_-(G \,\square\, G') \geq \mathrm{M}_-(G)\,\mathrm{M}_-(G') + 2m_2^-(G)m_2^-(G')$$

and

$$\mathrm{M}_0(G \,\square\, G') \geq \mathrm{M}_0(G)\,\mathrm{M}_0(G') + m_2^0(G)m_2^0(G').$$

In particular,

$$\mathrm{M}_-(G \,\square\, G') \geq \mathrm{M}_-(G)\,\mathrm{M}_-(G') + 2 \text{ and } \mathrm{M}_0(G \,\square\, G') \geq \mathrm{M}_0(G)\,\mathrm{M}_0(G') + 1.$$

COROLLARY 9.96. *Let G and G' be graphs. If $\mathrm{M}_-(G) = \mathrm{Z}_-(G)$ and $\mathrm{M}_-(G') = \mathrm{Z}_-(G')$, then $\mathrm{Z}_-(G \,\square\, G') \geq \mathrm{Z}_-(G)\,\mathrm{Z}_-(G') + 2$. If $\mathrm{M}_0(G) = \mathrm{Z}_-(G)$ and $\mathrm{M}_0(G') = \mathrm{Z}_-(G')$, then $\mathrm{Z}_-(G \,\square\, G') \geq \mathrm{Z}_-(G)\,\mathrm{Z}_-(G') + 1$.*

Recall that Kronecker products of matrices can be used to construct matrices for graph products (see Section 2.4). Here we describe constructions for tensor products, because hollow matrices are used.

OBSERVATION 9.97. **[44]** *Let G and G' be graphs. If $A \in \mathcal{S}_0(G)$ and $A' \in \mathcal{S}_0(G')$, then $A \otimes A' \in \mathcal{S}_0(G \times G')$. If $A \in \mathcal{S}_-(G)$ and $A' \in \mathcal{S}_-(G')$, then $A \otimes A' \in \mathcal{S}_0(G \times G')$. If $A \in \mathcal{S}_-(G)$ and $A' \in \mathcal{S}_0(G')$, or $A \in \mathcal{S}_0(G)$ and $A' \in \mathcal{S}_-(G')$, then $A \otimes A' \in \mathcal{S}_-(G \times G')$.*

If $\mathbf{x}$ is an eigenvector of A for eigenvalue λ and $\mathbf{x}'$ is an eigenvector of A' for eigenvalue λ', then $\mathbf{x} \otimes \mathbf{x}'$ is an eigenvector $A \otimes A'$ for eigenvalue $\lambda\lambda'$. Thus $\operatorname{rank}(A \otimes A') = \operatorname{rank} A \operatorname{rank} A'$.

PROPOSITION 9.98. *Let G and G' be graphs of orders n and n', respectively. Then*

(1) $\mathrm{mr}_0(G \times G') \leq \mathrm{mr}_0(G)\,\mathrm{mr}_0(G')$ *and*
$\mathrm{M}_0(G \times G') \geq n'\,\mathrm{M}_0(G) + n\,\mathrm{M}_0(G') - \mathrm{M}_0(G)\,\mathrm{M}_0(G')$.
(2) $\mathrm{mr}_0(G \times G') \leq \mathrm{mr}_-(G)\,\mathrm{mr}_-(G')$ *and*
$\mathrm{M}_0(G \times G') \geq n'\,\mathrm{M}_-(G) + n\,\mathrm{M}_-(G') - \mathrm{M}_-(G)\,\mathrm{M}_-(G')$.
(3) $\mathrm{mr}_-(G \times G') \leq \min\{\mathrm{mr}_0(G)\,\mathrm{mr}_-(G'), \mathrm{mr}_-(G)\,\mathrm{mr}_0(G')\}$,
$\mathrm{M}_-(G \times G') \geq n'\,\mathrm{M}_0(G) + n\,\mathrm{M}_-(G') - \mathrm{M}_0(G)\,\mathrm{M}_-(G')$, *and*
$\mathrm{M}_-(G \times G') \geq n'\,\mathrm{M}_-(G) + n\,\mathrm{M}_0(G') - \mathrm{M}_-(G)\,\mathrm{M}_0(G')$.

PROOF. (1): Choose matrices $A \in \mathcal{S}_0(G)$ and $A' \in \mathcal{S}_0(G')$ such that $\operatorname{rank} A = \operatorname{mr}_0(G)$ and $\operatorname{rank} A' = \operatorname{mr}_0(G')$. Then $A \otimes A' \in \mathcal{S}_0(G \times G')$ and $\operatorname{rank}(A \otimes A') = (\operatorname{rank} A)(\operatorname{rank} A')$ by Observation 9.97, so $\operatorname{mr}_0(G \times G') \leq \operatorname{mr}_0(G)\operatorname{mr}_0(G')$. The statement for maximum nullity follows by algebraic manipulation. The proofs of remaining statements are analogous. □

Skew forcing can be used to give a bound for standard zero forcing of tensor products.

THEOREM 9.99. [**38**] *Let G be a graph of order n and $r \geq 4$. Then*

$$\mathrm{Z}(G \times K_r) \leq (r-2)n + 2\,\mathrm{Z}_-(G)$$

and this bound is tight.

Families of Graphs. Theorem 9.100, Proposition 9.101, and Corollary 9.102 list values of $\mathrm{Z}_-(G)$, $\mathrm{M}_-(G)$, and $\mathrm{M}_0(G)$ for various families of graphs G. For Theorem 9.100, results that that concern $\mathrm{Z}_-(G)$ or $\mathrm{M}_-(G)$ can be found in [**199**] (or [**212**] for $\mathrm{Z}_-(G)$), whereas those concerning $\mathrm{M}_0(G)$ can be found in [**163**] (or [**97**] for the wheel).

THEOREM 9.100. [**97, 163, 199, 212**]

(1) *For $n \geq 1$, $\mathrm{Z}_-(P_n) = \mathrm{M}_-(P_n) = \mathrm{M}_0(P_n) = 0$ for even n, and $\mathrm{Z}_-(P_n) = \mathrm{M}_-(P_n) = \mathrm{M}_0(P_n) = 1$ for odd n.*
(2) *If T is a tree, then $\mathrm{Z}_-(T) = \mathrm{M}_-(T) = \mathrm{M}_0(T) = |V(T)| - 2\operatorname{match}(T)$.*
(3) *For $n \geq 2$, $\mathrm{Z}_-(K_n) = \mathrm{M}_-(K_n) = n-2$, and $\mathrm{M}_0(K_n) = n-3$ for $n \geq 3$.*
(4) *Let $n_1 \geq \cdots \geq n_r \geq 1$ and $n = n_1 + \cdots + n_r$. For $r \geq 2$, $\mathrm{Z}_-(K_{n_1,\ldots,n_r}) = \mathrm{M}_-(K_{n_1,\ldots,n_r}) = n-2$. For $r \geq 3$, $\mathrm{M}_0(K_{n_1,\ldots,n_r}) = n-3$, whereas $\mathrm{M}_0(K_{n_1,n_2}) = n-2$.*
(5) *For odd $n \geq 3$, $\mathrm{Z}_-(C_n) = \mathrm{M}_-(C_n) = 1$ and $\mathrm{M}_0(C_n) = 0$. For even $n \geq 4$, $\mathrm{Z}_-(C_n) = \mathrm{M}_-(C_n) = \mathrm{M}_0(C_n) = 2$.*
(6) *For even $n \geq 4$, $\mathrm{Z}_-(W_n) = \mathrm{M}_-(W_n) = 2$ and $\mathrm{M}_0(W_n) = 1$. For odd $n \geq 5$, $\mathrm{Z}_-(W_n) - \mathrm{M}_-(W_n) = 3$, $\mathrm{M}_0(W_n) = 3$ if $n \geq 7$, and $\mathrm{M}_0(W_5) = 2$.*
(7) *For $k \geq 2$, $\mathrm{Z}_-(Q_d) = \mathrm{M}_-(Q_d) = \mathrm{M}_0(Q_d) = 2^{d-1}$.*
(8) *For $s \geq 1$, $\mathrm{Z}_-(\mathrm{Half}_s) = \mathrm{M}_-(\mathrm{Half}_s) = \mathrm{M}_0(\mathrm{Half}_s) = 0$, where Half_s is the sth half-graph.*

PROPOSITION 9.101.

(1) *For $k \geq 1$, $\mathrm{Z}_-(F_k) = \mathrm{M}_-(F_k) = \mathrm{M}_0(F_k) = 1$, where F_k is the kth friendship graph.*
(2) *For $k \geq 1$, $\mathrm{Z}_-(B_k) = \mathrm{M}_-(B_k) = \mathrm{M}_0(B_k) = 2$, where B_k is the kth book graph.*
(3) *For $n \geq 4$, $\mathrm{Z}_-(\overline{P_n}) = n-4$.*
(4) *For $n \geq 4$, $\mathrm{Z}_-(\overline{C_n}) = n-4$.*
(5) *For $s \geq 2$, the complement $\overline{\mathrm{Half}_s}$ can be constructed from Half_{s-1} by adding an independent twin of the non-clique vertex that is adjacent to every clique vertex, and taking the disjoint union with a K_1. For $s \geq 2$, $\mathrm{Z}_-(\overline{\mathrm{Half}_s}) = 2$.*

PROOF. (1): $\mathrm{Z}_-(F_k) = 1$ because the vertex of degree $2k$ is a minimum skew forcing set and $\delta(F_k) = 2$. It follows from Theorems 9.82 and 9.83 that $\mathrm{M}_-(F_k) \neq 0$ and $\mathrm{M}_0(F_k) \neq 0$.

(2): Let the two spine vertices (of degree $k+1$) be numbered 1 and 2. Then $\mathrm{Z}_-(B_k) \leq 2$ because $\{1,2\}$ is a skew forcing set. The matrix $A \in \mathcal{S}_0(B_k)$ obtained from the adjacency matrix of B_k by setting $a_{1,2} = a_{2,1} = k$ has $\operatorname{null} A = 2$.

$\mathrm{M}_-(B_k) \geq 2$ because $\mathrm{mr}_-(B_k)$ must be even and $\mathrm{mr}_-(B_k) < |V(B_k)|$ since B_k has more than one perfect matching.

(3) and (4): Note that $\overline{C_4} = 2K_2$ so $\mathrm{Z}_-(\overline{C_4}) = 0$. Let G be $\overline{P_n}$ with $n \geq 4$ or $\overline{C_n}$ with $n \geq 5$. Since G has an induced P_4 subgraph and $\mathrm{Z}_-(P_4) = 0$, $\mathrm{Z}_-(G) \leq n - 4$ by Proposition 9.65, and $\mathrm{Z}_-(G) \geq n - 4 = \delta(G) - 1$ by Remark 9.63.

(5): Recall that Half_s is the graph constructed from a copy of K_s on vertices $\{a_1, \ldots, a_s\}$ and a copy of $\overline{K_s}$ on vertices $\{b_1, \ldots, b_s\}$ by adding all edges $\{a_i, b_j\}$ such that $i + j \leq s + 1$. Note that $\overline{\mathrm{Half}_s}[\{a_2, \ldots, a_s, b_2, \ldots, b_s\}]$ is isomorphic to Half_{s-1} using the vertices in the order $b_s, b_{s-1}, \ldots, b_2, a_s, a_{s-1}, \ldots, a_2$ (so $\{b_s, b_{s-1}, \ldots, b_2\}$ induces a K_{s-1}). Furthermore, b_1 is an independent twin of a_s in $\overline{\mathrm{Half}_s}$ (and a_s is the non-clique vertex adjacent to every clique vertex) and a_1 is an isolated vertex. This is illustrated in Example ZF.4-8 in [**183**]. Then $\mathrm{Z}_-(\overline{\mathrm{Half}_s}) = 2$ follows from Proposition 9.87. □

The next result follows from Propositions 9.98 and 9.101 and Theorem 9.99.

COROLLARY 9.102. *Let $r \geq 4$ and let G be a graph of order n such that $\mathrm{mr}_-(G) = \mathrm{mz}_-(G)$ or $\mathrm{mr}_0(G) = \mathrm{mz}_-(G)$. Then*

$$\mathrm{Z}_-(G \times K_r) = \mathrm{Z}(G \times K_r) = (r-2)n + 2\,\mathrm{Z}_-(G).$$

If $\mathrm{M}_-(G) = \mathrm{Z}_-(G)$, then $\mathrm{M}_0(G \times K_r) = \mathrm{M}(G \times K_r) = (r-2)n + 2\,\mathrm{Z}_-(G)$. If $\mathrm{M}_0(G) = \mathrm{Z}_-(G)$, then $\mathrm{M}_-(G \times K_r) = (r-2)n + 2\,\mathrm{Z}_-(G)$.

(1) *If T is a tree of order n, then $\mathrm{Z}(T \times K_r) = \mathrm{M}(T \times K_r) = \mathrm{Z}_-(T \times K_r) = \mathrm{M}_0(T \times K_r) = \mathrm{M}_-(T \times K_r) = rn - 4\,\mathrm{match}(T)$.*
(2) *For odd $n \geq 3$, $\mathrm{Z}(C_n \times K_r) = \mathrm{M}(C_n \times K_r) = \mathrm{Z}_-(C_n \times K_r) = \mathrm{M}_0(C_n \times K_r) = (r-2)n + 2$. For even $n \geq 4$, $\mathrm{Z}(C_n \times K_r) = \mathrm{M}(C_n \times K_r) = \mathrm{Z}_-(C_n \times K_r) = \mathrm{M}_0(C_n \times K_r) = \mathrm{M}_-(C_n \times K_r) = (r-2)n + 4$.*
(3) *For even $n \geq 4$, $\mathrm{Z}(W_n \times K_r) = \mathrm{M}(W_n \times K_r) = \mathrm{Z}_-(W_n \times K_r) = \mathrm{M}_0(W_n \times K_r) = (r-2)n + 4$. For odd $n \geq 5$, $\mathrm{Z}(W_n \times K_r) = \mathrm{M}(W_n \times K_r) = \mathrm{Z}_-(W_n \times K_r) = \mathrm{M}_0(W_n \times K_r) = (r-2)n + 6$.*
(4) *For $n \geq 2$, $\mathrm{Z}(K_n \times K_r) = \mathrm{M}(K_n \times K_r) = \mathrm{Z}_-(K_n \times K_r) = \mathrm{M}_0(K_n \times K_r) = rn - 4$.*
(5) *Let $n_1 \geq \cdots \geq n_r \geq 1$ and $n = n_1 + \cdots + n_r$. For $s \geq 2$, $\mathrm{Z}(K_{n_1,\ldots,n_r} \times K_r) = \mathrm{M}(K_{n_1,\ldots,n_r} \times K_r) = \mathrm{Z}_-(K_{n_1,\ldots,n_r} \times K_r) = \mathrm{M}_0(K_{n_1,\ldots,n_r} \times K_r) = rn - 4$.*
(6) *For $k \geq 2$, $\mathrm{Z}(Q_d \times K_r) = \mathrm{M}(Q_d \times K_r) = \mathrm{Z}_-(Q_d \times K_r) = \mathrm{M}_0(Q_d \times K_r) = \mathrm{M}_-(Q_d \times K_r) = (r-1)2^d$.*
(7) *For $s \geq 1$, $\mathrm{Z}(\mathrm{Half}_s \times K_r) = \mathrm{M}(\mathrm{Half}_s \times K_r) = \mathrm{Z}_-(\mathrm{Half}_s \times K_r) = \mathrm{M}_0(\mathrm{Half}_s \times K_r) = \mathrm{M}_-(\mathrm{Half}_s \times K_r) = (r-2)n$.*
(8) *For $k \geq 1$, $\mathrm{Z}(F_k \times K_r) = \mathrm{M}(F_k \times K_r) = \mathrm{Z}_-(F_k \times K_r) = \mathrm{M}_0(F_k \times K_r) = \mathrm{M}_-(F_k \times K_r) = (r-2)(2k+1) + 2$.*

PROOF. Suppose $\mathrm{M}_-(G) = \mathrm{Z}_-(G)$. As shown in [**198**], $\mathrm{mr}_-(K_r) = 2$. Thus $\mathrm{mr}_0(G \times K_r) \leq 2\,\mathrm{mr}_-(G)$ by Proposition 9.98. So

$$\begin{aligned}(r-2)n + 2\,\mathrm{Z}_-(G) &= (r-2)n + 2\,\mathrm{M}_-(G)) \leq \mathrm{M}_0(G \times K_r) \\ &\leq \mathrm{M}(G \times K_r) \leq \mathrm{Z}(G \times K_r) \leq (r-2)n + 2\,\mathrm{Z}_-(G)\end{aligned}$$

and $\mathrm{M}_0(G \times K_r) \leq \mathrm{Z}_-(G \times K_r) \leq \mathrm{Z}(G \times K_r)$. The proof of the case with $\mathrm{M}_0(G) = \mathrm{Z}_-(G)$ is analogous. The results for specific graphs follow from Theorem 9.100. □

Nordhaus-Gaddum Problems. The tight Nordhaus-Gaddum sum upper bound for skew zero forcing is almost the same as for standard and PSD zero forcing, as seen in the next result (the easy proof is omitted).

PROPOSITION 9.103. *Let G be a graph of order n. Then*

$$\mathrm{Z}_-(G) + \mathrm{Z}_-(\overline{G}) \leq 2n - 2$$

for $n \geq 2$, and $\mathrm{Z}_-(K_n) + \mathrm{Z}_-(\overline{K_n}) = 2n - 2$ for $n \geq 2$.

If each of the graphs G and $\overline{G}$ has an edge, then

$$\mathrm{Z}_-(G) + \mathrm{Z}_-(\overline{G}) \leq 2n - 4,$$

and $\mathrm{Z}_-(K_n - e) + \mathrm{Z}_-(\overline{K_n - e}) = 2n - 4$ for $n \geq 2$.

The situation is very different for the lower bound. Observe that $\overline{P_4} = P_4$ and $\mathrm{Z}_-(P_4) = 0$. To achieve a low sum for graphs of arbitrarily large order, consider the half-graph. The next result follows from Theorem 9.100(8) and Proposition 9.101(5).

PROPOSITION 9.104. *For $s \geq 2$, $\mathrm{Z}_-(\mathrm{Half}_s) + \mathrm{Z}_-(\overline{\mathrm{Half}_s}) = 2$.*

9.5. Connected and Total Zero Forcing

Both connected zero forcing, introduced by Brimkov and Hicks in [**63**], and total zero forcing, introduced by Davila and Hennings in [**103, 104**], use the standard color change rule but impose additional requirements. For a connected graph G, a *connected zero forcing set* is a standard zero forcing set B of a graph G such that $G[B]$ is connected, and the *connected zero forcing number* $\mathrm{Z}_c(G)$ is the minimum cardinality of a connected zero forcing set. For a graph G with no isolated vertices, a *total zero forcing set* is a standard zero forcing set B of a graph G such that $G[B]$ has no isolated vertices, and the *total zero forcing number* $\mathrm{Z}_t(G)$ is the minimum cardinality of a total zero forcing set. Connected zero forcing is motivated by the PMU placement problem for monitoring electric networks [**63**] (see Section 9.9), and also parallels the relationship between domination and connected domination [**105**]. Here we survey results for connected zero forcing and mention the obvious implications for total zero forcing; for more information on total zero forcing, see [**103, 104**].

OBSERVATION 9.105. *For any graph G for which the parameters are defined, $\mathrm{Z}(G) \leq \mathrm{Z}_c(G)$ and $\mathrm{Z}(G) \leq \mathrm{Z}_t(G)$. If G is connected and $G \neq P_n$, then $\mathrm{Z}_t(G) \leq \mathrm{Z}_c(G)$.*

Values of $\mathrm{Z}_t(G)$ and $\mathrm{Z}_c(G)$ are immediate for some well-known families, because some minimum zero forcing sets are connected:

OBSERVATION 9.106.

(1) $\mathrm{Z}_c(P_n) = \mathrm{Z}(P_n) = 1$, *whereas* $\mathrm{Z}_t(P_n) = 2$ *for* $n \geq 2$.
(2) $\mathrm{Z}_c(C_n) = \mathrm{Z}_t(C_n) = \mathrm{Z}(C_n) = 2$.
(3) $\mathrm{Z}_c(W_n) = \mathrm{Z}_t(W_n) = \mathrm{Z}(W_n) = 3$.
(4) $\mathrm{Z}_c(K_n) = \mathrm{Z}_t(K_n) = \mathrm{Z}(K_n) = n - 1$.
(5) *For* $2 \leq p \leq q$, $\mathrm{Z}_c(K_{p,q}) = \mathrm{Z}_t(K_{p,q}) = \mathrm{Z}(K_{p,q}) = p+q-2$, *whereas* $\mathrm{Z}_c(K_{1,n-1}) = \mathrm{Z}_t(K_{1,n-1}) = n - 1 < n - 2 = \mathrm{Z}(K_{1,n-1})$.

Extreme Values. Since $\mathrm{Z}_t(G) \geq 2$ for all G, and $\mathrm{Z}(G) \leq \mathrm{Z}_t(G) \leq \mathrm{Z}_c(G)$ if $G \neq P_n$, there is an immediate characterization of the lower bound whenever such a graph has a connected minimum zero forcing set.

OBSERVATION 9.107. *Let G be a graph. The following are equivalent:*
(1) *G is a path.*
(2) $\mathrm{Z}(G) = 1$.
(3) $\mathrm{Z}_c(G) = 1$.

REMARK 9.108. Let G be a 2-connected graph. Since any minimum standard zero forcing set in a polygonal path consists of two adjacent vertices, it follows from Theorem 9.12(3) that the following are equivalent:
(1) G is a polygonal path.
(2) $\mathrm{Z}(G) = 2$.
(3) $\mathrm{Z}_t(G) = 2$.
(4) $\mathrm{Z}_c(G) = 2$.

Davila et al. have characterized the highest (nontrivial) value of the connected zero forcing number.

THEOREM 9.109. [**105**] *For G be a connected graph of order $n \geq 2$, $\mathrm{Z}_c(G) = n - 1$ if and only if $G = K_n$ with $n \geq 2$ or $G = K_{1,n-1}$ with $n \geq 4$.*

Computation and Additional Bounds. Brimkov and Hicks [**63**] show that the computation of the connected zero forcing number is NP-complete. In that paper they present methods for computing $\mathrm{Z}_c(G)$ in $O(|G|)$ time when G is a tree or a unicylic graph.

Davila et al. present bounds on the connected zero forcing number that use the girth.

THEOREM 9.110. [**105**] *If G is a 2-connected graph of order n, then $\mathrm{Z}_c(G) \leq n - g(G) + 2$.*

THEOREM 9.111. [**105**] *If G is a connected graph with $g(G) \geq 3$ and $\delta(G) \geq 3$, then $\mathrm{Z}_c(G) \geq g(G) + \delta(G) - 3$, and this bound is tight.*

Theorem 9.29 implies the next result, although it was proved first for the connected zero forcing number.

THEOREM 9.112. [**105**] *If G is a connected graph, then $\mathrm{Z}_c(G) \geq b(G) + 1$, and this bound is tight.*

Graph Operations. There are no useful bounds on the change in connected zero forcing number for vertex deletion, edge deletion, and edge contraction:

PROPOSITION 9.113. [**63**] *For any $c_1 < 0$ and $c_2 > 0$, there exist graphs G_1 and G_2, vertices $v_1 \in V(G_1)$ and $v_2 \in V(G_2)$, and edges $e_1 \in E(G_1)$ and $e_2 \in E(G_2)$ such that $\mathrm{Z}_c(G_1) - \mathrm{Z}_c(G_1 - v_1) < c_1$, $\mathrm{Z}_c(G_2) - \mathrm{Z}_c(G_2 - v_2) > c_2$, $\mathrm{Z}_c(G_1) - \mathrm{Z}_c(G_1 - e_1) < c_1$, $\mathrm{Z}_c(G_2) - \mathrm{Z}_c(G_2 - e_2) > c_2$.*

For any $c_1 < 0$ and $c_2 > 0$, there exist graphs G_1 and G_2 and edges $e_1 \in E(G_1)$ and $e_2 \in E(G_2)$ such that $\mathrm{Z}_c(G_1) - \mathrm{Z}_c(G_1/e_1) < c_1$ and $\mathrm{Z}_c(G_2) - \mathrm{Z}_c(G_2/e_2) > c_2$.

An edge subdivision can raise the connected zero forcing number by an arbitrary amount but cannot lower it:

PROPOSITION 9.114. **[63]** *If G is a graph and $e \in E(G)$, then $\mathrm{Z}_c(G_e) \geq \mathrm{Z}_c(G)$. For any $c > 0$, there exists a graph G and edge $e \in E(G)$ such that $\mathrm{Z}_c(G_e) - \mathrm{Z}_c(G) > c$.*

Unlike standard zero forcing, it is possible a vertex is in every connected zero forcing set, as in the next result (note that **[63]** also contains more general results about cut-sets).

PROPOSITION 9.115. **[63]** *If G is a graph and v is a cut-vertex of G whose deletion leaves at least three components, then v is in every connected zero forcing set.*

PROPOSITION 9.116. **[63]** *Let G be a connected graph different from a path and let ℓ be a leaf adjacent to a vertex of degree 2. Then, ℓ does not belong to any minimum connected forcing set of G, and $\mathrm{Z}_c(G) = \mathrm{Z}_c(G - \ell)$.*

Nordhaus-Gaddum Problems. Of course, for connected zero forcing both G and $\overline{G}$ must be connected (we call this the *connectivity requirement*). The extremal graph for the lower sum Nordhaus-Gaddum problem for standard zero forcing, P_n, satisfies the connectivity requirement, and any lower bound of Z is a lower bound for Z_c. Thus $\mathrm{Z}_c(G) + \mathrm{Z}_c(\overline{G}) \geq n - 2$ and this bound is tight for graphs of order $n \geq 4$.

The extremal graph for the upper sum Nordhaus-Gaddum problem for standard zero forcing, K_n, does not satisfy the connectivity requirement. Furthermore, we must have the order $n \geq 4$ and each of G and $\overline{G}$ has an edge, so $\mathrm{Z}_c(G) + \mathrm{Z}_c(\overline{G}) \leq 2n - 2$. The next example comes close to showing this upper bound is tight (see **[63]** for more information about this and other examples).

EXAMPLE 9.117. For $n \geq 6$, define the graph H_n obtained by attaching a pair of pendent vertices to each end of a path P_{n-4}; H_8 is shown in Figure 9.6. Since H_n has a tree that is not $K_{1,n-1}$, $\mathrm{Z}(\overline{H_n}) = n - 3$ (cf. Theorem 9.5(23)). Furthermore, $\overline{H_n}$ has an induced P_4, and a set consisting of one endpoint of this P_4 and all the vertices not in this P_4 is a connected minimum zero forcing set. Thus $\mathrm{Z}_c(\overline{H_n}) = n - 3$. Any connected zero forcing set of H_n has a vertex of degree one at each end, and thus has at least $n - 2$ vertices, so $\mathrm{Z}_c(H_n) = n - 2$. Thus

$$\mathrm{Z}_c(H_n) + \mathrm{Z}_c(\overline{H_n}) = 2n - 5.$$

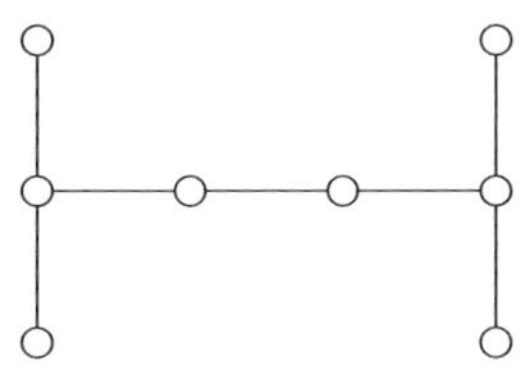

FIGURE 9.6. The graph H_8.

9.6. Additional Zero Forcing Parameters Related to the IEP-G

There are additional zero forcing parameters closely connected to the IEP-G. The enhanced zero forcing number $\widehat{\mathrm{Z}}(G)$ uses zero forcing on loop graphs (constraining the nonzero pattern of the diagonal) to improve the upper bound on the maximum nullity. There is also a family of parameters $\mathrm{Z}_q(G)$, each of which upper bounds the maximum nullity among symmetric matrices described by G that have at most q negative eigenvalues. We discuss these parameters in this section.

$\widehat{\mathrm{Z}}(G)$**.** The enhanced zero forcing number was introduced in [**19**]; it is based on the zero forcing number for loop graphs, which was defined in [**180**]. A more recent discussion of zero forcing and minimum rank for loop graphs can be found in [**56**], from which the notation used here is taken. To avoid confusion, $\mathfrak{G}$ is used to denote a loop graph, including a loop graph with no loops, whereas G continues to be used to denote a (simple) graph (which does not constrain the diagonal of the associated matrices). For a *loop graph* $\mathfrak{G} = (V(\mathfrak{G}), E(\mathfrak{G}))$, $V(\mathfrak{G})$ is the (finite nonempty) set of vertices and the set of edges $E(\mathfrak{G})$ is a set of two element multisets of vertices (i.e., the two vertices in an edge need not be distinct). A *loop* is an edge with two copies of one vertex. The edge $\{u, v\}$ is often denoted by uv (or by uu in the case of a loop $\{u, u\}$).

For a symmetric $n \times n$ real matrix A, the *loop graph* of A is $\mathfrak{G}(A) = (V, E)$ where $V = \{1, \dots, n\}$ and $E = \{uv : u \leq v \text{ and } a_{uv} \neq 0\}$. Let $\mathfrak{G} = (\{1, \dots, n\}, E)$ be a loop graph of order n. The set of real symmetric matrices described by $\mathfrak{G}$ is

$$\mathcal{S}(\mathfrak{G}) = \{A = [a_{ij}] \in S_n(\mathbb{R}) : \mathfrak{G}(A) = \mathfrak{G}\}.$$

The *minimum rank* and the *maximum nullity* of a loop graph $\mathfrak{G}$ are

$$\mathrm{mr}(\mathfrak{G}) = \min\{\operatorname{rank} A : A \in \mathcal{S}(\mathfrak{G})\} \quad \text{and} \quad \mathrm{M}(\mathfrak{G}) = \max\{\operatorname{null} A \mid A \in \mathcal{S}(\mathfrak{G})\}.$$

Clearly $\mathrm{mr}(\mathfrak{G}) + \mathrm{M}(\mathfrak{G}) = |\mathfrak{G}|$. Note that a loop graph $\mathfrak{G}$ constrains the zero-nonzero pattern of the main diagonal entries of matrices described by $\mathfrak{G}$. There is a distinction between a (simple) graph and a loop graph that has no loops – the loop graph with no loops forces the matrices to have zero diagonal, whereas the graph does not. The color change rule for loop graphs is the same as the skew color change rule for graphs (if exactly one neighbor w of u is unfilled, then fill w) but now a vertex is a neighbor of itself if it has a loop. Observe that white vertex forcing is permitted. Zero forcing sets and the *zero forcing number of a loop graph* $\mathfrak{G}$, denoted by $\mathrm{Z}(\mathfrak{G})$, are defined analogously to the definitions for graphs. The same proof that shows that $\mathrm{M}(G) \leq \mathrm{Z}(G)$ also shows $\mathrm{M}(\mathfrak{G}) \leq \mathrm{Z}(\mathfrak{G})$ [**180**].

The *underlying simple graph* of a loop graph $\mathfrak{G}$ is the graph G obtained from $\mathfrak{G}$ by deleting all loops. The *enhanced zero forcing number* of a graph G, denoted by $\widehat{\mathrm{Z}}(G)$, is the maximum of $\mathrm{Z}(\mathfrak{G})$ over all loop graphs $\mathfrak{G}$ such that the underlying simple graph of $\mathfrak{G}$ is G [**19**]. Since $\mathrm{Z}(\mathfrak{G}) \leq \mathrm{Z}(G)$ whenever G is the underlying graph of $\mathfrak{G}$, $\widehat{\mathrm{Z}}(G) \leq \mathrm{Z}(G)$. Furthermore, $\mathrm{M}(G) \leq \widehat{\mathrm{Z}}(G)$, because if $\operatorname{null} A = \mathrm{M}(G)$, then $\mathrm{M}(G) = \operatorname{null} A \leq \mathrm{M}(\mathfrak{G}(A)) \leq \mathrm{Z}(\mathfrak{G}(A)) \leq \widehat{\mathrm{Z}}(G)$.

EXAMPLE 9.118. Let $H_5 = C_5 \circ K_1$, commonly known as the *pentasun*. It was shown in [**25**] that $\mathrm{M}(H_5) = 2$ (by cut-vertex reduction) and the path cover number is $\mathrm{P}(H_5) = 3$. It is straightforward to verify that $\mathrm{Z}(H_5) = 3$, and it is shown in [**19**] that $\widehat{\mathrm{Z}}(H_5) = 2$.

The enhanced zero forcing number has been further extended. Lin introduced the odd cycle zero forcing number to further refine the zero forcing number and used loop graphs for define the enhanced odd cycle zero forcing number in [**220**].

$\mathrm{Z}_q(G)$. The PSD zero forcing number and the standard zero forcing number can be seen as the extreme values of a sequence of parameters $\mathrm{Z}_q(G)$ that upper bound a restricted maximum nullity where q denotes the number of negative eigenvalues permitted, so $\mathrm{Z}_+(G) = \mathrm{Z}_0(G)$ and $\mathrm{Z}(G) = \mathrm{Z}_n(G)$ for a graph G of order n. This family of parameters was introduced in [**73**] through a two-player game and studied further [**72**]. We follow the notation of [**72**] where the game is described as between a player and an oracle.

The Z_q*-forcing game* is defined as follows: All the vertices of a graph $G = (V, E)$ are initially unfilled. There is one player who has tokens and one oracle. The player will repeatedly apply one of the following three operations until all vertices are filled.

(1) For one token, any vertex can be changed from unfilled to filled.
(2) At no cost, the player can apply the standard color change rule.
(3) Let the vertices currently filled be denoted by B, and $W_1, \dots, W_k$ be the vertex sets of the connected components of $G[V \setminus B]$ (i.e., components of unfilled vertices). If $k \geq q+1$, the player can select at least $q+1$ of the W_i and announces the selection to the oracle. The oracle selects a nonempty subset of these components, $\{W_{i_1}, \dots, W_{i_\ell}\}$, and announces it back to the player. At no cost, the player can apply the filling rule on $G[B \cup W_{i_1} \cup \cdots \cup W_{i_\ell}]$.

The Z_q*-number* of a graph, denoted by $\mathrm{Z}_q(G)$, is the minimum number of tokens the player needs to ensure that all vertices can be filled regardless of the oracle's responses.

Note that unlike standard zero forcing and positive semidefinite zero forcing, there may not be an optimal strategy for the player that allows spending all the tokens up front. In other words, there are graphs where the player will vary the choice of where to spend tokens depending on the response of the oracle. An example of this phenomenon is given in [**73**]. Thus there are Z_q-forcing strategies rather than Z_q-sets.

From (3) in definition, $\mathrm{Z}_+(G) = \mathrm{Z}_0(G) \leq \mathrm{Z}_1(G) \leq \cdots \leq \mathrm{Z}_n(G) = \mathrm{Z}(G)$ for any graph G of order n [**73**]. The function `Zq_compute(G,q)` (where q is a nonnegative integer) computes $Z_q(G)$; its use is illustrated in Example ZF.7-1 in [**183**].

THEOREM 9.119. [**73**]. *For a graph G,*

$$\max\{\operatorname{null} A : A \in \mathcal{S}(G) \textit{ and } A \textit{ has at most } q \textit{ negative eigenvalues}\} \leq \mathrm{Z}_q(G).$$

The parameters $\mathrm{Z}_q(G)$ naturally give information about possible inertias of matrices in $\mathcal{S}(G)$; this is exploited in [**73**]. In [**73**] Butler, Grout, and Hall use loop graphs to enhance $\mathrm{Z}_q(G)$ to $\widehat{\mathrm{Z}}_qG)$ by analogy with $\widehat{\mathrm{Z}}(G)$.

THEOREM 9.120. [**72**] *For a graph G, $\mathrm{Z}_q(G) \leq q$ implies $\mathrm{Z}_q(G) = \mathrm{Z}(G)$.*

Graphs having $\mathrm{Z}_1(G) \leq 2$ are characterized in [**72**]. Efficient techniques for computing $\mathrm{Z}_1(T)$ when T is a tree are also presented there.

9.7. Rigid Linkage Zero Forcing

We now describe a forcing process on a graph G related to rigid linkages (see Chapter 7). At each step the vertices of G are filled or unfilled; additionally certain filled vertices are designated as active vertices and the remaining filled vertices are inactive vertices. An unfilled vertex is neither active nor inactive. The set of filled vertices, respectively active vertices, at the end of stage k is denoted by B_k, respectively B_k^a.

A step of the process changes one vertex from unfilled to filled by application of the *rigid linkage color change rule* (CCR-RL for short) given below. Step 0 consists of designating a set B_0 of filled vertices and initializing B_0^a to be B_0.

An application of CCR-RL to go from step k to step $k+1$ consists of

(1) Choose a connected component K of $G - B_k$ having no vertex adjacent to an inactive vertex.
(2) Select an active vertex u and a vertex w in K such that w is the only unfilled neighbor of u in K.
   - Fill w; that is, set $B_{k+1} = B_k \cup \{w\}$.
   - Inactivate u and activate w; that is, set $B_{k+1}^a = B_k^a \setminus \{u\} \cup \{w\}$.
(3) Step $k+1$ is completed once (1) and (2) have been performed.

The coloring of w by u in Stage 2 is called an RL-*force* and is denoted by $u \to w$.

A *rigid linkage forcing process* or RL-*forcing process* on G is a sequence of applications of the CCR-RL to an initial set of filled vertices. Note there may still exist RL-forces at the end of the process. The RL-forcing number is denoted by $Z^{\#}(G)$ and defined to be the minimum number of vertices in a B_0 for which there is a RL-forcing process that results in all vertices of G being filled.

As indicated by the name, the goal of RL-forcing is to create a rigid linkage. The next example shows why rule (1), which prevents RL-forcing into an unfilled component with an inactive filled neighbor, is necessary.

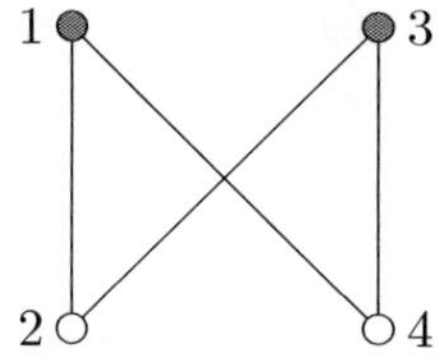

FIGURE 9.7. Graph and initial set showing need for (1) in the CCR-RL.

EXAMPLE 9.121. We will use the graph C_4 with initial filled vertices shown in Figure 9.7 to illustrate the need for the requirement that no vertex in K is adjacent to an inactive filled vertex in Step 2. Set $B_0 = \{1,3\}$. Then $B_0^a = \{1,3\}$. The components of $G \setminus B_0$ are the isolated vertices 2 and 4. We may perform a RL-force resulting in $1 \to 2$. This gives $B_1 = \{1,2,3\}$ and $B_1^a = \{2,3\}$. The only component of $G \setminus B_1$ is the isolated vertex 4. Vertex 4 is the unique unfilled vertex adjacent to the filled vertex 3. But 3 does not rigid linkage force 4, as 4 is adjacent to the non-active filled vertex 1. This is fortunate, as $\{1\text{–}2, 3\text{–}4\}$ is not a $(\{1,3\},\{2,4\})$ rigid linkage, while $\{1\text{–}2, 3\}$ is a rigid linkage.

The next example shows that making certain choices in an RL-forcing process can result in $B_r \neq V(G)$ even if B_0 is a standard zero forcing set and the rigid linkage forcing process is run until no more RL-forces are possible.

EXAMPLE 9.122. As illustrated in Figure 9.8, let $B_0 = \{2,3\}$ be the initial filled vertices in the paw graph. Then B_0 is a standard zero forcing set (with $3 \to 4$

and then $2 \to 1$). One option for rigid linkage forcing is to choose $2 \to 1$ as the first RL-force, and in this case then there is not another possible RL-force. Another option for rigid linkage zero forcing is to choose $3 \to 4$. After this one can use the rigid linkage forcing $2 \to 1$.

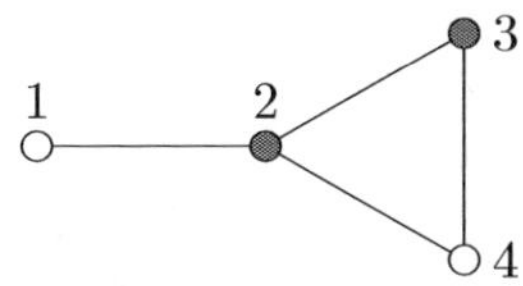

FIGURE 9.8. The paw graph with initial $B_0 = \{2,3\}$

Both the standard zero forcing and rigid linkage forcing processes allow one to choose which force may be performed at a given step. Thus, for a given initial set of filled vertices there may be many different forcing processes. For standard zero forcing, it is known **[6]** that there is a unique final set when the process is carried out until there are no Z-forces available. For rigid linkage forcing, choices may lead to different final sets of filled vertices. Example 9.122 gives a graph and an initial set for which there are final sets of filled vertices with different cardinalities.

The notions of RL-*forcing chains* and RL-*chain set* are defined in **[135]**, and are analogous to the definitions to forcing chains and forcing chain sets. It can be seen that any RL-chain set is a linkage. In **[135]**, graph theoretic arguments to show that being an RL-chain set is equivalent to being a rigid linkage. Here we summarize the results, without proof.

PROPOSITION 9.123. **[135]** *Any standard zero forcing process on a graph G is a rigid linkage forcing process.*

THEOREM 9.124. **[135]** *Let G be a graph and $\mathcal{P}$ a linkage in G. Then $\mathcal{P}$ is a rigid linkage if and only if $\mathcal{P}$ is an* RL-*chain set under some* RL-*forcing process.*

COROLLARY 9.125. **[135]** *Let G be a graph and $\mathcal{P}$ a spanning linkage in G. Then the following are equivalent:*

(1) *$\mathcal{P}$ is a rigid linkage.*
(2) *$\mathcal{P}$ is an* RL-*chain set.*
(3) *$\mathcal{P}$ is a* Z-*chain set.*

Therefore, $Z^{\#}(G) = \mathrm{Z}(G)$ for any graph G, and every Z-*chain set or* RL-*chain set is a vital linkage of the subgraph of G induced by the set of filled vertices at the end of the forcing process.*

We conclude this section with a new graph theoretic characterization of rigid-linkages of order t. Let $\mathcal{P} = (p_1, \dots, p_t)$ be an (α, β)-linkage of order t with the initial vertex of p_i being u_i and the terminal vertex of p_i being v_i for $i = 1, \dots, t$, so $\alpha = \{u_1, \dots u_t\}$ and $\beta = \{v_1, \dots, v_t\}$.

A *chord path* of $\mathcal{P}$ is a path of length at least 1 whose initial and terminal vertices lie on the same p_i for some i and if the path has length 1 its vertices are not consecutive vertices of p_i, and whose internal vertices lie outside of $V(\mathcal{P})$. For $k \in \{2, \dots, t\}$ an X^k-*subgraph* of G with respect to $\mathcal{P}$ is a subgraph of the form illustrated in Figure 9.9, where $i_1, \dots, i_t$ is a permutation of $1, \dots, t$, the vertical lines represent the paths in $\mathcal{P}$ in some order, the non-vertical lines represent paths in G whose internal vertices are disjoint from $V(\mathcal{P})$, the vertices w_i and x_{i-1} are distinct vertices, but not necessarily consecutive on p_i. The vertices u_{i_j} and w_j may be equal, as may the vertices x_{j-1} and v_{i_j}.

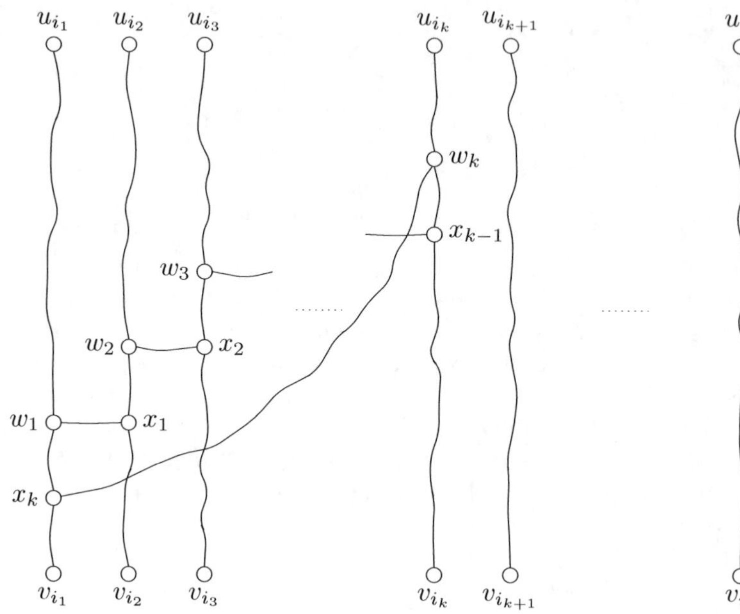

FIGURE 9.9. An X^k-subgraph of G with respect to a linkage $\mathcal{P}$.

THEOREM 9.126. *Let $\mathcal{P}$ be an order t (α, β)-linkage of the graph G. Then $\mathcal{P}$ is an (α, β)-rigid linkage if and only if $\mathcal{P}$ has no chord path, and contains no X^k-subgraph with respect to $\mathcal{P}$ for $k \in \{2, \ldots, t\}$.*

PROOF. Let $\mathcal{P} = (p_1, \ldots, p_t)$. Let $\alpha = \{u_1, \ldots u_t\}$ and $\beta = \{v_1, \ldots, v_t\}$ such that the initial vertex of p_i is u_i and the terminal vertex of p_i is v_i for $i = 1, \ldots, t$.

We prove both directions by contrapositive. First assume that G has a chord path. Let p_i be the path containing the terminal vertices of such a chord path. Then one can replace the path p_i by a path from u_i to v_i that uses the chord path and is disjoint from the other paths in $\mathcal{P}$. Hence $\mathcal{P}$ is not an (α, β)-rigid linkage. Next assume that G has an X^k-subgraph for some k. Then one can replace each path p_{i_ℓ} by the path that follows p_{i_ℓ} from u_{i_ℓ} to w_ℓ, then the path from w_ℓ to x_ℓ, and finally the subpath of $p_{i_{\ell+1}}$ from x_ℓ to $v_{i_{\ell+1}}$ (except that for $\ell = k$, use the subpath from x_k to v_{i_1}). This gives a different (α, β)-linkage. Hence, $\mathcal{P}$ is not rigid. This proves the "only if" part by contrapositive.

Now assume that $\mathcal{P}$ is not (α, β)-rigid. Then there exists another linkage $\mathcal{Q}$. Pick a i_1 with p_{i_1} not in $\mathcal{Q}$, and let w_1 be vertex before the first vertex exiting p_{i_1}, and x_1 the first vertex that comes back to a path in $\mathcal{P}$. If x_1 is on p_{i_1}, then $\mathcal{P}$ has a chord path. If not, then x_1 is on another path, say p_{i_2}. We may apply the same argument on p_{i_2}, p_{i_3}, and so on, until p_{i_b} is the same as some visited path p_{i_a}. Thus, G has an X^k-subgraph using the paths from p_{i_a} to p_{i_b} with $k = b - a + 1$. $\square$

We note that characterization of rigid linkages of order 2 was proven in [**135**], and is related to the work in [**251**] on graphs of two parallel paths and the work in [**227**] on vital linkages of order 2.

9.8. Minor Monotone Floors of Zero Forcing Parameters

Much of the power of the strong properties discussed in Part 2 comes from the fact that they have at least some version of the minor monotonicity property. Minor monotonicity is a very powerful property, and minor monotone floors (and ceilings) of graph parameters have been introduced for the purpose of creating a minor monotone graph parameter related to the original parameter. Although the terms "minor monotone floor" and "minor monotone ceiling" were not used, the minor monotone floor of the crossing number was introduced and studied in [**45**], as were the minor monotone ceilings of minimum degree and vertex connectivity in [**141**]. Minor monotone floors of zero forcing parameters and maximum nullity parameters were introduced in [**19**].

For any graph parameter taking values in the natural numbers, a direct conversion to a minor monotone parameter can be made in two possible ways, by taking a minimum over all graphs H that contain G as a minor, called the minor monotone floor, or by taking a maximum over all graphs H that are minors of G, called the minor monotone ceiling. Here we focus on the floor, because that is more useful for zero forcing parameters, but mention the ceiling at the end of this section. Let p be a graph parameter whose values are nonnegative integers. The *minor monotone floor* of p is $\lfloor p \rfloor(G) = \min\{p(H) : G \preceq H\}$.

OBSERVATION 9.127. [**19**] *Let p be a graph parameter with nonnegative integers values. The minor monotone floor $\lfloor p \rfloor$ is characterized by the three properties:*

(1) *If $G \preceq H$, then $\lfloor p \rfloor(G) \leq \lfloor p \rfloor(H)$.*
(2) *For any graph G, $\lfloor p \rfloor(G) \leq p(G)$.*
(3) *For any graph G, there exists a graph H such that $G \preceq H$ and $\lfloor p \rfloor(G) = p(H)$.*

OBSERVATION 9.128. [**19**] *If p and q are graph parameters such that for all graphs G, $q(G) \leq p(G)$, then $\lfloor q \rfloor(G) \leq \lfloor p \rfloor(G)$. If q is a minor monotone graph parameter, then $q(G) = \lfloor q \rfloor(G)$.*

For each of the zero forcing and maximum nullity parameters discussed, its minor monotone floor is of course specified by the minor monotone floor definition (and denoted by floor notation, e.g., $\lfloor \mathrm{Z} \rfloor(G)$ denotes the minor monotone floor of the standard zero forcing number, evaluated on G). For the standard and PSD zero forcing numbers (and some additional zero forcing parameters discussed in [**19**]), there is a color change rule for the minor monotone floor of the parameter involving the original color change rule and the concept of hopping. For these parameters, it is a theorem that the zero forcing number defined by the color change rule with hopping is the same as the minor monotone floor of the original parameter (see, for example, Theorem 9.129). *Hopping* is where a filled vertex u that has not yet performed a force but has all neighbors filled is allowed to force any one unfilled vertex w, and is analogous to adding the edge uw to the graph. In this case we say that *u hops to w* or *u forces w by a hop*. For u to hop, it is not required that u have any neighbors.

Recall that for the zero forcing parameters discussed previously except rigid linkage forcing $Z^{\#}$ (and the parameters $\widehat{\mathrm{Z}}$ and Z_q that not defined by color change rules), the final coloring is unique. This is not the case for monotone floors of zero forcing parameters, as illustrated in Example 9.132 below.

For the standard and PSD zero forcing numbers (and some other zero forcing parameters discussed in [**19**]), there is also a characterization in terms of a type of k-tree, and thus a relationship to tree-width, which is one of the most widely studied minor monotone graph parameters and plays an important role in the famous Graph Minor Theorem. A *k-tree* is constructed inductively by starting with a complete graph on $k+1$ vertices and connecting each new vertex to the vertices of an existing clique on k vertices. A k-tree is a k-connected chordal graph with maximum clique size $k+1$. Every clique in a k-tree is part of a maximal clique of order $k+1$. For a graph G, the *tree-width* $\text{tw}(G)$ is the minimum k such that G is a subgraph of a k-tree. The standard definition of tree-width uses a tree of bags, but that definition is equivalent to the definition we give here.

Minor Monotone Floor of Standard Zero Forcing. Let G be a graph. The *$\lfloor \mathrm{Z} \rfloor$-color change rule* is: If u is filled and w is the only unfilled neighbor of u, then fill w. If u is filled, all neighbors of u are filled, and u has not yet performed a force, then fill any one unfilled vertex w. That is, apply the standard color change rule with hopping. The next theorem says that the minor monotone floor of the standard zero forcing number is the zero forcing parameter defined by the $\lfloor \mathrm{Z} \rfloor$-color change rule.

THEOREM 9.129. [**19**] *For every graph G,*

$$\lfloor \mathrm{Z} \rfloor(G) = \min\{|B| : B \subseteq V(G) \text{ is a } \lfloor \mathrm{Z} \rfloor\text{-zero forcing set of } G\}.$$

Since hopping is analogous to adding edges to the graph, an immediate consequence of this is that only supergraphs of the same order are needed.

COROLLARY 9.130. *Let G be a graph. Then*

$$\lfloor \mathrm{Z} \rfloor(G) = \min\{\mathrm{Z}(H) : G \leq H \text{ and } V(H) = V(G)\}.$$

OBSERVATION 9.131. *Since a vertex v must be filled and have all but at most one neighbor filled in order for v to perform a $\lfloor \mathrm{Z} \rfloor$-color change, $\delta(G) \leq \lfloor \mathrm{Z} \rfloor(G)$ for every graph G.*

It is convenient to define an *active vertex* to be a filled vertex that has not performed a force; an *inactive vertex* is a filled vertex that has performed a force. Each vertex in the subset $B \subseteq V(G)$ that defines a coloring is initially active, and there are always $|B|$ active vertices. Only an active vertex can force another vertex using the $\lfloor \mathrm{Z} \rfloor$-color change rule, because a vertex can standard zero force at most one other vertex. When vertex u forces vertex w, u becomes inactive and w becomes active. Thus the number of active vertices remains constant.

As discussed above (see Observation 9.7), the standard final coloring of a given set of filled vertices is unique. However, a $\lfloor \mathrm{Z} \rfloor$-final coloring needs not be unique, as the next example shows.

EXAMPLE 9.132. [**19**] The graph G in Figure 9.10 has more than one $\lfloor \mathrm{Z} \rfloor$-final coloring for $B = \{1, 2, 3\}$: The set B is a $\lfloor \mathrm{Z} \rfloor$-zero forcing set of G, with the forces

2 forces 4 by a hop; 1, 3 and 4 are active;
4 forces 5; 1, 3 and 5 are active;
5 forces 6; 1, 3 and 6 are active;

6 forces 7.

But $\{1, 2, 3, 6\}$ is also a $\lfloor Z \rfloor$-final coloring for B, with the force

2 forces 6 by a hop; 1, 3 and 6 are active.

At this point no further forces are possible.

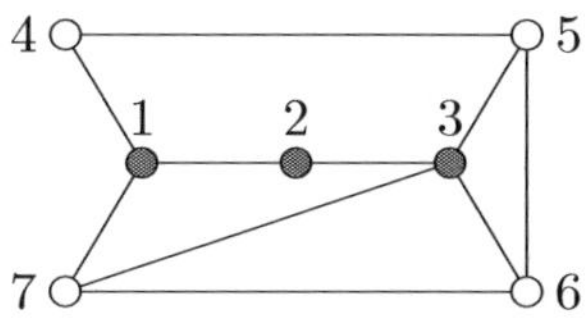

FIGURE 9.10. The graph G for Example 9.132.

The behavior of minor monotone floors on connected components is very different from that of the original zero forcing parameters, which sum across the components.

OBSERVATION 9.133. **[19]** *If G_i, $i = 1, \dots, k$ are the connected components of a graph G, then $\lfloor Z \rfloor(G) = \max_{i=1}^{k} \lfloor Z \rfloor(G_i)$.*

A *linear k-tree* is constructed iteratively by starting with K_{k+1} and connecting each new vertex to the vertices of an existing K_k that includes a vertex of degree k (it can be assumed that the new vertex is adjacent to the vertex just added); an example with $k = 2$ is shown in Figure 9.11(a). A linear k-tree is either K_{k+1} or a k-tree in which exactly two vertices have degree k. A *partial linear k-tree* is a subgraph of a linear k-tree. The *proper path-width* of a graph G, denoted by $\mathrm{ppw}(G)$, is the minimum k for which G is a partial linear k-tree. Proper path-width was introduced in **[259]**.

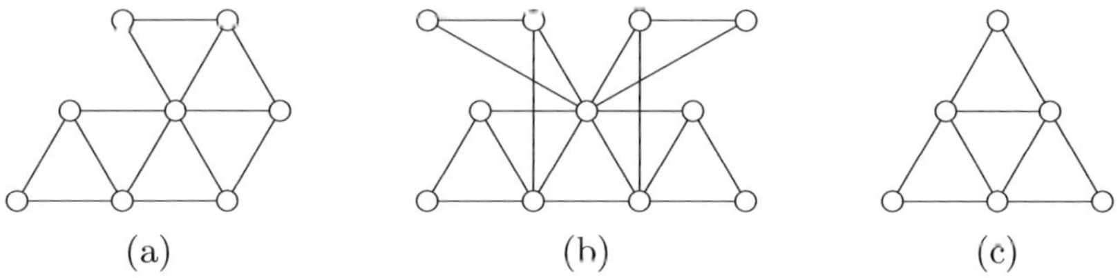

FIGURE 9.11. (a) a linear 2-tree, (b) a two-sided 2-tree that is not a linear 2-tree, (c) a 2-tree that is not a two-sided 2-tree (the supertriangle T_3)

THEOREM 9.134. **[19]** *For any graph G having at least one edge, $\lfloor Z \rfloor(G)$ is the minimum k for which G is a partial linear k-tree, i.e., $\lfloor Z \rfloor(G) = \mathrm{ppw}(G)$.*

Minor Monotone Floor of PSD Zero Forcing. Let G be a graph. The *$\lfloor Z_+ \rfloor$-color change rule* is: Let S be the set of filled vertices. Let $W_1, \dots, W_k$ be the sets of vertices of the k components of $G - S$ (it is possible that $k = 1$). For each component $1 \le i \le k$, let $C_i \subseteq S$ be the subset of filled vertices that are considered to be *active* with regard to that component. (Initially, each $C_i = S = B$, where B is the set of initially filled vertices.) If $u \in C_i$, $w \in W_i$, and u has no unfilled neighbors in $G[W_i \cup S] - w$, then fill w. (This allows for either a normal PSD force, or a hop from u to w.) To each connected component of $G[W_i] - w$, associate a new active set equal to $(C_i \setminus \{u\}) \cup \{w\}$.

THEOREM 9.135. **[19]** *For every graph G,*

$$\lfloor \mathrm{Z}_+ \rfloor(G) = \min\{|B| : B \subseteq V(G) \text{ is a } \lfloor \mathrm{Z}_+ \rfloor\text{-zero forcing set of } G\}.$$

Since hopping is analogous to adding edges to the graph, an immediate consequence of this is that only supergraphs of the same order are needed.

COROLLARY 9.136. *Let Gbe a graph. Then*

$$\lfloor \mathrm{Z}_+ \rfloor(G) = \min\{\mathrm{Z}_+(H) : G \leq H \text{ and } V(H) = V(G)\}.$$

OBSERVATION 9.137. **[19]** *Since* $\delta(G) \leq \mathrm{tw}(G)$, $\delta(G) \leq \lfloor \mathrm{Z}_+ \rfloor(G)$ *for every graph* G.

OBSERVATION 9.138. **[19]** *If* $G_i, i = 1, \ldots, k$ *are the connected components of a graph* G, *then* $\lfloor \mathrm{Z}_+ \rfloor(G) = \max_{i=1}^{k} \lfloor \mathrm{Z}_+ \rfloor(G_i)$.

A *two-sided k-tree* is constructed iteratively by starting with K_{k+1} and connecting each new vertex to the vertices of an existing K_k that either includes a vertex of degree k or is the same as the K_k to which some previous vertex was connected; an example with $k = 2$ is shown in Figure 9.11(b). Every tree is a two-sided 1-tree. A *partial two-sided k-tree* is a subgraph of a two sided k-tree. The *two-sided tree-width* of a graph G is the minimum k for which G is a partial two sided k-tree, and is denoted by $\mathrm{tstw}(G)$. It is known that the two-sided tree width is equal to largeur d'arborescence $\mathrm{la}(G)$, and thus that $\mathrm{tw}(G) \leq \mathrm{tstw}(G) \leq \mathrm{tw}(G) + 1$ (see **[191]**).

THEOREM 9.139. **[19]** *For any graph G having at least one edge,* $\lfloor \mathrm{Z}_+ \rfloor(G)$ *is the minimum k for which G is a partial two-sided k-tree, i.e.,* $\lfloor \mathrm{Z}_+ \rfloor(G) = \mathrm{tstw}(G)$.

Minor Monotone Floors of Other Zero Forcing Parameters. The minor monotone floor of the enhanced zero forcing number $\widehat{\mathrm{Z}}$, denoted by $\lfloor \widehat{\mathrm{Z}} \rfloor(G)$ for a graph G, is defined in **[19]**, but no color-change rule is presented and no k-tree characterization is given; note $\widehat{\mathrm{Z}}$ is not defined by a color change rule.

Unfortunately none of the other zero forcing parameters discussed previously except standard and positive semidefinite zero forcing has an interesting minor monotone floor.

The minor monotone floor of the skew forcing number is not a useful parameter because it is always zero: $\lfloor \mathrm{Z}_- \rfloor(G) = 0$ for every graph G because $G \prec G \circ K_1$ and $\mathrm{Z}_-(G \circ K_1) = 0$.

The minor monotone floor of the connected (or total) zero forcing number is not interesting because it is the same as the minor monotone floor of standard zero forcing number: Given a graph G, choose a $\lfloor \mathrm{Z} \rfloor$-zero forcing set B such that $|B| = \lfloor \mathrm{Z} \rfloor(G)$. Then create a supergraph H with $V(H) = V(G)$ by adding edges between the vertices of B as needed to connect $H[B]$ (one option is to make $H[B]$ a clique).

Minor Monotone Floors of Maximum Nullity Parameters. Much less is know about the minor monotone floors of maximum nullity parameters. The *minor monotone floor of maximum standard nullity* and the *minor monotone floor of maximum PSD nullity* are defined in **[19]** as

$$\lfloor \mathrm{M} \rfloor(G) = \min\{\mathrm{M}(H) : G \preceq H\} \text{ and } \lfloor \mathrm{M}_+ \rfloor(G) = \min\{\mathrm{M}_+(H) : G \preceq H\}.$$

Since ξ and ν are minor monotone, $\xi(G) \leq \lfloor M \rfloor(G)$ and $\nu(G) \leq \lfloor M_+ \rfloor(G)$ for every graph G. An example of a graph G such that $\xi(G) < \lfloor M \rfloor(G)$ is presented in [**19**]. No example of a graph G such that $\nu(G) < \lfloor M \rfloor(G)$ is known, but it is suggested in [**19**] that this is due to needing larger parameter values as well as larger order.

Minor Monotone Ceilings. Let p be a graph parameter whose values are nonnegative integers. The *minor monotone ceiling* of p is $\lceil p \rceil(G) = \max\{p(H) : H \preceq G\}$. There are observations analogous to Observations 9.127 and 9.128 for the minor monotone ceiling. The minor monotone ceiling is interesting for minimum degree and vertex connectivity (see [**19**]) but not for any of the zero forcing numbers discussed here because it is always the order of the graph: For example, $\lceil Z \rceil(G) = n$ for every graph G of order n because $\overline{K_n} \preceq G$ and $Z(\overline{K_n}) = n$.

9.9. Power Domination, $\gamma_P(G)$

The study of power domination on graphs was introduced by Haynes et al. in [**173**] to cost-effectively monitor an electric power network by modeling the monitoring capabilities of Phase Measurement Units (PMUs). A minimum power dominating set gives a placement of PMUs that observes the network using the minimum number of PMUs.

A set $S \subseteq V(G)$ *dominates* its closed neighborhood $N[S]$, and is a *dominating set* of G if every vertex of G is adjacent to a vertex in S or is in S, i.e., if $V(G) = N[S]$. The minimum cardinality of a dominating set is the *domination number* of G, and is denoted by $\gamma(G)$. Domination, and many of its variants, have been studied extensively (see, for example, [**174**] and the references therein).

Power domination can be thought of as a domination step followed by standard zero forcing, as is clear from the simplified observation rules introduced by Brueni and Heath in [**68**]: For a set S of vertices in a graph G (interpreted as the location of PMUs), define the set of observed vertices $PD(S) \subseteq V(G)$ iteratively as follows:

(1) $PD(S) = S \cup N(S) = N[S]$.
(2) While there exists $v \in PD(S)$ such that $|N(v) \setminus PD(S)| = 1$:
$PD(S) = PD(S) \cup N(v)$.

A set $S \subseteq V(G)$ is a *power dominating set* of a graph G if at the end of the process $PD(S) = V(G)$. The *power domination number* $\gamma_P(G)$ is the minimum cardinality of a set S such that $PD(S) = V(G)$.

OBSERVATION 9.140. *For any graph G and set $S \subseteq V(G)$, the final set of observed vertices $PD(S)$ is unique.*

Power domination takes its name from its application to electric power networks and the fact that Step (1) in the definition is a domination step. Step (2) is one application of the color change rule for standard zero forcing (where vertices in $PD(S)$ are considered filled). Note that zero forcing had not been defined when power domination was introduced, and power domination appears to be the first use of the Z-color change rule (although it did not refer to changing color and was originally stated in a more complicated form).

OBSERVATION 9.141. *For any graph G, $1 \leq \gamma_P(G) \leq \gamma(G)$ and $\gamma_P(G) \leq Z(G)$.*

OBSERVATION 9.142. *Let G be a graph and $S \subseteq V(G)$. If $N[S]$ is a standard zero forcing set for G, then S is a power dominating set.*

OBSERVATION 9.143. *If* $\gamma(G) = 1$, *then* $\gamma_P(G) = 1$, *so* $\gamma_P(G) = 1$ *for* $G \in \{K_n, W_n, F_k\}$. *Furthermore,* $\gamma_P(G) = 1$ *for* $G \in \{C_n, P_n, K_{2,n-2}\}$.

OBSERVATION 9.144. *If the connected components of* G *are* $G_1, \dots, G_c$, *then* $\gamma_P(G) = \sum_{i=1}^{c} \gamma_P(G_i)$.

REMARK 9.145. Suppose G is a graph of order n and $G' = G[W]$ is an induced subgraph of G of order n'. Observe that $\gamma_P(G) \leq n - n' + \gamma_P(G')$ by choosing $S = S' \cup (V(G) \setminus V(G'))$ where S' is a minimum power dominating set for G'. This is the analog of Remark 9.3, but it is a poor bound. Since it suffices to dominate the vertices not in G', $\gamma_P(G') + \gamma(G[V(G) \setminus W])$ is a better bound.

The next result describes a relationship between the power domination number and the zero forcing number.

THEOREM 9.146. [**38, 110**] *If* G *is a graph that has an edge, then* $\left\lceil \frac{Z(G)}{\Delta(G)} \right\rceil \leq \gamma_P(G)$, *and this bound is tight.*

Dean et al. applied this result to determine the power domination numbers of some hypercubes.

THEOREM 9.147. [**110**] *For every positive integer* d,

$$\frac{2^{d-1}}{d} \leq \gamma_P(Q_d) \leq 2^{d - \lfloor \log_2 d \rfloor - 1}.$$

When $d = 2^k$ *for a positive integer* k, *this determines the power domination number:* $\gamma_P(Q_{2^k}) = 2^{2^k - k - 1}$.

Recall that $Q_d = Q_{d-1} \,\square\, K_2$ and the set of vertices of one copy of Q_{d-1} is a minimum standard zero forcing set for Q_d. It is known that $\gamma(Q_{d-1}) = \frac{2^{d-1}}{d} = 2^{d-1-k}$ for $d = 2^k$ [**110, 172**], and it was conjectured in [**110**] that $\gamma_P(Q_d) = \gamma(Q_{d-1})$ for all d. This conjecture was disproved by Pai and Chiu in [**240**]; it was shown there that $\gamma_P(Q_7) = 10$, whereas $\gamma(Q_6) = 12$ [**172**]. The values of $\gamma_P(Q_d)$ for $d \leq 8$ in Table 9.1 were summarized in [**240**].

TABLE 9.1. Values of $\gamma_P(Q_d)$ for $d \leq 8$

d	1	2	3	4	5	6	7	8
$\gamma_P(Q_d)$	1	1	2	2	4	6	10	16

Computation. It is known that computing the power domination number is NP-hard [**173**]. *Sage* functions for finding a minimum power dominating set and the power domination number of a specific graph (that is not too large) are available in [**183**]. This code has several functions for power domination. The function `minPowerDominatingSet(G)` returns a minimum power dominating set. The function `PowerDom(G)`$= \gamma_P(G)$ returns the power domination number of G. These functions call the zero forcing set checking function for standard zero forcing. Examples for power domination appear in Section 10.1 of [**183**].

The Spider Cover Number and Forcing Spiders. Just as standard zero forcing creates forcing paths and PSD zero forcing creates forcing trees, power domination creates forcing spiders, and the spider cover number was introduced in [**173**]. Recall that a spider is a tree that has at most one high degree vertex (vertex of degree three or more). The *spider cover number* of G, denoted by $\operatorname{sp}(G)$, is the minimum number of vertex disjoint spiders occurring as subgraphs of G that cover all the vertices of G. Note that if T is tree, then every subgraph of T is induced, so the definition given here coincides with that in [**173**] (where the definition was restricted to trees but required the spiders to be induced).

OBSERVATION 9.148. *If H is a spanning subgraph of G, then* $\operatorname{sp}(G) \le \operatorname{sp}(H)$, *because any spider cover of H is a spider cover of G.*

THEOREM 9.149. [**173**] *If T is a tree, then* $\operatorname{sp}(T) = \gamma_P(T)$.

For a power dominating set $S = \{x_1, \dots, x_s\} \subseteq V(G)$, create a *power domination set of forces $\mathcal{F}$* of S as follows:

(0) $\mathcal{F} = \emptyset$.

(1) For $i = 1, \dots, s$:

$$\text{For } u \in N(x_i) \setminus \left(S \cup \bigcup_{j<i} N(x_j) \right):$$

$\mathcal{F} := \mathcal{F} \cup \{x_i \to u\}$.

(2) For each application of step (2) in the definition of power domination with $v \in PD(S)$ such that $|N(v) \setminus PD(S)| = 1$:
$\mathcal{F} := \mathcal{F} \cup \{v \to w\}$ where w is the one neighbor of v not already in $PDG(S)$.

Suppose S is a power dominating set of G and $\mathcal{F}$ is a power domination set of forces of S. For a vertex $x \in S$, define V_x to be the set of all vertices w such that there is a sequence of forces $x = v_1 \to v_2 \to \cdots \to v_k = w$ in $\mathcal{F}$; the empty sequence of forces is permitted, i.e., $x \in V_x$. The *forcing spider* S_x is the subgraph with vertex set V_x and all the edges of the form $\{v_{k-1}, v_k\}$ such that there is a sequence of forces $x = v_1 \to v_2 \to \cdots \to v_k$ in $\mathcal{F}$. The *forcing spider cover* (for a set of forces $\mathcal{F}$) is $\{S_x : x \in S\}$. These definitions and the next result are adapted from [**43**, Section 3.6] where the analogous result was established for trees, which was in turn inspired by the PSD forcing trees in [**122**].

PROPOSITION 9.150. *Suppose G is a graph, S is a power dominating set of G, and $\mathcal{F}$ is a power domination set of forces of S. Then*

(1) *S_x is a spider for $x \in S$.*
(2) *The forcing spider cover $\{S_x : x \in S\}$ is a spider cover of G.*
(3) $\operatorname{sp}(G) \le \gamma_P(G)$.

PROOF. For $x \in S$, $x \in V_y$ if and only if $y = x$. Thus $|\{S_x : x \in S\}| = |S|$. The sets V_x of vertices forced by distinct $x \in S$ are disjoint and include all vertices of G because by definition each vertex of G not in S is forced exactly once. Furthermore, $\deg_{S_x} v \le 2$ for $v \in S_x - \{x\}$. Thus x is the only possible high degree vertex of S_x. Also S_x does not contain a cycle because a vertex can perform at most one force (of a vertex not already forced) and only if it is in S or has been forced itself. Since S_x is connected, S_x is a spider. Since the sets V_x partition the vertices of

G, $\{S_x : x \in S\}$ is a spider cover of G, and $\mathrm{sp}(G) \leq \gamma_P(G)$ by choosing S to be a minimum power dominating set. □

Unlike a forcing path for Z or a forcing tree for Z_+, a forcing spider need not be induced. For $k \geq 3$ and $r \geq 1$, the *generalized wheel* $GW(k,r)$ is the graph of order $kr+1$ constructed from $C_k \,\square\, P_r$ by adding a new vertex 0 adjacent to every vertex in one end copy of C_k; Figure 9.12 shows $GW(6,3)$. Note that the generalized wheel $GW(k,1)$ is the wheel W_{k+1}. In $GW(k,r)$, standard notation denotes the vertices of the $C_k \,\square\, P_r$ by (i,j) with $i = 1,\ldots,k$ and $j = 1,\ldots,r$; the vertex 0 is adjacent to $(i,1), i = 1,\ldots,k$.

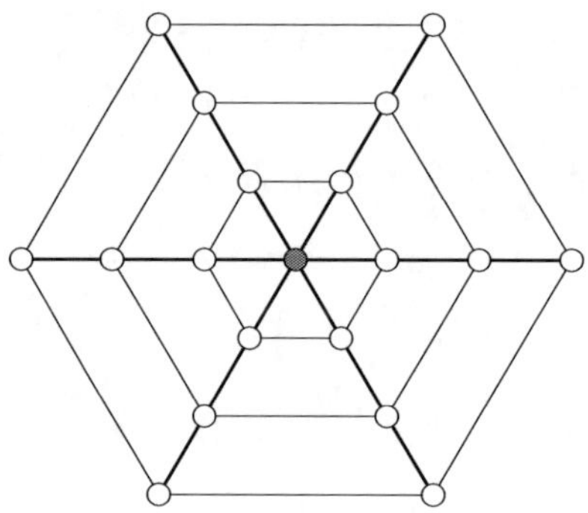

FIGURE 9.12. The generalized wheel $GW(6,3)$ with its forcing spider edges shown as thick lines.

EXAMPLE 9.151. The set $S = \{0\}$ is a power dominating set for $GW(s,r)$, because S dominates $(i,1), i = 1,\ldots,k$, i.e., an end copy of C_k in $C_k \,\square\, P_r$, and forcing proceeds along the "path" edges from one copy of C_k to the next: For $j = 1,\ldots,r-1$, for $i = 1,\ldots,k$, $(i,j) \to (i,j+1)$. The forcing spider has edges $\{0,(i,1)\}$ for $i = 1,\ldots,k$ and $\{(i,j),(i,j+1)\}$ for $i = 1,\ldots,k\ j = 1,\ldots,r-1$. Example 10.1-1 in [**183**] illustrates $\gamma_P(GW(6,3)) = 1$.

Just as the path cover number can be much less than the standard zero forcing number, the spider cover number can be much less than the power domination number.

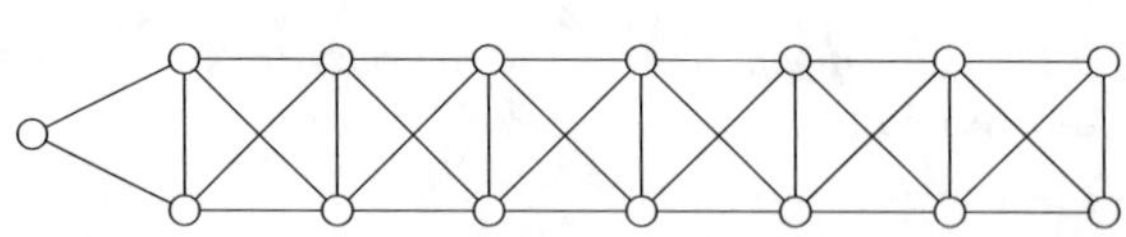

FIGURE 9.13. The graph G_7 has $\mathrm{sp}(G) = 1$ and $\gamma_P(G) = 3$

EXAMPLE 9.152. Denote the vertices of the strong product of paths $P_2 \boxtimes P_r$ by (i,j) with $i = 1,2$ and $j = 1,\ldots,r$ and add a new vertex 0 is adjacent to $(1,1)$ and $(2,1)$ to create a graph G_r of order $2r+1$; Figure 9.13 shows G_7. All the vertices can be covered by one spider that has edges $\{0,(1,1)\}$, $\{0,(2,1)\}$, and $\{(1,j),(1,j+1)\}$, $\{(2,j),(2,j+1)\}$ for $j = 1,\ldots,r-1$. Observe that $\Delta(P_2 \boxtimes P_r) = 5$ and recall that $\mathrm{Z}(P_2 \boxtimes P_r) = r+1$ [**6**], which implies $\Delta(G_r) = 5$ and $\mathrm{Z}(G_r) = r+1$. By Theorem

9.146, $\gamma_P(G_r) \geq \left\lceil \frac{Z(G_r)}{\Delta(G_r)} \right\rceil = \left\lceil \frac{r+1}{5} \right\rceil$, so $\gamma_P(G_r)$ grows arbitrarily large, whereas $\operatorname{sp}(G_r) = 1$. Example 10.1-2 in [**183**] shows $\gamma_P(G_7) = 3 > 2 = \left\lceil \frac{8}{5} \right\rceil = \left\lceil \frac{Z(G_7)}{\Delta(G_7)} \right\rceil$.

Graph Operations.

THEOREM 9.153. [**116**] *Let G be a graph, $v \in V(G)$, and $e \in E(G)$. Then*
(1) $\gamma_P(G) - 1 \leq \gamma_P(G - v)$. *There is no useful upper bound for* $\gamma_P(G - v)$.
(2) $\gamma_P(G) - 1 \leq \gamma_P(G - e) \leq \gamma_P(G) + 1$.
(3) $\gamma_P(G) - 1 \leq \gamma_P(G/e) \leq \gamma_P(G) + 1$.
All the bounds are tight.

COROLLARY 9.154. *Let G be a graph and $e \in E(G)$. Then*

$$\gamma_P(G) \leq \gamma_P(G_e) \leq \gamma_P(G) + 1$$

and the bounds are tight.

PROOF. Let $e = \{u, w\}$ and let v_e be the new vertex adjacent to u and w in edge subdivision. The upper bound $\gamma_P(G_e) \leq \gamma_P(G) + 1$ is immediate from Theorem 9.153 because $G = (G_e)/\{u, v_e\}$. The graph G in Figure 9.14 shows the upper bound is tight.

For the lower bound $\gamma_P(G) \leq \gamma_P(G_e)$, choose a minimum power dominating set $S \subseteq V(G_e)$ and construct a propagating set of forces $\mathcal{F}$. Suppose first that v_e performs a force; without loss of generality $v_e \to w \in \mathcal{F}$. If $v_e \in S$, then let $B' = B \setminus \{v_e\} \cup \{w\}$. If $v_e \notin B$, then $u \to v_e \in \mathcal{F}$; define $S' = B$. In all cases, S' is a power dominating set for G and $\gamma_P(G) \leq \gamma_P(G_e)$.

Finally, consider the case where v_e does not perform a force; without loss of generality, w is observed before or at the same time as u. If $v_e \in S$, define $S' = S \setminus \{v_e\} \cup \{u\}$. If $v_e \notin S$, define $S' = S$. In all cases, S' is a power dominating set for G and $\gamma_P(G) \leq \gamma_P(G_e)$.

A path shows the lower bound is tight. □

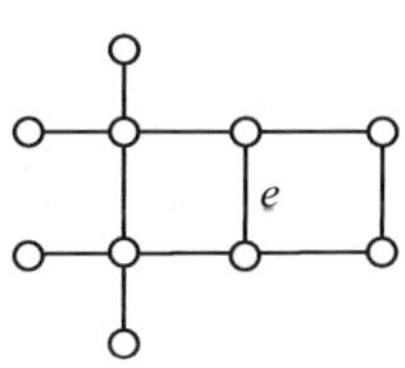

FIGURE 9.14. A graph G with $\gamma_P(G_e) = \gamma_P(G) + 1$.

Extreme values. High power domination numbers have been characterized. For a graph H, the corona $H \circ \overline{K_2}$ is constructed from H by adding two leaves to each vertex of H and the corona $H \circ K_2$ can be constructed from $H \circ \overline{K_2}$ by adding an edge between each pair of leaves both adjacent to one vertex of H. The set $\mathfrak{T}(H) = \{G : H \circ \overline{K_2} \leq G \leq H \circ K_2\}$ consists of graphs constructed from H by adding two leaves to each vertex and then either adding the edge between these leaves or not adding this edge (choosing independently for each vertex of H). Observe that $V(H)$ is a minimum power dominating set for any $G \in \mathfrak{T}(H)$. Define $\mathfrak{T} = \cup_H \mathfrak{T}(H)$ where the union is take over connected graphs H. Whereas the standard zero forcing number can be close to the order (e.g., $Z(K_n) = n - 1$), the power domination number cannot.

THEOREM 9.155. [**269**] *For any connected graph G of order $n \geq 3$, $\gamma_P(G) \leq \frac{n}{3}$ with equality if and only if $G = K_{3,3}$ or $G \in \mathfrak{T}$.*

For low power domination number, power domination number equal to one is not yet characterized.

REMARK 9.156. If $\gamma_P(G) = 1$, then G has a spider that is a spanning subgraph by Proposition 9.150. If T is a tree, then $\gamma_P(G) = 1$ if and only if T is a spider by Theorem 9.149.

Nordhaus-Gaddum Problems. Nordhaus Gaddum problems for power domination are studied in [**39**]. The upper bound $\gamma_P(G) + \gamma_P(\overline{G}) \leq n + 1$ follows from the analogous bound for domination number, which is realized by K_n. The trivial lower bound is $2 \leq \gamma_P(G) + \gamma_P(\overline{G})$ (since $1 \leq \gamma_P(H)$ for all H), and this bound is realized by P_n for $n \geq 4$ [**39**]. The upper bound is less immediate when both G and $\overline{G}$ are required to be connected.

THEOREM 9.157. [**39**] *Suppose G is a graph of order n such that both G and $\overline{G}$ are connected. Then for $n \neq 12, 13, 14, 15, 16, 17, 18, 20, 21, 24$,*

$$\gamma_P(G) + \gamma_P(\overline{G}) \leq \left\lceil \frac{n}{3} \right\rceil + 1.$$

This bound is attained for arbitrarily large n by every $G \in \mathfrak{T}$ of order $n \geq 6$.

9.10. Cops and Robbers, $c(G)$

Cops and Robbers is a two-player pursuit-evasion game played on a graph. The game has two players, one who places and moves a set of cops and the other who places and moves one robber. The goal for the cops is to catch the robber by occupying the same vertex the robber occupies, which is called a *capture*. The goal of the robber is to avoid capture. Initially the cops are placed on a multiset S of vertices (more than one cop can occupy a single vertex), and then the robber is placed. The game is played in a sequence of rounds during which the players take turns, with both playing in a single round, starting with the cops: The team of cops takes a turn by allowing each cop one chance to move to an adjacent vertex or decide to stay in place. Then the robber takes a turn by moving to an adjacent vertex or deciding to stay in place. The cops win the game if a cop captures the robber after a finite number of rounds. If the robber has a strategy to evade the cops indefinitely, the robber wins. If choosing the set S results in the cops winning regardless of the robber's strategy, then S is called a *capture set*. In the literature, a reflexive graph, i.e., a graph in which each vertex has a loop, is often used, but this is equivalent to using a simple graph in which staying in place is considered a move. The information in this section is very limited; see [**48**] for more information on Cops and Robbers.

The connection of Cops and Robbers to zero forcing is not as clear as that of power domination. Both Cops and Robbers and zero forcing can be considered forms of graph searching. Zero forcing is viewed as fast-mixed graph searching in [**267**]. See also the discussion of tree-width as a variant of zero forcing modeled by a variant of Cops and Robbers in [**19**]. Connections have also been built through the study of throttling (see Chapter 11).

A graph G is *cop-win* if one cop can capture the robber, and Cops and Robbers was introduced in [**237, 248**] focused on determining whether a graph was cop-win. Any graph that has a universal vertex is cop-win, but so are many additional graphs. Aigner and Fromme [**5**] defined the *cop number* $c(G)$ of a graph G to be the minimum cardinality of a capture set. Subsequently much of the study of the game of Cops and Robbers has focused on determining the cop number.

REMARK 9.158. One fundamental difference between Cops and Robbers and all variants of zero forcing (including power domination) is that only the number of cops $|S|$ affects whether S is a capture set, whereas which vertices are filled (or have PMUs in power domination) matters for all forms of zero forcing. That is, if S is a capture set and $|S'| = |S|$, then S' is a capture set for Cops and Robbers. Of course, the number rounds needed to capture the robber depends on the initial placement of the cops. For a color change rule X and most graphs G, there exist sets $S, S' \subset V(G)$ such that S is an X-forcing set, S' is not an X-forcing set, and $|S| = |S'|$.

OBSERVATION 9.159. *Let G be a graph. If the connected components of G are $G_1, \dots, G_t$, then $c(G) = \sum_{i=1}^{t} c(G_i)$.*

REMARK 9.160. Suppose G is a graph of order n and $G' = G[W]$ is an induced subgraph of G of order n'. Observe that $c(G) \leq n - n' + c(G')$ by choosing $S = S' \cup (V(G) \setminus V(G'))$ where S' is a minimum capture set for G'. This is the analog of Remark 9.3, but it is a very poor bound. Since it suffices to dominate the vertices not in G', $c(G') + \gamma(G[V(G) \setminus W])$ is a better bound.

The cop numbers of some graphs are listed in the next result.

PROPOSITION 9.161. [**48**]

(1) $c(K_n) = 1$.
(2) $c(C_n) = 2$.
(3) $c(K_{r,n-r}) = 2$ *for* $2 \leq r \leq n-2$.
(4) $c(H) = 1$ *for any chordal graph* H.

Computation. The k-cop problem is to determine for any given graph G and positive integer k whether $c(G) \leq k$. It is known that the k-cop problem is NP-hard (see [**142**] or [**48**, Theorem 5.5]). However, polynomial-time algorithms for each fixed k to determine whether a graph is k-cop-win are known [**87**] (and the relational characterization of k-cop-win graphs also implies polynomial-time algorithms for each fixed k to determine whether a graph is k-cop-win).

Meyniel's Conjecture and bounds on cop number. While it is easy to see that the cop number of a graph can be greater than one, it is not known how large it can be in terms of order. Since a dominating set is a capture set, it is immediate that $c(G) \leq \gamma(G)$, and $\gamma(G) \leq \lfloor \frac{n}{2} \rfloor$ for a graph of order n having no isolated vertices [**174**, Theorem 2.1]. *Meyniel's Conjecture* is that $c(G) = O(\sqrt{n})$ [**144**]. Projective planes can be used to construct *Meyniel extremal* graphs with $c(G) \geq \sqrt{\frac{n}{2}}$ for unbounded order n, as described in the next example.

EXAMPLE 9.162. [**48**, Theorem 1.5] A *projective plane* is a set of points and a set of lines that satisfy the following conditions: (1) Every pair of distinct lines intersect in a unique point. (2) There is exactly one line incident to every pair of distinct points. (3) There are four points such that no line is incident to more than two of them. A projective plane $\mathbb{P}_q$ of *order* q has q^2+q+1 points (and the same number of lines). Each line contains $q+1$ points and each point is on $q+1$ lines. It is known that there is a projective plane of order q whenever q is a prime power.

The *incidence graph* $G(\mathbb{P}_q)$ of the projective plane $\mathbb{P}_q$ is the bipartite graph of order $2(q^2+q+1)$ where one partite set is points, one is lines, and two vertices in different parts are adjacent if they are incident. Prałat showed in [**246**] that $c(G(\mathbb{P}_q)) = q+1 > \sqrt{\frac{2(q^2+q+1)}{2}}$ (see also Proposition 10.70).

It was shown in the study of throttling that the PSD zero forcing number is an upper bound for the cop number.

PROPOSITION 9.163. [**57**] *For every graph* G, $c(G) \le \mathrm{Z}_+(G)$.

The full result (Theorem 11.63) from which Proposition 9.163 is abstracted is useful in the study throttling for sparse graphs. However, for most graphs it is a poor bound, since it is known that almost all graphs have PSD zero forcing number close to the order (see Section 9.11). There are known sublinear upper bounds on the cop number (see [**48**, Theorem 1.6] and additional discussion therein).

The next result provides a lower bound on cop number (and can be used to show $c(G(\mathbb{P}_q)) \ge q+1$ [**48**].

THEOREM 9.164. [**48**, Theorem 1.3]. *if* G *is a graph with girth* $g(G) \ge 5$, *then* $c(G) \ge \delta(G)$.

Observe that $g(K_{3,3}) = 4$, $\delta(K_{3,3}) = 3$, and $c(K_{r,n-r}) = 2$ for $2 \le r \le n-2$, so they hypothesis $g(G) \ge 5$ necessary in Theorem 9.164.

Graph Operations. A vertex u is a *corner* if $N[u] \subseteq N[v]$ for some vertex $v \ne u$. Corners play an important role in the study of Cops and Robbers. A graph is *dismantlable* if there is a sequence of corners whose sequential deletion results a single vertex.

THEOREM 9.165. [**48**, Theorems 1.9, 2.3] *Let* G *be a graph.*

(1) *If* u *is a corner of* G, *then* $c(G-u) \le c(G)$.
(2) $c(G) = 1$ *if and only if* G *is dismantlable.*

A few of the bounds on the effects of graph operations that work for zero forcing also work for cop number.

PROPOSITION 9.166. *Let* G *be a graph.*

(1) *For any vertex* v *of* G, $c(G) - c(G-v) \le 1$.
(2) *For any edge* e *of* G, $c(G) - c(G-e) \le 1$.

All bounds are tight.

PROOF. (1) is immediate from Remark 9.160. Let $e = uw$. Choose a minimum capture set for $G-e$, place an additional cop on u, and have that cop remain in place. Then the capture process for $G-e$ works for G because the robber can never use e without being captured. Deleting a vertex or edge from a cycle C_n shows the bounds are tight. □

PROPOSITION 9.167. *Let* G *be a graph.*

(1) *If v and w are independent twins in G, then $c(G-v) \le c(G) \le c(G-v)+1$ and both bounds are tight.*
(2) *If v and w are adjacent twins in G, then $c(G-v) = c(G)$.*

PROOF. First assume v and w are independent twins in G. The upper bound follows from Proposition 9.166. Let S be a capture set for G. If any cops are on v in S, then move them to w to create S'; if no cops are on v then $S' = S$. Then S' is a capture set for $G - v$ using the same capture process (with v replaced by w as needed for the cops) except the robber no longer has the option of going to v. Consider $K_{2,n-2}$ with $n \ge 5$. Since $c(K_{1,p-1}) = 1$ and $c(K_{2,p-2}) = 2$ for $p \ge 4$, choosing v in the partite set of two vertices shows that the upper bound is tight and choosing v in the partite set of $n-2$ vertices shows that the lower bound is tight.

Now assume v and w are adjacent twins in G. Then $c(G-v) \le c(G)$ by Theorem 9.165 since v is a corner. To show $c(G) \le c(G-v)$, we claim that any capture set for $G - v$ is still a capture set for G: If there is a cop on any vertex in $N[w]$ and the robber moves to v, the robber is caught in the next round. If there is no cop in $N[w]$ and the robber moves to v, having the robber move to w instead has no effect on the outcome (including the number of rounds needed to catch the robber). □

REMARK 9.168. As for PSD zero forcing and power domination, there is no useful lower bound for $c(G) - c(G-v)$. For cop number and power domination, this is due to the ability of one vertex to dominate a graph. Let $G = K_1 \vee G(\mathbb{P}_q)$. Then $c(G) = 1$ and $c(G-v) = c(G(\mathbb{P}_q)) = q + 1 \approx \sqrt{|V(G)|}$.

The ability to move multiple cops across a single edge is a fundamental difference between Cops and Robbers and variants of zero forcing (including power domination). It is a basic property of all forms of zero forcing (except minor monotone floors, which allow hopping) that only one 'fill' can be transmitted across an edge in one direction (skew forcing can transmit one each way, resulting in two). This is formalized in Propositions 9.25, 9.54, and 9.93. To see that there is no useful lower bound on $c(G) - c(G-e)$, let G be constructed from two disjoint copies of $G(\mathbb{P}_q)$ by adding an edge $e = uw$ between the two, so $G - e = G(\mathbb{P}_q) \sqcup G(\mathbb{P}_q)$. We see that $c(G) \le q + 2$ by leaving a cop on u and moving the remaining $q+1$ cops into the copy of $G(\mathbb{P}_q)$ that contains the robber. Clearly $c(G-e) = 2(q+1)$, so $c(G-e) - c(G) \ge q \approx \sqrt{\frac{|V(G)|}{2}}$.

PROPOSITION 9.169. *Let G be a graph and $e \in E(G)$. Then:*
(1) *$c(G/e) \le c(G)$.*
(2) *$c(G) \le c(G_e)$.*

Both these bounds are tight.

PROOF. Let $e = uw$, let v_e be the new vertex that results from contraction in G/e. If S is a capture set for G, then S is also a capture set for G/e, using the same capture process except that whenever a cop occupies u or w in G, that cop occupies v_e in G/e. Thus $c(G/e) \le c(G)$. The bound for subdivision follows from that for contraction, because $(G_e)/(uv_e) = G$. A path shows both bounds are tight. □

9.11. Average Values and Random Graphs

It is interesting to consider the "average" behavior of a graph parameter $\zeta(G)$. The *average of* ζ over all labeled graphs of order n is

$$\overline{\zeta}(n) = \frac{\sum_{|V(G)|=n} \zeta(G)}{2^{\binom{n}{2}}}.$$

Note that $\overline{\zeta}(n) = E[\zeta(G(n, \frac{1}{2}))]$ where $G(n,p)$ is the Erdős-Rényi random graph with edge probability p and $E[x]$ denotes the expected value of x. Here we summarize what is known about the value of $\overline{\zeta}(n)$ for ζ being one of the variants of zero forcing or one of the related parameters discussed in this chapter, and also for tree-width, domination number, and the motivating maximum nullity parameters M, M_+, and M_0.

It is well known that $\mathrm{tw}(G) \leq \mathrm{Z}_+(G) \leq \widehat{\mathrm{Z}}(G) \leq \mathrm{Z}(G) \leq n$ and thus $\mathrm{tw}(G) \leq \lfloor \mathrm{Z}_+ \rfloor(H) \leq \lfloor \widehat{\mathrm{Z}} \rfloor(G) \leq \lfloor \mathrm{Z} \rfloor(G) \leq n$ for all graphs of order n [**19**]. Since $\mathrm{tw}(G(n, \frac{1}{2})) = n - o(n)$ [**242**], $\overline{\mathrm{Z}_+}(n) = \overline{\widehat{\mathrm{Z}}}(n) = \overline{\mathrm{Z}}(n) = \overline{\lfloor \mathrm{Z}_+ \rfloor}(n) = \overline{\lfloor \widehat{\mathrm{Z}} \rfloor}(n) = \overline{\lfloor \mathrm{Z} \rfloor}(n) = n - o(n)$. Since $\mathrm{Z}_-(G) \leq \mathrm{Z}(G)$ for all G, $\overline{\mathrm{Z}_-}(n) \leq \overline{\mathrm{Z}}(n)$.

The nullity parameters behave differently from the zero forcing parameters. Hall et al. showed in [**170**] that $\overline{\mathrm{mr}}(n) \geq 0.146907n$ for n sufficiently large by using results concerning the number of possible zero patterns of a sequence of polynomials. Thus $\overline{\mathrm{M}_+}(n) \leq \overline{\mathrm{M}}(n) \leq 0.86n$. Since $\mathrm{M}_0(n) \leq \mathrm{M}(n)$, this also implies $\overline{\mathrm{M}_0}(n) \leq 0.86n$. It is also shown there that $\overline{\nu}(n) \leq \overline{\xi}(n) \leq \sqrt{2}n(1 + o(1))$.

It is also shown in [**170**] that $\overline{\kappa}(n) = 0.5n(1 \pm o(1))$. Then $\kappa(G) \leq \nu(G) \leq \xi(G) \leq \mathrm{M}(G)$ and $\nu(G) \leq \mathrm{M}_+(G)$ are applied to conclude the expected value of $\overline{\zeta}(n) \geq 0.5n(1 - o(1))$ for $\zeta \in \{\nu, \xi, \mathrm{M}_+, \mathrm{M}\}$.

The power domination number and the cop number behave very differently from zero forcing numbers because they have the domination number as an upper bound. For any fixed $\epsilon > 0$, $(1 - \epsilon)\log_2 n \leq \gamma(G(n, \frac{1}{2})) \leq (1 + \epsilon)\log_2 n$ with probability going to 1 as $n \to \infty$ [**235, 265**]. Thus $\gamma(G(n, \frac{1}{2})) = (\log_2 n)(1 \pm o(1))$.

Table 9.2 summarizes what is known about average value of various parameters.

TABLE 9.2. Summary of values or bounds for $\overline{\zeta}(G) = E[\zeta(G(n, \frac{1}{2}))]$ for n sufficiently large for parameters related to zero forcing

ζ	$\overline{\zeta}(G(n)$
Z, $\widehat{\mathrm{Z}}$, Z_+, $\lfloor \mathrm{Z} \rfloor$, $\lfloor \widehat{\mathrm{Z}} \rfloor$, $\lfloor \mathrm{Z}_+ \rfloor$, tw	$= n(1 - o(1))$
Z_-	$\leq n(1 - o(1))$
M, M_+	$\geq 0.5n(1 - o(1))$ and $\leq .86n$
ξ, ν	$\geq 0.5n(1 - o(1))$ and $\leq 0.71n(1 + o(1))$
M_0	$\leq 0.86n$
γ	$= (\log_2 n)(1 \pm o(1))$
γ_P, c	$\leq (\log_2 n)(1 \pm o(1))$

9.12. Topics Not Covered

We have not discussed k-power domination or k-forcing. Chang et al. introduced *generalized power domination* or *k-power domination* in [**82**]. This allows a vertex with at most k unobserved neighbors to observe them during the iterative (forcing) part of the power domination process. This paper has led to further study of k-power domination (see, for example, citations on MathSciNet). In a similar spirit, Amos et al. introduced *k-forcing* in [**13**], which allows a vertex with at most k unfilled neighbors to fill them. They derived bounds that apply to the zero forcing number (see Proposition 9.24). This paper has also led to further study of k-forcing (see, for example, citations on MathSciNet).

The (standard) *Grundy domination number* is the maximum number of vertices in a sequence $(v_1, \dots, v_g)$ such that $N[v_i] \setminus \cup_{j=1}^{i-1} N[v_j] \neq \emptyset$. Brešar et al. introduced a variant of Grundy domination as the complementary parameter for zero forcing in [**58**] (complementary graph parameters sum to the order of the graph, e.g., minimum rank and maximum nullity). In [**222**] Lin established complementary relations between additional variants of zero forcing and variants of Grundy domination (including standard Grundy domination); relations with variants of minimum rank and maximum nullity are also discussed there.

CHAPTER 10

Propagation Time and Capture Time

Propagation time measures the number of time steps (also called rounds) needed to fill every vertex of the graph (or observe all vertices) when performing in one time step all possible forces that can be done independently of each other; capture time measures the number of rounds needed to catch the robber with both sides playing optimally. Capture time was introduced in 2009 by Bonato et al. Propagation time for standard zero forcing was introduced in 2012 by Hogben et al. [**187**] and Chilakamarri et al. [**85**]; subsequently propagation time for other variants of zero forcing was studied.

In this chapter and in Chapter 11 we begin with the universal definitions of propagation (see Section 10.1) and throttling for various kinds of zero forcing rather than the original definitions, in order to more efficiently discuss multiple variants while emphasizing their similarities and differences. Discussion of propagation time for each of the three forcing rules Z, Z_+, and Z_- follows in Sections 10.2, 10.3, and 10.4. Propagation time for power domination is not explicitly discussed in Section 10.1, but the material there adapts readily to power domination (see Section 10.5). Capture time for Cops and Robbers is discussed in Section 10.6.

10.1. Universal Definitions for Forcing Propagation Time

When discussing propagation time for zero forcing, it assumed that a fixed X-color change rule is being used (think Z, Z_+, or Z_-). For the forcing rules discussed here, the final coloring is unique, and this allows the use of simpler definitions than those in [**77**], where Carlson introduced a universal perspective on propagation and throttling.

Starting with $B \subseteq V(G)$, we define two sequences of sets, the set $B^{(i)}$ of vertices that are filled at time step i and the set $B^{[i]}$ of vertices that are filled after time step i. Thus $B^{[0]} = B^{(0)} = B$ is the set of initial filled vertices and $B^{[i+1]} = B^{[i]} \cup B^{(i+1)}$. Assume $B^{(i)}$ and $B^{[i]}$ have been constructed. Then

$$B^{(i+1)} = \{w : \ w \text{ can be } X\text{-forced by some } v \text{ (given } B^{[i]} \text{ filled)}\}.$$

The *X-propagation time* of $B \subseteq V(G)$, denoted by $\mathrm{pt}_X(G; B)$, is the least t such that $B^{[t]} = V(G)$ or infinity if B is not a X-forcing set of G. The *X-propagation time of G*, $\mathrm{pt}_X(G)$, also known as the *minimum X-propagation time of G*, is

$$\mathrm{pt}_X(G) = \min\{\mathrm{pt}(G; B) : B \text{ is a minimum } X\text{-forcing set}\}.$$

In Figure 10.1 we illustrate the Z-propagation process and filled vertex sets $B^{[i]}$ for a tree T_1 and an initial set B such that $\mathrm{pt}_{\mathrm{Z}}(G) = \mathrm{pt}_{\mathrm{Z}}(G; B)$. The tree T_1 appears in examples throughout this chapter and the next. Additional examples are presented in the sections devoted to specific color change rules.

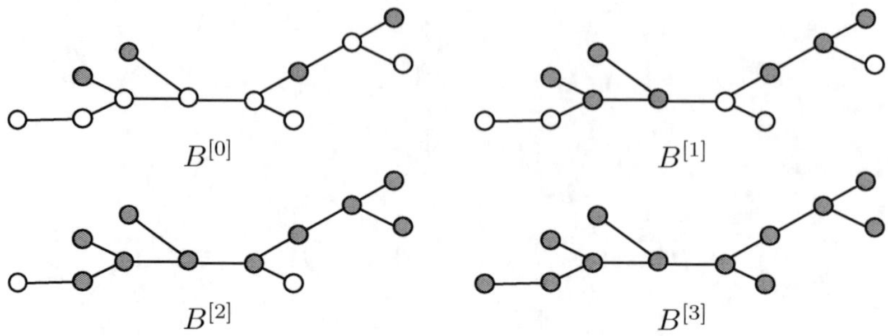

FIGURE 10.1. Standard propagation on the tree T_1

OBSERVATION 10.1. *For a graph G of order n, $\operatorname{pt}_X(G;B) \le n - |B|$ because at least one force must be performed at each time step. This implies*

$$\operatorname{pt}_X(G) \le n - X(G).$$

Define the *k-propagation time of G* by

$$\operatorname{pt}_X(G,k) = \min_{|B|=k} \operatorname{pt}_X(G;B).$$

Then $\operatorname{pt}_X(G) = \operatorname{pt}_X(G, X(G))$. The parameter k-propagation time generalizes the k-capture time introduced in the study of Cops and Robbers (see Section 10.6). An X-set B is an *optimal X-set for k-propagation time* of G if $\operatorname{pt}_X(G;B) = \operatorname{pt}_X(G,k)$.

When studying the propagation time of an X-forcing set B, we focus on a *propagating chronological list of forces* (and associated *propagating set of forces*), in which all vertices of $B^{[i-1]}$ are filled before any vertex in $B^{(i)}$ is filled. That is, a propagating chronological list of forces for B is one that carries out the propagation process of constructing $B^{(i)}$ for $i = 0, \ldots, \operatorname{pt}_X(G;B)$ in order. A subset B of vertices of G is an *efficient X-forcing set* for G if B is a minimum X-forcing set of G and $\operatorname{pt}_X(G;B) = \operatorname{pt}_X(G)$. Thus the set B in Figure 10.1 is an efficient Z-forcing set for the tree T_1.

Two X-forcing sets B_1 and B_2 of a graph G are *isomorphic* if there is a graph automorphism $\psi(G)$ of G such that $\psi(B_1) = B_2$.

OBSERVATION 10.2. *If B_1 and B_2 are isomorphic X-forcing sets, then*

$$\operatorname{pt}_X(G;B_1) = \operatorname{pt}_X(G;B_2).$$

However, non-isomorphic zero forcing sets may also have the same propagation time.

A *color change rule requires adjacency* if v and w must be adjacent for v to force w. Most color change rules, including the Z-, Z_+-, and Z_--color change rules, require adjacency (however, the color change rules for minor monotone floors of zero forcing parameters do not; see Section 9.8). If X is a zero forcing color change rule that requires adjacency and that does not allow unfilled vertex forcing (i.e., v must be filled in order for v to X-force w), distance plays a fundamental role as a lower bound for propagation time and throttling. (The rules Z and Z_+ do not allow unfilled vertex forcing, but Z_- does allow unfilled vertex forcing.) The *distance* from a set U to a vertex v is $\operatorname{dist}(U,v) = \min_{u\in U} \operatorname{dist}(u,v)$. The *eccentricity* of a set U of vertices in a graph G is defined by

$$\operatorname{ecc}(U) = \max_{v \in V(G)} \operatorname{dist}(U, v).$$

The *k-radius* of G is

$$\operatorname{rad}_k(G) = \min_{S \subseteq V, |S|=k} \operatorname{ecc}(S).$$

The 1-radius is called the *radius* and denoted by $\operatorname{rad}(G)$. An X-forcing set B of G has *full X-forcing* if for all $i = 1, \dots, \operatorname{pt}_X(G; B)$, every unfilled neighbor of a filled vertex is forced at the next time step, i.e., $v \in B^{[t]}$ implies $N(v) \subseteq B^{[t+1]}$.

REMARK 10.3. Let X be a color change rule that requires adjacency and does not allow unfilled vertex forcing. If G is a graph and $B \subset V(G)$, then

$$\operatorname{ecc}(B) \leq \operatorname{pt}_X(G; B)$$

because each time step can advance at most distance one. If B has full forcing, then $\operatorname{pt}_X(G; B) = \operatorname{ecc}(B)$. Furthermore,

$$\operatorname{rad}_k(G) \leq \operatorname{pt}_X(G, k) \text{ and } \operatorname{rad}_{X(G)}(G) \leq \operatorname{pt}(G).$$

We can also consider maximum propagation time. The *maximum X-propagation time* of G is defined to be

$$\operatorname{PT}_X(G) = \max\{\operatorname{pt}(G; B) : B \text{ is a minimum } X\text{-forcing set}\}.$$

The *X-propagation time interval* of G is defined as

$$[\operatorname{pt}_X(G), \operatorname{PT}_X(G)] = \{\operatorname{pt}_X(G), \operatorname{pt}_X(G) + 1, \dots, \operatorname{PT}_X(G) - 1, \operatorname{PT}_X(G)\}.$$

The graph G has a *full X-propagation time interval* if every integer between $\operatorname{pt}_X(G)$ and $\operatorname{PT}_X(G)$ is realized by some minimum X-forcing set. The use of 'full' with two different meanings for propagation (full forcing and full propagation time interval) is unfortunate, because there is no connection between the two uses (both of which appear in the literature).

As with zero forcing, a universal vertex plays a special role. Due to the similarity across variants of zero forcing, we state and prove a version for several variants, which is similar to the proofs for zero forcing.

PROPOSITION 10.4. *Let X be one of the* Z*-color rule, the* Z_+*-color rule, or the* Z_-*-color rule. Let H be a graph of order n with no isolated vertices and let $G = H \vee K_1$ where $V(K_1) = \{u\}$, and let k be such that $X(G) \leq k \leq n+1$. Then* $\operatorname{pt}_X(G, k) = \operatorname{pt}_X(H, k-1)$. *Furthermore, there is an X-set $B \subseteq V(G)$ that contains u such that B is optimal for k-propagation time of G and $B \setminus \{u\}$ is optimal for $(k-1)$-propagation time of H.*

PROOF. It is immediate that $\operatorname{pt}_X(G, k) \leq \operatorname{pt}_X(H, k-1) + 1$ by choosing an X-set that is optimal for $(k-1)$-propagation time of H and adding u. To prove $\operatorname{pt}_X(H, k-1) \leq \operatorname{pt}_X(G, k) - 1$, let $B \subseteq V(G)$ be an X-set that is optimal for k-propagation time of G and let $\mathcal{F}$ be a propagating set of X-forces of B.

Suppose first that $u \notin B$. If u performs a force at the first time step, then $X = Z_-$ and only one vertex of H is unfilled, so we can replace B by any set of vertices that contains u and omits exactly two adjacent vertices of H without changing the propagation time (which is one). So assume u does not force at the first time step. No vertex of H can a force another vertex of H until u is filled. So there must be a vertex $x \in B$ such $N_H[x] \subseteq B$ and $B^{(1)} = \{u\}$. Since $\deg_H(x) \geq 1$, there is a vertex $y \in N_H(x)$. Let $B' = B \setminus \{y\} \cup \{u\}$, so $|B'| = |B|$, $x \to y$ in the first time step using B', and $\mathrm{pt}_X(G; B') \le \mathrm{pt}_X(G; B)$. Thus there is an X-set that contains u that is optimal for k-propagation time of G.

Let $\hat{B} = B \setminus \{u\}$. If u does not force, then $\hat{B}$ is a forcing set for H with $\mathcal{F}$ and same propagation time. Suppose $u \to w \in \mathcal{F}$. This must be the last force performed in the component H' of H that contains w, because every other vertex of H' except w must be filled in order for u to force. Since there is a vertex v of H' adjacent to w, we can replace $u \to w$ by $v \to w$ without changing the propagation time, so $\mathrm{pt}_X(H, k-1) \le \mathrm{pt}_X(H; \hat{B}) = \mathrm{pt}_X(G; B) = \mathrm{pt}_X(G, k)$. □

10.2. Z-Propagation Time

The Z-*propagation time* is denoted by $\mathrm{pt}(G; B)$ and $\mathrm{pt}(G)$ (in addition to $\mathrm{pt}_Z(G; B)$ and $\mathrm{pt}_Z(G)$). The Z-propagation process and filled vertex sets $B^{[i]}$ for the tree T_1 and a set $B \subset V(T_1)$ were illustrated in Figure 10.1.

OBSERVATION 10.5. *Observations 10.1, 10.2, and Remark 10.3 apply to Z-propagation time.*

In the next example we compute sets $B^{(i)}$ and $B^{[i]}$ for two different sets $B = B^{[0]}$ to determine Z-propagation time for the graph $K_2 \,\square\, P_5$.

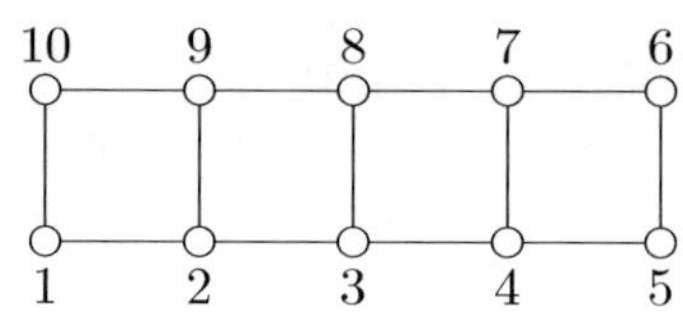

FIGURE 10.2. The graph $K_2 \,\square\, P_5$

EXAMPLE 10.6. The graph $K_2 \,\square\, P_5$ is shown in Figure 10.2. First we consider $B_1 = \{5, 6\}$. Then

i	0	1	2	3	4
$B_1^{(i)}$	$\{5,6\}$	$\{4,7\}$	$\{3,8\}$	$\{2,9\}$	$\{1,10\}$
$B_1^{[i]}$	$\{5,6\}$	$\{4,5,6,7\}$	$\{3,4,5,6,7,8\}$	$\{2,3,4,5,6,7,8,9\}$	$V(K_2 \,\square\, P_5)$

so $\mathrm{pt}(K_2 \,\square\, P_5; B_1) = 4$.

Next consider $B_2 = \{4, 5\}$. Then

i	0	1	2	3	4	5
$B_2^{(i)}$	$\{4,5\}$	$\{6\}$	$\{7\}$	$\{3,8\}$	$\{2,9\}$	$\{1,10\}$
$B_2^{[i]}$	$\{4,5\}$	$\{4,5,6\}$	$\{4,5,6,7\}$	$\{3,4,5,6,7,8\}$	$\{2,3,4,5,6,7,8,9\}$	$V(K_2 \,\square\, P_5)$

so $\mathrm{pt}(K_2 \,\square\, P_5; B_2) = 5$.

Since every other minimum standard zero forcing set is isomorphic to one of B_1 or B_2, $\mathrm{pt}(G) = 4$.

REMARK 10.7. [**85, 187**] Let G be a graph. Then

$$\frac{|V(G)| - \mathrm{Z}(G)}{\mathrm{Z}(G)} \le \mathrm{pt}(G)$$

because when using minimum standard zero forcing set, at most $\mathrm{Z}(G)$ forces can be performed at each time step. This extends to larger standard zero forcing sets:

$$\frac{|V(G)|-|B|}{|B|} \leq \mathrm{pt}(G; B) \tag{10.1}$$

PROPOSITION 10.8. [**187**] *Let G be a graph. Every connected graph of order at least two has more than one efficient zero forcing set. However, there are connected graphs of arbitrarily large order in which the intersection of all efficient standard zero forcing sets is nonempty.*

EXAMPLE 10.9. The wheel W_n with n odd is an example in which the intersection of all efficient standard zero forcing sets is nonempty, because for $n \geq 5$ there is a unique universal vertex u and every efficient standard zero forcing set contains u: There are only two minimum standard zero forcing sets up to isomorphism, B_1 consisting of two consecutive cycle vertices and u, and B_2 consisting of three consecutive cycle vertices. Observe that $\mathrm{pt}(W_{2k+1}; B_1) = k - 1$, $\mathrm{pt}(W_{2k+1}; B_2) = k$, and u is in the intersection of all sets of the form B_1. Note that $\mathrm{pt}(W_{2k}; B_1) = k - 1 = \mathrm{pt}(W_{2k+1}; B_2)$, so when n even the intersection of the efficient standard zero forcing sets is empty. This is illustrated in Example PT.2-1 in [**184**] for $n = 7$ and $n = 8$. See Remark 10.11 for the propagation time of the wheel.

Computation and Additional Bounds. The function `pt(G)` $= \mathrm{pt}(G)$ is available in [**184**] for computing the propagation time of a specific graph (that is not too large). The value of $\mathrm{pt}(G)$ is computed as $\mathrm{pt}(G) =$ `ptk(G,Z(G))` where `ptk(G,k)` $= \mathrm{pt}(G, k)$ is the minimum of $\mathrm{pt}(G; B)$ over all sets B such that $|B| = k$. The function `ptz(G,B)` $= \mathrm{pt}(G; B)$ is also available. The use of these functions is illustrated in Section 10.2 of [**184**].

In general, the diameter and the propagation time of a graph are not comparable (examples are presented in [**187**]). But for trees there is a relationship:

THEOREM 10.10. [**187**] *Let T be a tree and B be a minimum zero forcing set of T. Then $\mathrm{pt}(T, B) \leq \mathrm{diam}(T)$. Hence, $\mathrm{pt}(T) \leq \mathrm{diam}(T)$.*

Families of Graphs. The next remark lists the propagation times of some families of graphs.

REMARK 10.11. Most of the following appeared in [**85**] or [**187**]; all are immediate from the known minimum zero forcing sets, but for several we provide simple explanations.

(1) For $n \geq 2$, $\mathrm{pt}(P_n) = n - 1$ because only an endpoint x is a minimum zero forcing set and $\mathrm{pt}(P_n, \{x\}) = n - 1$.
(2) For $n \geq 3$, $\mathrm{pt}(C_n) = \lceil \frac{n-2}{2} \rceil$ because $\mathrm{Z}(C_n) = 2$, a set two of vertices is a zero forcing set if and only if the vertices are adjacent, and for two adjacent vertices x and y, $\mathrm{pt}(C_n, \{x, y\}) = \lceil \frac{n-2}{2} \rceil$.
(3) For $n \geq 4$, $\mathrm{pt}(W_n) = \lceil \frac{n-3}{2} \rceil$.
(4) For $n \geq 2$, $\mathrm{pt}(K_n) = 1$ because $\mathrm{Z}(K_n) = n - 1$.
(5) For $n \geq 3$, $\mathrm{pt}(K_{1,n-1}) = 2$.
(6) For $2 \leq r \leq n - 2$, $\mathrm{pt}(K_{r,n-r}) = 1$.
(7) For $d \geq 1$, $\mathrm{pt}(Q_d) = 1$ because $\mathrm{Z}(Q_d) = 2^{d-1}$ and if B is the vertices of one copy of Q_{d-1} in the construction $Q_d = Q_{d-1} \square K_2$, then $\mathrm{pt}(Q_d; B) = 1$.

(8) For $k \geq 1$, $\text{pt}(F_k) = 1$ because $\text{Z}(F_k) = k + 1$ and if B is an independent set of k vertices of degree 2 together with the vertex of degree $2k$, then $\text{pt}(F_k; B) = 1$.

PROPOSITION 10.12. *Let $n \geq 5$.*

(1) For $k \geq 3$, $\text{pt}(B_k) = 3$.

(2) For $n \geq 5$, $\text{pt}(\overline{C_n}) = \begin{cases} 2 & \text{unless } n = 6 \\ 1 & \text{if } n = 6 \end{cases}$.

(3) For $n \geq 5$, $\text{pt}(\overline{P_n}) = 2$.

PROOF. For the book graph, recall that $\text{Z}(B_k) = k$. There are k pairs of adjacent degree two vertices (called page vertices) and two vertices of degree $k+1$, called spine vertices. Any zero forcing set B must contain at least one vertex from at least $k - 1$ of the page vertices. It must also contain a spine vertex adjacent to one or more of the page vertices in B, or both page vertices on one page. A zero forcing set B that contains both page vertices on one page or a spine vertex and all the page vertices in B are adjacent to this spine vertex, then $\text{pt}(B_k; B) = 3$; otherwise $\text{pt}(B_k; B) = 4$. This is illustrated in Example PT.2-2 [**184**].

For the complements of a cycle or path, we first consider the special cases $n = 5, 6$. For $n = 5$, $\overline{C_5} = C_5$ and $\text{pt}(C_5) = 2$. For $n = 6$, $\overline{C_6} = K_3 \,\square\, P_2$, $\text{Z}(K_3 \,\square\, P_2) = 3$ and $\text{pt}(K_3 \,\square\, P_2) = 1$ using one copy of K_3. For $\overline{P_5}$ and $\overline{P_6}$, an extra edge is added. This does not change the propagation time for $\overline{P_5}$, but $\text{pt}(\overline{P_6}) = 2$.

Now assume $n \geq 7$. Note that $\text{Z}(\overline{C_n}) = n-3 = \text{Z}(\overline{P_n})$. Let $V = \{0, 1, \ldots, n-1\}$ be the set of vertices of C_n in cycle order and perform arithmetic modulo n; for $\overline{P_n}$ the edge $\{0, n-1\}$ is added. Let $B = V \setminus \{1, 3, 5\}$ and note that $\text{pt}(\overline{C_n}; B) = 2 = \text{pt}(\overline{P_n}; B)$. To complete the proof, we show there is no set of $n - 3$ vertices that can force $\overline{C_n}$ (or $\overline{P_n}$) in one time step. In $\overline{C_n}$, vertex k is adjacent to every vertex except vertices $k - 1$ and $k + 1$. In order for vertex k to perform the first force with a set B of filled vertices having $|B| = n - 3$, necessarily $B = V \setminus \{k - 1, k + 1, \ell\}$ where $\ell \notin \{k - 1, k, k + 1\}$. We claim that with such a B, at most one filled vertex different from k can perform a force in the first time step. Every vertex except $k, k - 2, k + 2$ is adjacent to both $k - 1$ and $k + 1$ in $\overline{C_n}$ and so cannot force. Since $k-2$ is adjacent to the unfilled vertex $k+1$ in $\overline{C_n}$, in order for the force $k-2 \to k+1$ to occur in the first step, $k - 2$ is not adjacent to ℓ. This implies $\ell = k - 3$. A similar analysis for $k + 2$ shows that in order for the force $k + 2 \to k - 1$ to occur in the first step, $\ell = k + 3$. Since $k - 3 = k + 3$ implies $n = 6$, this is not possible. Thus an additional step is needed and $\text{pt}(\overline{C_n}) = 2$. The extra edge in $\overline{P_n}$ does not help and $\text{pt}(\overline{P_n}) = 2$. □

Reversals. Recall that a reversal of a standard zero forcing set B is the set of last vertices of the zero forcing chains of a chronological list of forces, i.e., the vertices that do not perform forces. A *propagating reversal* $Rev(B)$ of a standard zero forcing set B is the set of last vertices of the maximal zero forcing chains of a propagating chronological list of forces. Note that $Rev(B)$ need not be unique, as shown in the next example.

EXAMPLE 10.13. Consider the graph shown in Figure 10.3. The minimum Z-forcing set $B = \{5, 6\}$ has two propagating reversals, $\{1, 3\}$ and $\{1, 8\}$, depending on whether $8 \to 2$ or $3 \to 2$ at time step 3.

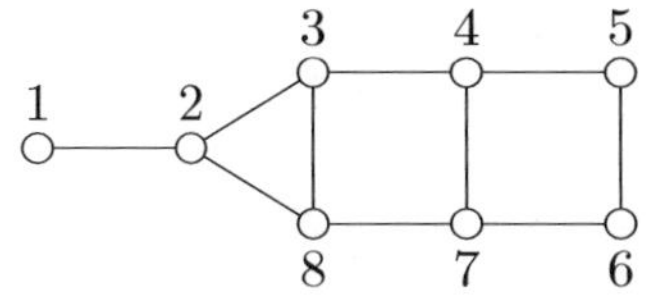

FIGURE 10.3. The graph in Example 10.13

EXAMPLE 10.14. Recall the graph $K_2 \square P_5$ and Z-forcing sets B_1 and B_2 in Example 10.6. Then $Rev(B_1) = \{1, 10\}$ and $Rev(B_2) = \{1, 10\}$ also (there are other possible reversals using non-propagating chronological lists of forces, but they are not of interest here). Note that $\mathrm{pt}(K_2 \square P_5, \{1, 10\}) = 4$, so the propagation time of a propagating reversal can be less than the propagation time of the original. Note that $Rev(\{1, 10\}) = \{5, 6\}$. In fact, $\{4, 5\}$ is not a propagating reversal of any minimum standard zero forcing set.

THEOREM 10.15. [**187**] *Let G be a graph, B a standard zero forcing set of G, and $Rev(B)$ be a propagating reversal of B. Then*

$$\mathrm{pt}(G; Rev(B)) \leq \mathrm{pt}(G; B).$$

If B is efficient, then $\mathrm{pt}(G; Rev(B)) = \mathrm{pt}(G) = \mathrm{pt}(G; B)$.

Extreme Values. Here we summarize some of what is known about extreme propagation time.

PROPOSITION 10.16. *Let G be a graph.*

(1) $\mathrm{pt}(G) = 0$ *if and only if G has no edges.*
(2) $\mathrm{pt}(G) = |V(G)| - 1$ *if and only if G is a path.*
(3) Graphs G having $\mathrm{pt}(G) = |V(G)| - 2$ *are characterized in* [**187**].

OBSERVATION 10.17. *If G is a graph with* $\mathrm{pt}(G) = 1$*, then* $\mathrm{Z}(G) \geq \frac{|V(G)|}{2}$ *(by the bound in Remark 10.7).*

Let $H_1 = (V_1, E_1)$ and $H_2 = (V_2, E_2)$ be graphs of equal order and let M be a matching between V_1 and V_2. The *M-sum of H_1 and H_2*, denoted by $H_1M^+H_2$, has $V(H_1M^+H_2) = H_1 \cup H_2$ and $E(H_1M^+H_2) = E(H_1) \cup E(H_2) \cup M$. Any graph that can be expressed in the form $H_1M^+H_2$ is called a *matched-sum graph*. This terminology was introduced by Georges and Mauro in [**157**]. An equivalent definition was given in [**187**] (without reference to a matched-sum graph): For a bijection μ from V_1 to V_2, the *matching graph* (H_1, H_2, μ) is $H_1M^+H_2$ when M is the set of edges $\{v\mu(v) : v \in V_1\}$.

PROPOSITION 10.18. [**187**] *Let G be a graph of order n. Any two of the following conditions imply the third.*

(1) $n = 2\,\mathrm{Z}(G)$.
(2) $\mathrm{pt}(G) = 1$.
(3) G is a matched-sum graph.

When these conditions are true, B has full forcing for any efficient Z*-forcing set.*

Graph Operations. Because deletion of a vertex or edge or subdivision or contraction of an edge of a graph G can change the zero forcing number, there is no relationship between $\mathrm{pt}(G)$ and $\mathrm{pt}(G-v)$, $\mathrm{pt}(G-e)$, $\mathrm{pt}(G_e)$, or $\mathrm{pt}(G/e)$. For the path P_n, deleting a vertex or an edge near the middle reduces propagation time by about $\frac{n}{2}$. For a cycle C_n, deleting a vertex or edge increases propagation time by about $\frac{n}{2}$. The next two examples present families of graphs where an edge subdivision causes an unbounded decrease or increase in the propagation time. This automatically covers edge contraction also, because $(G_e)/uv_e = G$ when $e = uw$ and v_e is the new vertex.

EXAMPLE 10.19. Let $G(k)$ be the graph of order $2k+3$ constructed from $P_{2k} \sqcup P_3$ with vertices $(1, 2, \dots, 2k)$ and $(2k+1, 2k+2, 2k+3)$ by adding edge $e = \{k, 2k+2\}$ (see Figure 10.4, which shows both $G(k)$ and $G(k)_e$). Then $\mathrm{Z}(G(k)) = 2$ and $\mathrm{pt}(G(k)) = 2k-1$, whereas $\mathrm{Z}(G(k)_e) = 3$ and $\mathrm{pt}(G(k)_e) = k+1$ (efficient standard zero forcing sets are illustrated in Figure 10.4 and computations are provided in Example PT.2-3 in [**184**] for $k = 5$). So $\mathrm{pt}(G(k)) - \mathrm{pt}(G(k)_e) = k-2$ and the difference between the propagation time of the original graph and a subdivision can be arbitrarily large.

$k = 5$ $\quad$ $\mathrm{Z}(G(k)) = 2$ & $\mathrm{pt}(G(k)) = 2k-1$ $\quad$ $\mathrm{Z}(G(k)_e) = 3$ & $\mathrm{pt}(G(k)_e) = k+1$

FIGURE 10.4. Subdivision (respectively, contraction) can lower (respectively, raise) propagation time substantially

EXAMPLE 10.20. Let $H(k)$ be the graph of order $2k$ constructed from $P_k \sqcup P_k$ with vertices $(1, 2, \dots, k)$ and $(k+1, k+2, \dots, 2k)$ by adding edges $e = \{k-2, 2k-2\}$ and $\{k-1, 2k-1\}$ (see Figure 10.5, which shows both $H(k)$ and $H(k)_e$). Then $\mathrm{Z}(H(k)) = 2$ and $\mathrm{pt}(H(k)) = k-1$, whereas $\mathrm{Z}(H(k)_e) = 2$ and $\mathrm{pt}(H(k)_e) = 2k-4$ (efficient standard zero forcing sets are illustrated Figure 10.5 and computations are provided in Example PT.2-4 in [**184**] for $k = 7$). So $\mathrm{pt}(H(k)) - \mathrm{pt}(H(k)_e) = -k+3$ and the difference between the propagation time of a subdivision and the original graph can be arbitrarily large.

$k = 7$ $\quad$ $\mathrm{Z}(H(k)) = 2$ & $\mathrm{pt}(H(k)) = k-1$ $\quad$ $\mathrm{Z}(H(k)_e) = 2$ & $\mathrm{pt}(H(k)_e) = 2k-4$

FIGURE 10.5. Subdivision (respectively, contraction) can raise (respectively, lower) propagation time substantially

A universal vertex whose deletion does not result in isolated vertices is a special case where the propagation times are related, as established in Proposition 10.4.

COROLLARY 10.21. *Let H be a graph of order n with no isolated vertices and let $G = H \vee K_1$ where $V(K_1) = \{u\}$, and let k be such that $\mathrm{Z}(G) \le k \le n+1$. Then $\mathrm{pt}(G, k) = \mathrm{pt}(H, k-1)$. Furthermore, there is a zero forcing set $B \subseteq V(G)$ that contains u such that B is optimal for k-propagation time of G and $B \setminus \{u\}$ is optimal for $(k-1)$-propagation time of H.*

Nordhaus-Gaddum Problems. The Nordhaus-Gaddum sum problem for propagation time is to establish tight lower and upper bounds on $\operatorname{pt}(G) + \operatorname{pt}(\overline{G})$.

PROPOSITION 10.22. *For every graph G of order n,*

$$1 \leq \operatorname{pt}(G) + \operatorname{pt}(\overline{G}) \leq n + 2$$

and the lower bound is tight.

PROOF. By Observation 10.1, $\operatorname{pt}(G) + \operatorname{pt}(\overline{G}) \leq (n - \operatorname{Z}(G)) + (n - \operatorname{Z}(\overline{G}))$. By Theorem 9.30, $\operatorname{Z}(G) + \operatorname{Z}(\overline{G}) \geq n - 2$. Thus $\operatorname{pt}(G) + \operatorname{pt}(\overline{G}) \leq n + 2$. If G has an edge, then $\operatorname{pt}(G) \geq 1$, so $1 \leq \operatorname{pt}(G) + \operatorname{pt}(\overline{G})$. Note that $\operatorname{pt}(K_n) = 1$ and $\operatorname{pt}(\overline{K_n}) = 0$. □

While the upper bound $n + 2$ is realized by P_4 (which is self-complementary), a tight Nordhaus-Gaddum bound is expected to be realized by arbitrarily large n. The largest value of $\operatorname{pt}(G) + \operatorname{pt}(\overline{G})$ realized by arbitrarily large n (that we are aware of) is $n + 1$ for the path, which follows from Remark 10.11 and Proposition 10.12.

COROLLARY 10.23. *For every* $n \geq 5$, $\operatorname{pt}(P_n) + \operatorname{pt}(\overline{P_n}) = n + 1$.

Full Forcing.

REMARK 10.24. Let G be a graph and let B be a minimum Z-forcing set. If $\frac{|V(G)| - \operatorname{Z}(G)}{\operatorname{Z}(G)} = \operatorname{pt}(G; B)$, then B has full Z-forcing, because each standard zero forcing chain must have length $\frac{|V(G)| - \operatorname{Z}(G)}{\operatorname{Z}(G)}$ vertices and a force must be performed in each chain at every time step. The next example shows the converse is false.

EXAMPLE 10.25. Let H be shown in Figure 10.6 and let $B = \{1, 6\}$. Observe that $\operatorname{Z}(H) = 2$, $\operatorname{pt}(H) = \operatorname{pt}(H; B) = 4$, B has full Z-forcing, whereas $\frac{|V(G)| - \operatorname{Z}(G)}{\operatorname{Z}(G)} = 3$.

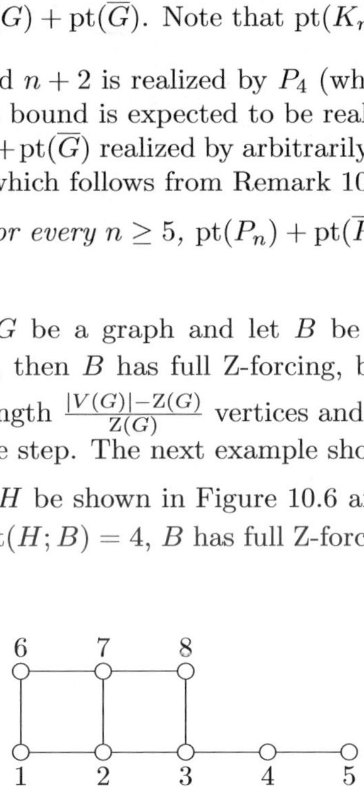

FIGURE 10.6. A graph H that has full forcing

Maximum Propagation Time and the Propagation Time Interval. Maximum Z-propagation time of a graph G, which is the maximum propagation tome of a minimum standard zero forcing set, is denoted by $\operatorname{PT}(G)$. There are examples of graphs that have full Z-propagation time intervals (e.g., P_n, C_n, $K_{r,n-r}$), based on their minimum standard zero forcing sets. The next example shows that a graph need not have a full Z-propagation time interval (another example is given in [**187**]).

EXAMPLE 10.26. Let $H(k)$ be the graph in Example 10.20 (see Figure 10.5, which shows both $H(k)$ and $H(k)_e$). Up to isomorphism, $H(k)$ has only two minimum standard zero forcing sets: $B_1 = \{1, k + 1\}$ and $B_2 = \{1, k\}$. Since $\operatorname{pt}(H(k); B_1) = k - 1$ and $\operatorname{pt}(H(k); B_2) = 2k - 5$, the propagation time interval is $\{k - 1, 2k - 5\}$, which is not full for $k \geq 6$. Computations for $k = 7$ are illustrated in Example PT.2-5 in [**184**].

10.3. Z_+-Propagation Time

The study of *PSD propagation time* or Z_+*-propagation time*, denoted by $\mathrm{pt}_+(G; B)$ and $\mathrm{pt}_+(G)$, was introduced by Warnberg in [**261**]. Figure 10.7 illustrates the PSD propagation process and sets for an efficient Z_+-forcing set for the tree T_1 in Figure 10.1.

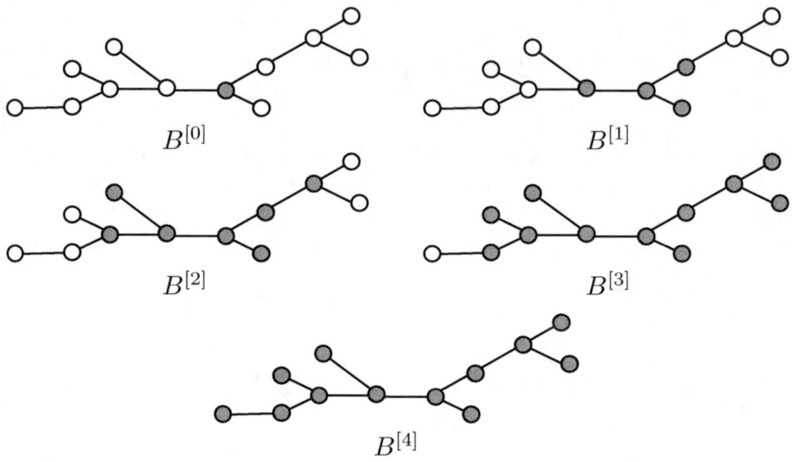

FIGURE 10.7. PSD propagation on the tree T_1

REMARK 10.27. Let G be a graph and $B \subseteq V(G)$. Since any Z-force is a valid Z_+-force, $\mathrm{pt}_+(G; B) \leq \mathrm{pt}(G; B)$. However, there is no relationship between $\mathrm{pt}(G)$ and $\mathrm{pt}_+(G)$ because $\mathrm{pt}_+(G)$ may use a smaller (minimum) forcing set.

OBSERVATION 10.28. *Observations 10.1, 10.2, and Remark 10.3 apply to* Z_+*-propagation time.*

The lower bound $\frac{|V(G)|-Z(G)}{Z(G)} \leq \mathrm{pt}(G)$ in Remark 10.7 is not valid for PSD propagation time due to the ability of one vertex to force many vertices. For example, $Z_+(K_{1,n-1}) = 1$ and $\mathrm{pt}_+(K_{1,n-1}) = 1$.

Families of Graphs. The values of $\mathrm{pt}_+(G)$ of many graphs are straightforward from the known minimum PSD zero forcing sets.

THEOREM 10.29. [**261**].

(1) *For* $n \geq 1$, $\mathrm{pt}_+(P_n) = \left\lceil \frac{n-1}{2} \right\rceil$.
(2) *For* $n \geq 3$, $\mathrm{pt}_+(C_n) = \left\lceil \frac{n-2}{4} \right\rceil$.
(3) *For* $n \geq 4$, $\mathrm{pt}_+(W_n) = \left\lceil \frac{n-3}{4} \right\rceil$.
(4) *For* $n \geq 2$, $\mathrm{pt}_+(K_n) = 1$.
(5) *For* $r \geq 2$ *and* $n_1 \geq \cdots \geq n_r \geq 1$, $\mathrm{pt}_+(K_{n_1,\dots,n_r}) = 1$.
(6) *For* $d \geq 1$, $\mathrm{pt}_+(Q_d) = 1$.
(7) *If* T *is a tree, then* $\mathrm{pt}_+(T) = \mathrm{rad}(T) = \left\lceil \frac{\mathrm{diam}(T)}{2} \right\rceil$.

PROPOSITION 10.30.

(1) *For* $k \geq 1$, $\mathrm{pt}_+(F_k) = 1$.
(2) *For* $k \geq 1$, $\mathrm{pt}_+(B_k) = 1$.

(3) $\text{pt}_+(K_{1,n-1} \square K_r) = 1$.
(4) *For* $n \geq 5$, $\text{pt}_+(\overline{P_n}) = 1$.
(5) *For* $n \geq 5$, $\text{pt}_+(\overline{C_n}) = 1$.

PROOF. (1) and (2) are immediate by choosing appropriate minimum PSD forcing sets (recall that $Z_+(F_k) = k + 1$ and $Z_+(B_k) = 2$ by Theorem 9.44).

(3): Recall that $Z_+(T \square K_r) = r$ (Example 9.56). Let c be the center of $K_{1,n-1}$, let $V(K_r) = \{x_1, \dots, x_r\}$, and let $B = \{(c, x_1), \dots, (c, x_r)\}$. Then $\text{pt}_+(K_{1,n-1} \square K_r; B) = 1$.

(4) and (5): Note that $Z_+(\overline{C_n}) = n - 3 = Z_+(\overline{P_n})$. Let $V = \{0, 1, \dots, n-1\}$ be the set of vertices of C_n in cycle order and perform arithmetic modulo n; for $\overline{P_n}$ the edge $\{0, n-1\}$ is added. Let $B = V \setminus \{1, 2, 3\}$ and note that $\text{pt}_+(\overline{C_n}; B) = 1 = \text{pt}_+(\overline{P_n}; B)$. □

Computation. The function `ptp(G)` $= \text{pt}_+(G)$ is available in [**184**] for computing the PSD propagation time of a specific graph (that is not too large). The value of $\text{pt}_+(G)$ is computed as $\text{pt}_+(G) =$ `ptpk(G, Zplus(G))` where `ptpk(G,k)` $= \text{pt}_+(G, k)$ is the minimum of $\text{pt}_+(G; B)$ over all sets B such that $|B| = k$. The function `pt_plus(G;B)` $= \text{pt}_+(G; B)$ is also available. The use of these functions is illustrated in Section 10.3 of [**184**].

Full Forcing. The concept of full forcing was introduced in [**261**] in the context of PSD forcing. Any graph G with $\text{pt}_+(G) = 1$ has full PSD forcing for every efficient PSD forcing set. A graph has *universal full PSD forcing* if every minimum PSD forcing set exhibits full forcing. Trees, cycles, and complete graphs have universal full PSD forcing. Not every graph has universal full PSD forcing or even full forcing for any minimum PSD forcing set.

EXAMPLE 10.31. Consider the wheel W_n with $n \geq 8$. Let u be the universal vertex. Note that $Z_+(W_n) = 3$ and $\delta(G) = 3$, so any minimum PSD forcing set must be a cut-set or contain a vertex of degree 3 and two of its neighbors. Let B be any minimum PSD forcing set B. If $u \in B$, then u has unfilled neighbors that are not forced in the first time step. If $u \not\in B$, then B must consist of three consecutive cycle vertices, and u is the only vertex forced in the first time step, so B does not have full PSD forcing. Thus no minimum PSD forcing set exhibits full forcing. Furthermore, every efficient PSD forcing set must contain u. Computation of $\text{pt}_+(W_8)$ is illustrated in Example PT.3-1 in [**184**].

Extreme Values. The next result summarizes some early results about extreme PSD propagation time.

PROPOSITION 10.32. [**261**] *Let G be a graph.*

(1) $\text{pt}_+(G) = |V(G)| - 1$ *if and only if* $G = P_2$.
(2) $\text{pt}_+(G) = |V(G)| - 2$ *if and only if G is one of* P_3, P_4, C_3, *or* $P_2 \sqcup P_1$.
(3) $\text{pt}_+(G) = 0$ *if and only if G has no edges.*
(4) *If G has an edge, then* $\text{pt}_+(G) = 1$ *if and only if there exists a minimum PSD forcing set for G that is a dominating set and exhibits full forcing. If G is a tree, then* $\text{pt}_+(G) = 1$ *if and only if* $G = K_{1,n-1}$.

It has recently been shown by a group from the AMS Mathematics Research Community *Finding Needles in Haystacks: Approaches to Inverse Problems using Combinatorics and Linear Algebra* that for fixed k, the number of graphs with PSD propagation time bounded below by $|V(G)| - k$ is finite, as a consequence of a stronger result limiting PSD propagation time to half the order.

THEOREM 10.33. [**186**] *Let G be a graph of order n. Then*

$$\mathrm{pt}_+(G) \leq \left\lceil \frac{n - \mathrm{Z}_+(G)}{2} \right\rceil \leq \left\lceil \frac{n-1}{2} \right\rceil$$

and these bounds are tight.

COROLLARY 10.34. [**186**] *Let k be a fixed positive integer. If $\mathrm{pt}_+(G) \geq |V(G)| - k$, then $n \leq 2k$ and thus the number of graphs G having $\mathrm{pt}_+(G) \geq |V(G)| - k$ is finite.*

Graph Operations. As with standard propagation time, there is no relation between $\mathrm{pt}_+(G)$ and $\mathrm{pt}_+(G-v)$ or $\mathrm{pt}_+(G-e)$. For the path P_n, deleting a vertex or an edge near the middle reduces PSD propagation time by about $\frac{n}{2}$. For a cycle C_n, deleting a vertex or edge increases PSD propagation time by about $\frac{n}{4}$. The next results establish the effect on PSD propagation time of some specific types of vertex deletion.

PROPOSITION 10.35. [**261**] *If v and w are adjacent twins in a connected graph G of order at least three, then $\mathrm{pt}_+(G) = \mathrm{pt}_+(G - v)$.*

COROLLARY 10.36. *Let H be a graph of order n with no isolated vertices and let $G = H \vee K_1$ where $V(K_1) = \{u\}$, and let k be such that $\mathrm{Z}_+(G) \leq k \leq n+1$. Then $\mathrm{pt}_+(G, k) = \mathrm{pt}_+(H, k-1)$. Furthermore, there is a PSD forcing set $B \subseteq V(G)$ that contains u such that B is optimal for k-PSD propagation time of G and $B \setminus \{u\}$ is optimal for $(k-1)$-PSD propagation time of H.*

The special case $\mathrm{pt}_+(G) = \mathrm{pt}_+(H)$ of Corollary 10.36 was established in [**261**]. Since $\mathrm{Z}_+(G_e) = \mathrm{Z}_+(G)$ (see Proposition 9.51), we are able to bound $\mathrm{pt}_+(G_e) - \mathrm{pt}_+(G)$.

PROPOSITION 10.37. *Let $e = \{u, w\}$ be an edge of a graph G. Then*

$$\mathrm{pt}_+(G) \leq \mathrm{pt}_+(G_e) \leq \mathrm{pt}_+(G) + 1$$

and these bounds are tight.

PROOF. Let v_e denote the vertex added to G in the creation of G_e.

$\mathrm{pt}_+(G_e) \leq \mathrm{pt}_+(G) + 1$: The proof that $\mathrm{Z}_+(G_e) = \mathrm{Z}_+(G)$ presented in [**122**] shows that in addition to keeping the same B, we can use almost the same propagating set of PSD forces. Specifically, choose an efficient PSD forcing set B with propagating set of PSD forces $\mathcal{F}$. Without loss of generality, either $u \to w$ or neither forces the other in $\mathcal{F}$. In G_e, fill the vertices in B. If $u \to w$ is in $\mathcal{F}$, let $\mathcal{F}'$ be obtained from $\mathcal{F}$ by replacing $u \to w$ by $u \to v_e$, $v_e \to w$; this can increase the propagation time by at most one. If $u \to w$ is not in $\mathcal{F}$, then $\mathcal{F}' = \mathcal{F}$; when u and w are both filled, either can force v_e, and again the propagation time can increase by at most one. Thus $\mathrm{pt}_+(G_e) \leq \mathrm{pt}_+(G_e; B') \leq \mathrm{pt}_+(G) + 1$.

$\mathrm{pt}_+(G) \leq \mathrm{pt}_+(G_e)$: Now let B be an efficient PSD forcing set for G_e with $u \in B$ and let $\mathcal{F}$ be a propagating set of PSD forces. If $v_e \in B$, then set $B' = B \setminus \{v_e\} \cup \{w\}$.

Then B' is a positive semidefinite zero forcing set for G, $v_e \to w$ at time step one, and every other force in $\mathcal{F}$ can be performed in G no later than in G_e. Otherwise $B' = B$ is a positive semidefinite zero forcing set for G, and every force in $\mathcal{F}$ except $u \to v_e$ (or $w \to v_e$) can be performed in G no later than in G_e. Thus $\mathrm{pt}_+(G) \leq \mathrm{pt}_+(G; B') \leq \mathrm{pt}_+(G_e)$.

Even paths and odd paths show that the lower and upper bounds are tight. □

Edge contraction can substantially decrease Z_+-propagation time, as seen in the next example.

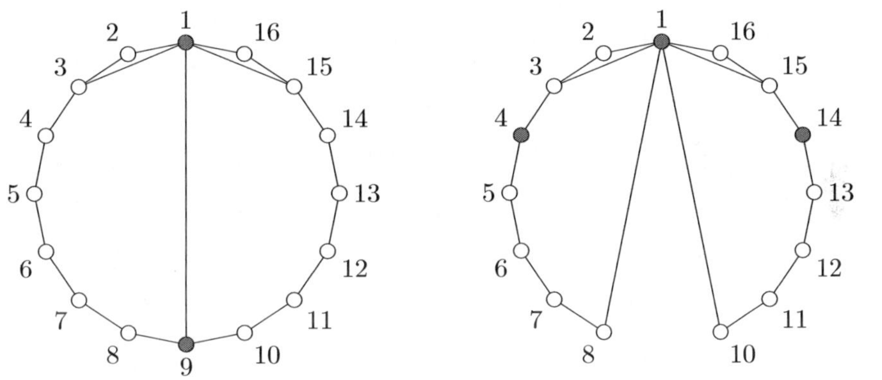

$k = 8$ $k = 8$

$\mathrm{Z}_+(L(k)) = 2$ & $\mathrm{pt}_+(L(k)) = k - 1$ $\mathrm{Z}_+(L(k)_e) = 3$ & $\mathrm{pt}_+(L(k)_e) = \lceil \frac{k-2}{3} \rceil$

FIGURE 10.8. Edge contraction can decrease Z_+-propagation time substantially

EXAMPLE 10.38. Let $L(k)$ be the graph of order $2k$ constructed from C_{2k} with vertices $(1, 2, \ldots, 2k)$ by adding edges $e = \{1, k+1\}$, $\{1, 3\}$, and $\{1, 2k-1\}$ (see Figure 10.8, which shows both $L(k)$ and $L(k)_e$). Then $\mathrm{Z}(L(k)) = 2$ and $\mathrm{pt}_+(L(k)) = k-1$, whereas $\mathrm{Z}(L(k)_e) = 3$ and $\mathrm{pt}_+(L(k)_e) = \lceil \frac{k-2}{3} \rceil$; efficient standard zero forcing sets are illustrated in Figure 10.5 and computations are provided in Example PT.3-2 in **[184]** for $k = 8$. So $\mathrm{pt}(L(k)) - \mathrm{pt}(L(k)_e) = k - 1 - \lceil \frac{k-2}{3} \rceil = \lfloor \frac{2k-1}{3} \rfloor$ and the difference between the propagation time of the original graph and a contraction can be arbitrarily large.

Maximum Propagation Time and the Propagation Interval. The next theorem lists maximum propagation time of families of graphs, all of which have full propagation time intervals.

THEOREM 10.39. **[261]** *Each of the following graphs has the maximum propagation time listed and has a full PSD propagation time interval.*

(1) If T is a tree, then $\mathrm{PT}_+(T) = \mathrm{diam}(T)$.
(2) $\mathrm{PT}_+(C_n) = \lceil \frac{n-2}{2} \rceil$.
(3) $\mathrm{PT}_+(W_n) = \lceil \frac{n-2}{2} \rceil$.
(4) $\mathrm{PT}_+(K_n) = 1$.

In contrast to the (standard) propagation interval, it was not known whether a PSD propagation time interval can have a gap. Numerous other graphs had been checked and determined to have a full propagation time interval, including all graphs of order at most ten [**261**]; Warnberg conjectured [**261**, Conjecture 3.15] that every PSD propagation time interval is full. In 2020, as part of the virtual preparation for the AMS Mathematics Research Community *Finding Needles in Haystacks: Approaches to Inverse Problems using Combinatorics and Linear Algebra*, Ben Small presented a counterexample to this conjecture, i.e., an example of a graph that does not have a full PSD propagation time interval [**256**].

Nordhaus-Gaddum Problems. The upper bound in the next result follows from Theorems 10.33 and 9.60.

PROPOSITION 10.40. [**186**] *For every graph G of order n,*

$$1 \leq \text{pt}_+(G) + \text{pt}_+(\overline{G}) \leq n + 2$$

and both bounds are tight.

10.4. Z$_-$-Propagation Time

The study of *skew propagation time* or *Z$_-$-propagation time*, denoted by $\text{pt}_-(G; B)$ and $\text{pt}_-(G)$, was introduced in was introduced by Kingsley in [**212**]. Figure 10.9 illustrates the skew propagation process and filled sets for the tree T_1 as in Figure 10.1. Note than many unfilled vertex forces are performed (by leaves).

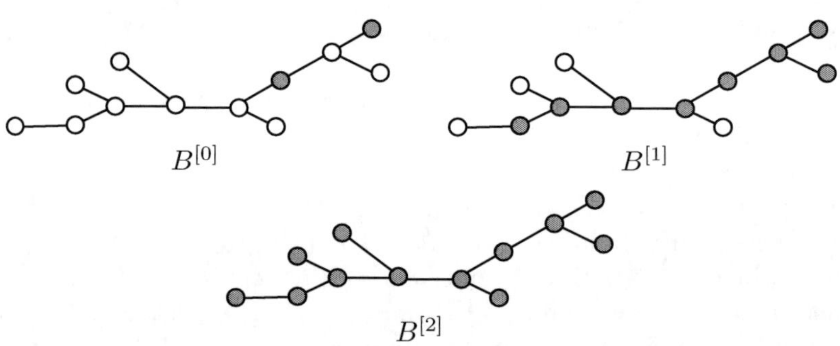

FIGURE 10.9. Skew propagation on the tree T_1

REMARK 10.41. Let G be a graph and $B \subseteq V(G)$. Since any Z-force is a valid Z$_-$-force, $\text{pt}_-(G; B) \leq \text{pt}(G; B)$. However, there is no relationship between $\text{pt}(G)$ and $\text{pt}_-(G)$ because $\text{pt}_-(G)$ may use a smaller (minimum) forcing set.

OBSERVATION 10.42. *Observations 10.1 and 10.2 apply to Z$_-$-propagation time.*

Most lower bounds for standard propagation time are not valid for skew propagation time due to the possibility of unfilled vertex forcing. The lower bound $\frac{|V(G)|-\text{Z}(G)}{\text{Z}(G)} \leq \text{pt}(G)$ in Remark 10.7 cannot apply because the denominator can be zero (and it is easy to find counterexamples with positive denominators). For the lower bound involving eccentricity in Remark 10.3, let P_{4k+1} be the path $1–2–\cdots–(4k)–(4k{+}1)$. Then $\text{Z}_-(P_{4k+1}) = 1$, $\text{pt}(P_{4k+1}, \{2k{+}1\}) = k$, and $\text{ecc}(\{2k{+}1\}) = 2k$. This is illustrated in Example PT.4-1 in [**184**].

Computation. The function `ptsk(G,k)` $= \mathrm{pt}_-(G,k)$, which is the minimum of $\mathrm{pt}_-(G;B)$ over all sets B such that $|B| = k$, and the function `pts(G)` $=$ `pts(G,Zskew(G))` $= \mathrm{pt}_-(G)$ are available in [**184**] for computing the skew propagation time of a specific graph (that is not too large). The skew propagation time of B can be computed by the function `prop_time_unlooped(G;B)` $= \mathrm{pt}_-(G;B)$. The use of these functions is illustrated in Section 10.4 of [**184**].

Families of Graphs. The next result lists the propagation times of some families of graphs. Many are immediate from the known minimum skew forcing sets, but here we point out a quirk. For n even, $\mathrm{Z}_-(P_n) = 0$, so forcing takes place from each end (two forces at each time step) and the skew propagation time is about half the order. For n odd, $\mathrm{Z}_-(P_n) = 1$ and by placing the one filled vertex near the middle (while ensuring that each of the two paths of unfilled vertices is even) allows four forces simultaneously, and the skew propagation time is about one quarter of the order. As is often the case, larger skew forcing number results in smaller skew propagation time. The same ideas apply to cycles.

PROPOSITION 10.43.

(1) [**212**] *For* $n \geq 2$, $\mathrm{pt}_-(K_n) = 1$.
(2) [**212**] *For* $r \geq 2$ *and* $n_1 \geq \cdots \geq n_r \geq 1$, $\mathrm{pt}_-(K_{n_1,\dots,n_r}) = 1$.
(3) [**212**] $\mathrm{pt}_-(P_n) = \frac{n}{2}$ *for even* n *and* $\mathrm{pt}_-(P_n) = \left\lfloor \frac{n+1}{4} \right\rfloor$ *for odd* n.
(4) [**212**] $\mathrm{pt}_-(C_n) = \frac{n-1}{2}$ *for odd* n *and* $\mathrm{pt}_-(C_n) = \left\lceil \frac{n-2}{4} \right\rceil$ *for even* n.
(5) [**212**] $\mathrm{pt}_-(W_n) = \frac{n-2}{2}$ *for even* n *and* $\mathrm{pt}_-(W_n) = \left\lceil \frac{n-3}{4} \right\rceil$ *for odd* n.
(6) [**212**] $\mathrm{pt}_-(Q_k) = 1$.
(7) [**212**] $\mathrm{pt}_-(\mathrm{Half}_s) = 2s - 1$.
(8) *For* $k \geq 1$, $\mathrm{pt}_-(F_k) = 1$.
(9) *For* $k \geq 1$, $\mathrm{pt}_-(B_k) = 1$.
(10) *For* $n \geq 4$, $\mathrm{pt}_-(\overline{P_n}) = 2$.
(11) *For* $n \geq 5$, $\mathrm{pt}_-(\overline{C_n}) = 2$.

PROOF. (8) is immediate since filling the center vertex means each unfilled K_2 can force itself in one step. (9) is immediate since filling the two spine vertices means each unfilled K_2 can force itself in one step.

(10) and (11): Assume $n \geq 5$ because $\overline{P_4} = P_4$ and $\mathrm{pt}_-(P_4) = 2$. Let $G = \overline{C_n}$ or $G = \overline{P_n}$, and note that G has an induced P_4. Since $\mathrm{Z}_-(G) = n - 4$, $\mathrm{pt}_-(G) \leq 2$ by choosing $V(G) \setminus V(P_4)$ as the skew zero forcing set.

Now suppose B is a skew zero forcing set of $\overline{C_4}$ with $|B| = n - 4$. Let $V = \{0, 1, \dots, n-1\}$ be the set of vertices of C_n in cycle order and perform arithmetic modulo n. In $\overline{C_n}$, vertex k is adjacent to every vertex except vertices $k-1$ and $k+1$. In order for vertex k to perform the first force with a set B of filled vertices having $|B| = n-4$, necessarily $B = V \setminus \{k-1, k, k+1, \ell\}$ where $\ell \notin \{k-1, k, k+1\}$. For such a B, a vertex different from k can perform a force in the first time step only if $\ell = k-2$ or $\ell = k+2$ (and $k+2 \neq k-2$ since $n \geq 5$). Thus at most two vertices can perform forces in the first time step, and so at least two unfilled vertices remain, so $\mathrm{pt}_-(\overline{C_n}) = 2$. For $\overline{P_n}$, use the same vertex set, the edge $\{0, n-1\}$ is added to the edges of $\overline{C_n}$, and the argument is unchanged except some vertices cannot force at time one. $\square$

Extreme Values. The next two results summarize some results about extreme propagation time, very low and very high.

PROPOSITION 10.44. [**212**] *Let G be a graph.*

(1) $\mathrm{pt}_-(G) = 0$ *if and only if G has no edges.*
(2) $\mathrm{Z}_-(G) = 0$ *and* $\mathrm{pt}_-(G) = 1$ *if and only if* $G = rK_2$.
(3) If $|V(G)| = 2\,\mathrm{Z}_-(G)$ *and G is a matching graph, then* $\mathrm{pt}_-(G) = 1$.

Condition (3) is analogous to one of the implications in Proposition 10.18, but the analogs of the other two are false [**212**].

PROPOSITION 10.45. *Let G be a connected graph. Then* $\mathrm{Z}_-(G) = 0$ *and* $\mathrm{pt}_-(G) = 2$ *if and only if* $G = H \circ K_1$ *for some connected graph H of order at least two.*

PROOF. If $G = H \circ K_1$ for some connected graph H of order at least two, then it is immediate that $\mathrm{Z}_-(G) = 0$ and $\mathrm{pt}_-(G) = 2$. If the order n of G is less than four then it is not possible to have $\mathrm{pt}_-(G) = 2$. So assume $n \geq 4$, $\mathrm{Z}_-(G) = 0$, and $\mathrm{pt}_-(G) = 2$. Every vertex forced in the first time step is the neighbor of a leaf. If there is any unfilled vertex that is not a leaf, it cannot be forced in the second time step because every filled vertex has a (unfilled) leaf neighbor. So G consists of leaves and their neighbors; denote the set of neighbors by N. If vertex in N is adjacent to more than one leaf, it cannot force the leaves adjacent to it, contradicting $\mathrm{Z}_-(G) = 0$. Thus $G = G[N] \circ K_1$. □

Recall that for $s \geq 2$, the sth half-graph Half_s is the graph of order $2s$ constructed from a copy of K_s on vertices $\{a_1, \ldots, a_s\}$ and a copy of $\overline{K_s}$ on vertices $\{b_1, \ldots, b_s\}$ by adding all edges $\{a_i, b_j\}$ such that $i + j \leq s + 1$. The (k,p)*-lollipop*, denoted by $L_{k,p}$, is the graph constructed by connecting a K_k to a P_p by a bridge from an endpoint of the P_p to a vertex of the K_k. The $(3, 2s-3)$-lollipop within Half_s is formed from the K_3 with vertices b_1, a_{s-1}, a_s and the path b_2–a_{s-2}–b_3– $\cdots$ –a_2–b_{s-1}–a_1–b_s by connecting b_2 and a_{s-1}. The graph Half_4 and its $(3,5)$-lollipop is shown in Figure 10.10.

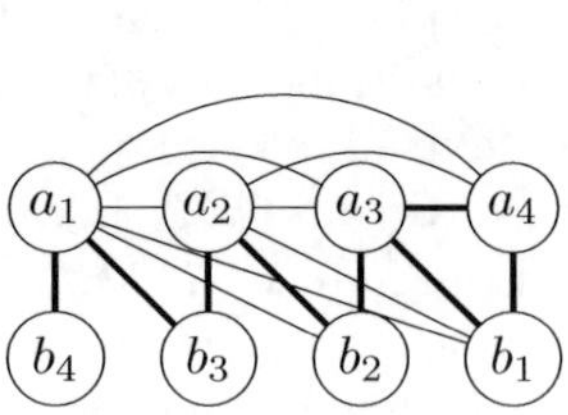

FIGURE 10.10. Half_4 and its $(3,5)$-lollipop, where the lollipop edges are thick

THEOREM 10.46. [**212**] *There is no graph for which* $\mathrm{pt}_-(G) = |V(G)|$. *If* $\mathrm{Z}_-(G) = 0$ *and* $\mathrm{pt}_-(G) = |V(G)| - 1$, *then G is a spanning subgraph of* Half_s *that contains the* $(3, 2s-3)$*-lollipop.*

Graphs with $\mathrm{Z}_-(G) = 0$. We continue an examination of skew propagation time when $\mathrm{Z}_-(G) = 0$.

OBSERVATION 10.47. *If* $\mathrm{Z}_-(G) = 0$ *then* $\emptyset$ *is the only minimum zero forcing set and* $\mathrm{pt}_-(G) = \mathrm{PT}_-(G)$.

Recall that $Z_-(G) = 0$ if and only if Algorithm 9.79 produces the empty set. The next algorithm builds on this to check whether $Z_-(G) = 0$ and when it does to compute $\text{pt}_-(G)$ by simultaneously stripping leaves. A different simultaneous leaf stripping algorithm appeared in [**212**, Algorithm 2.5.12], where it was used to give an upper bound on $\text{pt}_-(T)$ for a tree T.

ALGORITHM 10.48 (Simultaneous Leaf-Stripping Algorithm).
Input: *Graph G.*
Output: *$p(G)$ where $p(G) = \text{pt}_-(G)$ if $Z_-(G) = 0$ and $p(G) = -1$ if $Z_-(G) \neq 0$.*

$\tilde{G} := G$.
$k := 0$.
Repeat:

(1) $k := k + 1$.
(2) $L :=$ *the set of leaves of* $\tilde{G}$.
(3) $N :=$ *the set of neighbors of leaves in* $\tilde{G}$.
(4) **If** $\delta(\tilde{G}) = 0$:
  Return $p(G) = -1$.
(5) **If** $L = \emptyset$:
  Return $p(G) = -1$.
(6) **If** $|N| \neq |L|$:
  Return $p(G) = -1$.
(7) **If** $\Delta(\tilde{G}) = 1$:
  Return $p(G) = 2k - 1$.
(8) $\tilde{G} := \tilde{G} - (L \cup N)$.
(9) **If** $\tilde{G} = \emptyset$:
  Return $p(G) = 2k$.

Algorithm 10.48 is illustrated in the Figure 10.11, which shows the various iterations of $\tilde{G}$ produced, and proved in Theorem 10.49. The function `leafstripsimult(G)` in [**184**] implements it, and its use on the graph in Figure 10.11 is illustrated in Example PT.4-2.

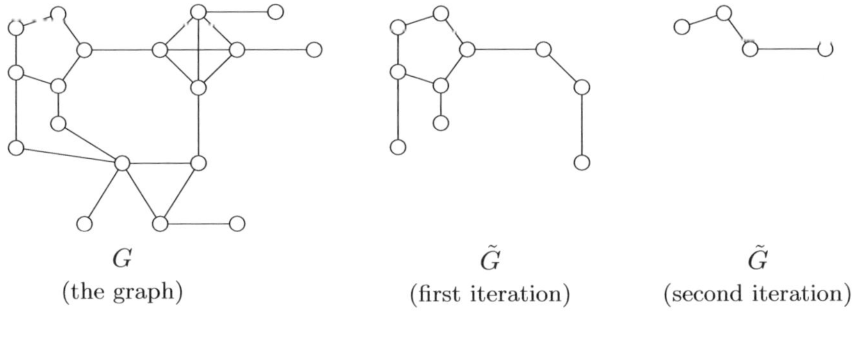

FIGURE 10.11. A graph G, $\tilde{G}$ after the first iteration of simultaneous leaf stripping, and $\tilde{G}$ after the second iteration of simultaneous leaf stripping ($\tilde{G} = \emptyset$ after the third iteration)

THEOREM 10.49. *Let G be a graph and let $p(G)$ be returned by Algorithm 10.48.*

(1) *If* $p(G) \geq 1$, *then* $\mathrm{Z}_-(G) = 0$ *and* $\mathrm{pt}_-(G) = p(G)$.
(2) *If* $p(G) = -1$, *then* $\mathrm{Z}_-(G) \neq 0$

Furthermore, when $\mathrm{Z}_-(G) = 0$ *and a propagation process is used for the empty set, every vertex forced at the last time step is a leaf.*

PROOF. The algorithm repeats until it is terminated by failure with $\mathrm{Z}_-(G) \neq 0$ or it succeeds with $\mathrm{Z}_-(G) = 0$. It fails when it finds an isolated vertex (4), the graph has no leaves (5), or the number of leaves is greater than the number of neighbors (6), indicating it is impossible to do a leaf-neighbor perfect match. If no failure happens then Algorithm would return $\tilde{G} = \emptyset$ and $\mathrm{Z}_-(G) = 0$. It remains to show that in this case the skew propagation time is returned.

So assume $\mathrm{Z}_-(G) = 0$, the algorithm has been run and $p(G) \geq 1$ has been returned. Let $k(G)$ be the value of the iteration counter after applying the algorithm to G. We show by induction that the empty set skew forces G in exactly $p(G)$ time steps and every vertex forced in the last time step of propagation of the empty set is a leaf of G. The proof is by induction on the number of iterations $k(G)$. The base cases $k = 1$ and $k = 2$ are covered by Propositions 10.44(2) and 10.45. So assume $\mathrm{pt}_-(H) = p(H)$ and every vertex forced in the last time step of propagation of the empty set is a leaf for every graph H for which $k(H) = k$. Suppose $k(G) = k + 1$ iterations. Let L demote the set of leaves of G and N the set of neighbors of leaves (called leaf-neighbors). In the first time step of skew propagation every leaf forces its neighbor. Now forcing can proceed on $H = G - (L \cup N)$. Observe that $k(H) = k(G) - 1 = k$ so $\mathrm{pt}_-(H) = p(H)$ and every vertex forced in the last time step of propagation of the empty set is a leaf of H. Then $\mathrm{pt}_-(G) \leq p(G)$ because the leaves of G can be forced in one step after H is filled, so $\mathrm{pt}_-(G) \leq \mathrm{pt}_-(H)+2 = p(H) + 2 = p(G)$. Let u be a vertex of H forced in the last time step when forcing H, so u is a leaf of H. Since u is not a leaf of G and not a leaf-neighbor of G, u can't be adjacent to a leaf of G, so u must be adjacent to some vertex $v \in N$. Then v cannot force the leaf of G it is adjacent to until after H is filled. Thus $\mathrm{pt}_-(G) \geq \mathrm{pt}_-(H) + 2 = p(G)$. And every vertex forced in the last time step is a leaf of G, because N is forced first and H is forced before the last time step. □

Note that step (6), checking whether it is impossible to do a leaf-neighbor perfect match is unnecessary in the original leaf-stripping algorithm, where a single leaf-neighbor is deleted at each step, but necessary here since we delete all leaves at once).

Graph Operations. As is the case for standard propagation time and PSD propagation time, there is no constant bound on the difference between $\mathrm{pt}_-(G)$ and $\mathrm{pt}_-(G-v)$, $\mathrm{pt}_-(G-e)$, or $\mathrm{pt}_-(G_e)$. The path P_n provides examples: With n even, deleting a vertex or edge at the end or subdividing an edge reduces propagation time by about $\frac{n}{4}$. With n odd, deleting a vertex or edge at the end or subdividing an edge increases propagation time by about $\frac{n}{4}$. However, independent twins and universal vertices are sometimes more manageable. The next result is illustrated in Example PT.4-3 in [**184**].

PROPOSITION 10.50. *Suppose* v *and* w *are independent twins in* G *and* $\mathrm{Z}_-(G - v) = 0$. *Then* $\mathrm{pt}_-(G) = \mathrm{pt}_-(G - v)$.

PROOF. By Proposition 9.87, $\mathrm{Z}_-(G) = 1$ since $\mathrm{Z}_-(G - v) = 0$, and one of v or w is in any skew forcing set. Thus the only minimum skew forcing sets are

$\{v\}$ and $\{w\}$. Suppose $B = \{v\}$, and note that $\text{pt}_-(G) = \text{pt}_-(G; B)$. Then $\text{pt}_-(G; B) \leq \text{pt}_-(G - v)$ using the same set of skew forces for the empty set in $G - v$, i.e., v does not perform a force. Now consider an efficient set of forces for B. If v does not perform a force, then the same set of forces works for $G - v$, so $\text{pt}_-(G - v) \leq \text{pt}_-(G; B)$. Suppose $v \to u$ at time i. Thus $N(v) \setminus \{u\} \subset B^{[i-1]}$. Since $N(v) \setminus \{u\} = N(w) \setminus \{u\}$, at time i $w \to u$ in $G - v$, and so $\text{pt}_-(G - v) \leq \text{pt}_-(G; B)$. Thus $\text{pt}_-(G) = \text{pt}_-(G - v)$. □

COROLLARY 10.51. *Let H be a graph of order n with no isolated vertices and let $G = H \vee K_1$ where $V(K_1) = \{u\}$, and let k be such that $\text{Z}_-(G) \leq k \leq n + 1$. Then $\text{pt}_-(G, k) = \text{pt}_-(H, k - 1)$. Furthermore, there is a skew forcing set $B \subseteq V(G)$ that contains u such that B is optimal for k-skew propagation time of G and $B \setminus \{u\}$ is optimal for $(k - 1)$-skew propagation time of H.*

Nordhaus-Gaddum Problems. The Nordhaus-Gaddum sum problem for skew propagation time is to establish tight lower and upper bounds on $\text{pt}_-(G) + \text{pt}_-(\overline{G})$, where $\overline{G}$ is the complement of G. The Nordhaus-Gaddum sum lower bound is 1, realized by $G = K_n$.

Unlike the case of standard propagation time and PSD propagation time, the upper bound can be close to twice the order.

PROPOSITION 10.52. *For $s \geq 2$,*

$$\text{pt}_-(\text{Half}_s) + \text{pt}_-(\overline{\text{Half}_s}) = 4s - 4 = 2|V(\text{Half}_s)| - 4.$$

PROOF. By Theorem 10.46, $\text{pt}_-(\text{Half}_s) = 2s - 1$. From Proposition 9.103 with $s \geq 2$, the complement $\overline{\text{Half}_s}$ of the s-half graph can be constructed from Half_{s-1} by adding an independent twin of the non-clique vertex that is adjacent to every clique vertex and taking the disjoint union with a K_1, and $\text{Z}_-(\overline{\text{Half}_s}) = 2$. Clearly the isolated vertex has no effect on skew propagation time. By Proposition 10.50, $\text{pt}_-(\overline{\text{Half}_s}) = \text{pt}_-(\text{Half}_{s-1}) = 2s - 3$. □

Since $\text{pt}_-(G) \leq n - 1$ for every graph G of order n, $\text{pt}_-(G) + \text{pt}_-(\overline{G}) \leq 2n - 2$, but this is unlikely to be realized and we expect that the half graph produces the largest Nordhaus-Gaddum sum.

Propagation Interval. The skew propagation time interval need not be full, as seen in the next example.

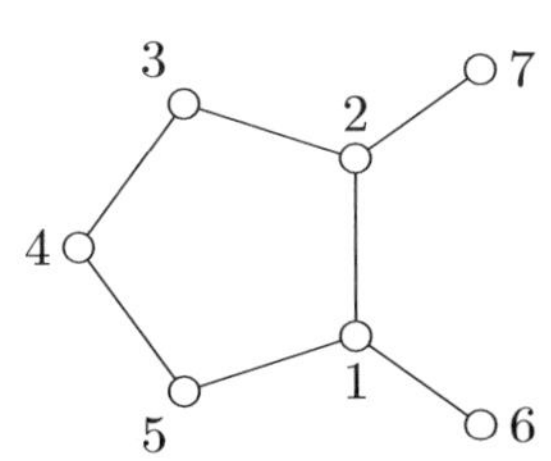

FIGURE 10.12. A graph H that does not have a full skew propagation time interval

EXAMPLE 10.53. [**212**] Let H be the graph shown in Figure 10.12. Since $\emptyset$ is not a zero forcing set and $\{3\}$ is, $\mathrm{Z}_{-}(H) = 1$. Observe that $\mathrm{pt}(H; \{3\}) = \mathrm{pt}(H; \{5\}) = 2$ and $\mathrm{pt}(H; \{6\}) = \mathrm{pt}(H; \{7\}) = 4$, whereas none of $\{1\}$, $\{2\}$ or $\{5\}$ is a zero forcing set. These values are computed in Example PT.4-4 in [**184**].

10.5. Propagation Time for Power Domination

Although the universal definition of propagation time for zero forcing does not immediately apply to power domination (because there is an initial domination step), it is easily adapted to power domination. Starting with $S \subseteq V(G)$, we define two sequences of sets, the set $S^{(i)}$ of vertices that become observed at time step i and the set $S^{[i]}$ of vertices that have been observed after time step i. Thus $S^{[0]} = S^{(0)} = S$ is the set of PMU sites, $S^{[1]} = N[S]$ is the closed neighborhood of S, $S^{(1)} = N[S] \setminus S$, and $S^{[i+1]} = S^{[i]} \cup S^{(i+1)}$ for $i \geq 1$. Assume $S^{(i)}$ and $S^{[i]}$ have been constructed and $S^{(i)} \neq \emptyset$. Then

$$S^{(i+1)} = \{w : \exists v,\ v \in S^{[i]} \text{ and } N(v) \cap \overline{S^{[i]}} = \{w\}\},$$

i.e., w is the only unfilled neighbor of the filled vertex v.

The *power propagation time* or *γ_P-propagation time* of $S \subseteq V(G)$, denoted by $\mathrm{pt}_{\gamma_P}(G; S)$, is the least t such that $S^{[t]} = V(G)$ (or infinity if S is not a power dominating set of G). The *power propagation time of G* or *γ_P-propagation time of G*, also known as the *minimum γ_P-propagation time*, is

$$\mathrm{pt}_{\gamma_P}(G) = \min\{\mathrm{pt}_{\gamma_P}(G; S) : S \text{ is a minimum power dominating set}\}.$$

The notation $\mathrm{ppt}(G)$ is sometimes used for $\mathrm{pt}_{\gamma_P}(G)$ in the literature. Define

$$\mathrm{pt}_{\gamma_P}(G, k) = \min_{|S|=k} \mathrm{pt}_{\gamma_P}(G; S).$$

Then $\mathrm{pt}_{\gamma_P}(G) = \mathrm{pt}_{\gamma_P}(G, \gamma_P(G))$.

The study of power propagation time was introduced in [**136**]. Figure 10.13 illustrates the power propagation process and observed sets (shown in gray) for the tree T_1 in Figure 10.1.

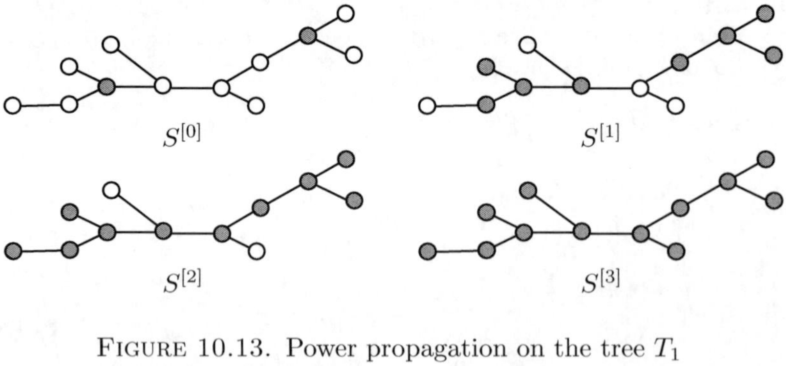

FIGURE 10.13. Power propagation on the tree T_1

A subset S of vertices of G is an *efficient power dominating set* for G if S is a minimum power dominating of G and $\mathrm{pt}_{\gamma_P}(G; S) = \mathrm{pt}_{\gamma_P}(G)$. The set S in Figure 10.13 is an efficient power dominating set for the tree T_1 (see [**184**, Example PT.5-1]). When studying the propagation time of a power dominating set S, a *propagating set of forces*), in which all vertices of $S^{[i-1]}$ are observed before any vertex in $S^{(i)}$ is observed. That is, a propagating set of forces for S is one that carries out the propagation process of constructing $S^{(i)}$ for $i = 0, \dots, \mathrm{pt}_{\gamma_P}(G; S)$ in order.

The definition presented here is the one used in [**54, 136**], and under this definition $\mathrm{pt}_{\gamma_P}(\overline{K_n}) = 0$. Another perspective is that the domination step always takes place, so $\mathrm{pt}_{\gamma_P}(G) \geq 1$ for all graphs G.

Many of the observations for propagation times of variants of zero forcing remain true for power propagation time.

OBSERVATION 10.54. *For a graph G of order n, $\mathrm{pt}_{\gamma_P}(G; S) \leq n - |S|$ because at least one observation must be performed at each time step. This implies*

$$\mathrm{pt}_{\gamma_P}(G) \leq n - \gamma_P(G).$$

Furthermore, $\mathrm{pt}_{\gamma_P}(G; S) \leq n - |N[S]| + 1$.

Two power dominating sets S_1 and S_2 of a graph G are *isomorphic* if there is a graph automorphism $\psi(G)$ of G such that $\psi(S_1) = S_2$.

OBSERVATION 10.55. *If B_1 and B_2 are isomorphic power dominating sets, then*

$$\mathrm{pt}_{\gamma_P}(G; S_1) = \mathrm{pt}_{\gamma_P}(G; S_2).$$

However, non-isomorphic power dominating sets may also have the same power propagation time.

OBSERVATION 10.56. *If G is a graph and $S \subset V(G)$, then*

$$\mathrm{ecc}(S) \leq \mathrm{pt}_{\gamma_P}(G; S).$$

It was claimed in [**219**] that $\mathrm{pt}_{\gamma_P}(G) \leq \mathrm{diam}(G)$ (where $\mathrm{pt}_{\gamma_P}(G)$ was described as the number of rounds needed to complete the propagation process). However that is not correct and a counterexample H_Δ with $\mathrm{pt}_{\gamma_P}(H_\Delta) > \mathrm{diam}(H_\Delta)$ is presented in [**136**]. Diameter is an upper bound for power propagation time for trees.

THEOREM 10.57. *If T is a tree with at least 3 vertices, then*

$$\mathrm{pt}_{\gamma_P}(T) \leq \mathrm{diam}(T) - 1$$

and T has a minimum power dominating set S with $\mathrm{pt}_{\gamma_P}(T; S) = \mathrm{pt}_{\gamma_P}(T)$ and $\deg_T u \geq 2$ for all $u \in S$.

The upper bound on power propagation time in Theorem 10.57 is tight for a star. It was observed by Haynes et al. in [**173**] that if $\Delta(G) \geq 3$, then G has a minimum power dominating set in which every vertex has degree at least three. By Theorem 10.57, every tree must have an efficient power dominating set in which every vertex has degree at least two. However, it is not true that there must be an efficient power dominating set in which every vertex has degree at least three.

EXAMPLE 10.58. Let T be the tree shown in Figure 10.14. Note that $\gamma_P(T) = 2$ and $\mathrm{pt}_{\gamma_P}(T) = 2$ using the set of two filled degree-2 vertices shown on the right. In each diagram, the number by a vertex shows the time at which it becomes observed. Note that the power propagation time using the two degree-3 vertices is three.

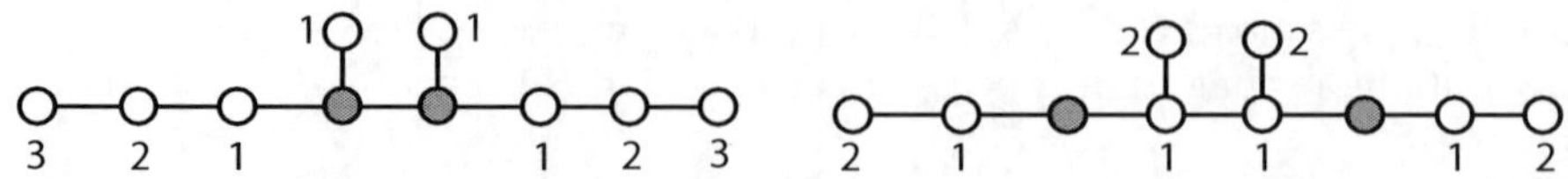

FIGURE 10.14. The tree T and power propagation times for two choices of a minimum power dominating set.

Computation. The function `ppt(G)` $= \mathrm{pt}_{\gamma_P}(G)$ is available in [**184**] for computing the power propagation time of a specific graph (that is not too large). The value of $\mathrm{pt}_{\gamma_P}(G)$ is computed by `ppt(G)` = `pptk(G,Z(G))` where `pptk(G,k)` $= \mathrm{pt}_{\gamma_P}(G,k)$ is the minimum of $\mathrm{pt}_{\gamma_P}(G;S)$ over all sets S such that $|B| = k$. The function `pptS(G,S)` $= \mathrm{pt}_{\gamma_P}(G;S)$ is also available. The use of these functions is illustrated in Section 10.5 of [**184**].

Bozeman made a study of power propagation time in [**53, 54**]. We summarize some of her results here.

Families of Graphs.

PROPOSITION 10.59. [**54**]

(1) *For* $n \geq 1$, $\gamma_P(K_n) = 1$ *and* $\mathrm{pt}_{\gamma_P}(K_n) = 1$.
(2) *For* $n \geq 1$, $\gamma_P(P_n) = 1$ *and* $\mathrm{pt}_{\gamma_P}(P_n) = \lfloor \frac{n}{2} \rfloor$.
(3) *For* $n \geq 3$, $\gamma_P(C_n) = 1$ *and* $\mathrm{pt}_{\gamma_P}(C_n) = \lfloor \frac{n}{2} \rfloor$.
(4) *For* $s, t \geq 3$, $\gamma_P(K_{s,t}) = 2$ *and* $\mathrm{pt}_{\gamma_P}(K_{s,t}) = 1$.
For $s \geq 2$, $\gamma_P(K_{s,2}) = 1$ *and* $\mathrm{pt}_{\gamma_P}(K_{s,2}) = 2$.
For $s \geq 1$, $\gamma_P(K_{s,1}) = 1$ *and* $\mathrm{pt}_{\gamma_P}(K_{s,1}) = 1$.

Graph Operations. As is the case for other types of propagation time, there is no constant bound on the difference between $\mathrm{pt}_{\gamma_P}(G)$ and $\mathrm{pt}_{\gamma_P}(G-v)$, $\mathrm{pt}_{\gamma_P}(G-e)$, $\mathrm{pt}_{\gamma_P}(G_e)$, or $\mathrm{pt}_{\gamma_P}(G/e)$. Examples: For the path P_n, deleting a vertex or an edge near the middle reduces propagation time by about $\frac{n}{2}$. Deleting a universal vertex can increase power propagation time from 1 to $\lfloor \frac{n-1}{2} \rfloor$ when G is the graph obtained from P_{n-1} by adding a universal vertex. For examples related to edge subdivision and contraction, see [**54**].

Extreme Values.

OBSERVATION 10.60. *A graph G has* $\mathrm{pt}_{\gamma_P}(G) = 0$ *if and only if* $G = \overline{K_n}$ *and* $\mathrm{pt}_{\gamma_P}(G) = 1$ *if and only if* $\gamma_P(G) = \gamma(G)$.

THEOREM 10.61. [**54**] *Let G be a graph of order n.*

- $\mathrm{pt}_{\gamma_P}(G) = n-1$ *if and only if* $G \in \{K_1, K_2\}$.
- $\mathrm{pt}_{\gamma_P}(G) = n-2$ *if and only if* $G \in \{K_1 \sqcup K_1, K_1 \sqcup K_2, P_3, P_4, C_3, C_4\}$.

A characterization of graphs G of order n with $\mathrm{pt}_{\gamma_P}(G) = n-3$ where G is a tree or $\gamma_P(G) = 2$ or $\gamma_P(G) = 3$ is given in [**54**]. As part of the characterization, infinite families of graphs having $\mathrm{pt}_{\gamma_P}(G) = n-3$ are presented, including the spider $S(1,1,n-3)$.

THEOREM 10.62. [**53**] *Let G be a connected graph on n vertices that is not planar. Then* $\mathrm{pt}_{\gamma_P}(G) \leq n-4$ *and this bound is tight. Furthermore, if G is connected and* $\mathrm{pt}_{\gamma_P}(G) = n-4$, *then* $\gamma_P(G) = 1$.

Nordhaus-Gaddum Problems. The Nordhaus-Gaddum sum problem for a power propagation time is to establish tight lower and upper bounds on $\mathrm{pt}_{\gamma_P}(G) + \mathrm{pt}_{\gamma_P}(\overline{G})$, where $\overline{G}$ is the complement of G. The Nordhaus-Gaddum sum lower bound is 1, realized by $G = K_n$; if both G and $\overline{G}$ are required to have edges, then the Nordhaus-Gaddum sum lower bound is 2, realized by $G = K_{n,n}$ and $n \geq 3$.

PROPOSITION 10.63. [**54**] *For every graph G of order n,*

$$\mathrm{pt}_{\gamma_P}(G) + \mathrm{pt}_{\gamma_P}(\overline{G}) \leq n + 2.$$

Bozeman did not find any graphs realizing $\mathrm{pt}_{\gamma_P}(G)+\mathrm{pt}_{\gamma_P}(\overline{G}) = n+2$ (or $n+1$). She found a family of graphs realizing $\mathrm{pt}_{\gamma_P}(G) + \mathrm{pt}_{\gamma_P}(\overline{G}) = n$, conjectured that $\mathrm{pt}_{\gamma_P}(G) + \mathrm{pt}_{\gamma_P}(\overline{G}) \leq n$ for all graphs of order n, and established the conjecture when the graph G satisfies an additional condition, such as G has a leaf, $g(G) \geq 5$, or $\gamma_P(G) + \gamma_P(\overline{G}) \geq 4$ [**54**].

10.6. Capture Time for Cops and Robbers

The *k-capture time* $\mathrm{capt}_k(G)$ is the minimum number of rounds needed for $k \geq c(G)$ cops to capture the robber on G over all possible games (when both the cops and the robber play optimally); $\mathrm{capt}_k(G) = \infty$ for $k < c(G)$. The *capture time* is the $c(G)$-capture time and is denoted by $\mathrm{capt}(G)$. The study of capture time was introduced in [**47**]. The study of k-capture time with $k > c(G)$ was introduced in [**49**]. Figure 10.15 illustrates the capture process for the tree T_1 in Figure 10.1. In this example (and for trees in general) the robber does not move; however, that is not true in general.

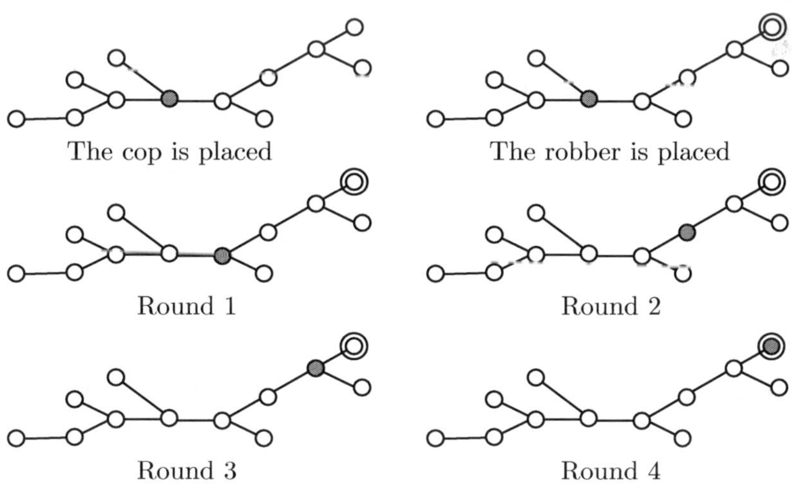

FIGURE 10.15. Capturing the robber on the tree T_1. The vertex occupied by the cop is gray and the vertex occupied by the robber is circled.

OBSERVATION 10.64. *If G has a universal vertex, then* $\mathrm{capt}(G) = 1$.

OBSERVATION 10.65. *Observation 10.2 and Remark 10.3 apply to capture time. In particular,* $\mathrm{rad}_{c(G)}(G) \leq \mathrm{capt}(G)$ *and* $\mathrm{rad}_k(G) \leq \mathrm{capt}_k(G)$.

Observation 10.1, i.e., $\mathrm{pt}_X(G) \le n - X(G)$, is not true for capture time when the cop number is greater than one.

THEOREM 10.66. *Let G be a graph of order n. Then* $\mathrm{capt}(G) \le O(n^{k+1})$ *and for fixed $k \ge 2$ there are examples of graphs of arbitrarily large order n with $c(G) = k$ and* $\mathrm{capt}(G) \ge \left(\frac{n}{40k^4}\right)^{k+1}$.

When a graph is cop-win, i.e., $c(G) = 1$, the situation is more like propagation time.

THEOREM 10.67.

(1) *Let G be a cop-win graph of order n.*
  - [**213**] *If $n \ge 7$, then* $\mathrm{capt}(G) \le n - 4$.
  - [**47**] *If $n \ge 5$, then* $\mathrm{capt}(G) \le n - 3$.
  - *If $n = 3, 4$, then* $\mathrm{capt}(G) \le n - 2$.
  - *If $n = 1, 2$, then* $\mathrm{capt}(G) \le n - 1$.

(2) [**47**] *For $n \ge 7$ there is a family of cop-win planar graphs $H(n)$ of order n having* $\mathrm{capt}(H(n)) = n - 4$ *($H(10)$ is shown in Figure 10.16).*

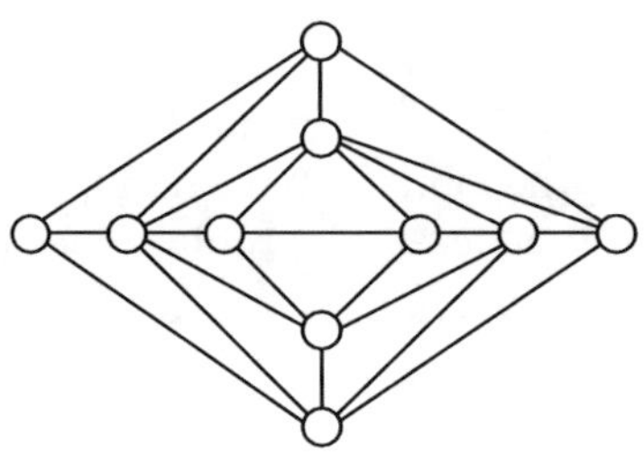

FIGURE 10.16. The graph $H(10)$.

Breen et al. established a relationship between capture time and PSD propagation time.

THEOREM 10.68. [**57**] *If $S \subseteq V(G)$ is a PSD zero forcing set, then S is a capture set,* $\mathrm{capt}(G; S) \le \mathrm{pt}_+(G; S)$, *and* $c(G) \le \mathrm{Z}_+(G)$. *Furthermore, if G is a tree or C_n with $n \ge 4$, then* $\mathrm{capt}(G; S) = \mathrm{pt}_+(G; S)$ *for $S \subseteq V(G)$ and* $\mathrm{capt}(G) = \mathrm{pt}_+(G)$.

Families of Graphs. Values of $\mathrm{capt}(G)$ in the next result for which no source is listed are straightforward.

THEOREM 10.69.

(1) [**49**] *If T is a tree, then* $\mathrm{pt}_+(T) = \mathrm{rad}(T) = \left\lceil \frac{\mathrm{diam}(T)}{2} \right\rceil$.
(2) [**49**] *For $n \ge 1$,* $\mathrm{capt}(P_n) = \left\lceil \frac{n-1}{2} \right\rceil$.
(3) [**57**] *For $n \ge 3$,* $\mathrm{capt}(C_n) = \left\lceil \frac{n-2}{4} \right\rceil$.
(4) *For $n \ge 4$,* $\mathrm{capt}(W_n) = 1$.
(5) *For $n \ge 2$,* $\mathrm{capt}(K_n) = 1$.
(6) *For $k \ge 1$,* $\mathrm{capt}(F_k) = 1$.

The next result is implicit in the proof in [**246**] that $c(G(\mathbb{P}_q)) = q + 1$.

PROPOSITION 10.70. *Let $G(\mathbb{P}_q)$ be the incidence graph of the projective plane $\mathbb{P}_q$. Then* $\operatorname{capt}(G(\mathbb{P}_q)) = 2$.

PROOF. In $G(\mathbb{P}_q)$, let $S = \{p_1, \dots, p_{q+1}\}$ be the set of all the points on line ℓ_1. We show that $\operatorname{capt}(G; S) = 2$. Since any two points are on exactly one line, the only line containing both p_i and p_j is ℓ_1 (for $i \neq j$). For $i = 1, \dots, q+1$, p_i is on q lines that are not ℓ_1. Together this accounts for $(q+1)q + 1 = q^2 + q + 1$ lines. Thus every line contains at least one point in S. If the robber is placed on a vertex of $G(\mathbb{P}_q)$ that is a line, then the robber is caught in Round 1. So assume the robber is placed on a point p_s that is not on line ℓ_1. Each pair of points p_s, p_i determines a line ℓ_{k_i} for $i = 1, \dots, q+1$. In round 1 the cop on p_i moves to line ℓ_{k_i}. Then every neighbor of the robber is occupied by a cop so the robber is caught in round 2. Since $\gamma(G(\mathbb{P}_q)) > q + 1$, and $c(G(\mathbb{P}_q)) = q + 1$ (see Example 9.162), this establishes $\operatorname{capt}(G(\mathbb{P}_q)) = 2$. □

10.7. Expected Propagation Time for Probabilistic Zero Forcing

Kang and Yi [**210**] introduced a variant of zero forcing called *probabilistic zero forcing* that is defined as follows: In one *round*, each filled vertex u attempts to *force* (fill) each of its unfilled neighbors w independently with probability

$$\mathbf{Pr}(u \to w) = \frac{|N[u] \cap B|}{\deg u},$$

where B denotes the set of filled vertices. This rule is known as the *probabilistic color change rule*, and *probabilistic zero forcing* refers to the process of filling every vertex of a graph by repeated application of the probabilistic color change rule. The study of probabilistic zero forcing therefore produces a discrete dynamical system that can be used to describes applications such as modeling the spread of a rumor through a social network, the spread of an infectious disease in a population, or the dissemination of a computer virus in a network.

For traditional zero forcing the minimum number of vertices required to fill the entire graph is a parameter of primary interest. However, in probabilistic zero forcing one filled vertex per connected component is necessary and sufficient to eventually fill the entire graph. Therefore finding a minimum probabilistic zero forcing set is not an interesting problem. However, expected propagation time, introduced by Geneson and Hogben in [**153**], is a natural parameter to study. Expected propagation time has also been studied in subsequent work, including [**81**] and [**124**].

Suppose that G is a connected graph with the vertices in $B \neq \emptyset$ filled and all other vertices unfilled. The *probabilistic propagation time* of B, denoted by $\operatorname{pt}_{pzf}(G; B)$, is defined as the random variable equal to the number of the round in which the last vertex is filled when applying the probabilistic color change rule [**153**]. For a connected graph G and a set $B \subseteq V(G)$ of vertices, the *expected propagation time of B* is the expected value of the probabilistic propagation time of B [**153**], i.e.,

$$\operatorname{ept}(G; B) = \mathbf{E}[\operatorname{pt}_{pzf}(G, B)].$$

The *expected propagation time* of a connected graph G is the minimum of the expected propagation time of B over all one-vertex sets B of G [**153**], i.e.,

$$\operatorname{ept}(G) = \min\{\operatorname{ept}(G, \{v\}) : v \in V(G)\}.$$

The use of Markov chains for probabilistic zero forcing was introduced in [**210**] and studied further in [**153**] and [**81**]. Let G be a graph and $B \subset V(G)$ be nonempty. A *simple state for* B is a set of filled vertices that it is possible to reach by starting with exactly the vertices in B filled, and then applying the probabilistic color change rule iteratively. It is often convenient to combine simple states that behave analogously into one *composite state*. An *ordered state list for* B, denoted by $\mathfrak{S} = (S_1, \ldots, S_s)$, is an ordered list of all states (simple or composite) for B in which S_1 is the initial state (where exactly the vertices in B are filled), S_s is the final state (where all vertices are filled), and the states $S_k, k = 2, \ldots, s-1$ are in some chosen order. A graph G and an ordered state list $\mathfrak{S}$ determine the Markov transition matrix for the process, which is denoted by $M(G, \mathfrak{S})$. Reordering the states $S_2, \ldots, S_{s-1}$ results in a Markov transition matrix that is obtained from $M(G, \mathfrak{S})$ by a permutation similarity. If $\mathfrak{S}$ is an ordered state list with s states, then the probability that all vertices are filled after round r is the $(1, s)$-entry of M^r, i.e., $(M^r)_{1s}$. Thus the probability that the propagation time is r is $\left(M^r - M^{r-1}\right)_{1s}$, and the expected propagation time is

$$\operatorname{ept}(G; B) = \sum_{r=1}^{\infty} r \left(M^r - M^{r-1}\right)_{1s}$$

[**153**]. The next result gives an exact formula for computing $\operatorname{ept}(G; B)$.

THEOREM 10.71. [**81**] *Suppose that G is a graph, $B \subset V(G)$ is nonempty, $\mathfrak{S}$ is an ordered state list for B with s states, and $M = M(G, \mathfrak{S})$. Then*

$$\operatorname{ept}(G, B) = ((M - \mathbb{1}\mathbf{e}_s{}^\top - I)^{-1})_{1s} + 1,$$

where $\mathbb{1} = [1, \ldots, 1]^\top$ and $\mathbf{e}_s = [0, \ldots, 0, 1]^\top$.

Expected propagation times have been determined exactly for a few families of graphs and asymptotically for some others.

THEOREM 10.72. [**153**] *For a path of order $n > 2$,*

$$\operatorname{ept}(P_n) = \begin{cases} \frac{n}{2} + \frac{2}{3} & \text{if } n \text{ is even} \\ \frac{n}{2} + \frac{1}{2} & \text{if } n \text{ is odd} \end{cases}.$$

For a cycle of order $n > 2$,

$$\operatorname{ept}(C_n) = \begin{cases} \frac{n}{2} + \frac{1}{3} & \text{if } n \text{ is even} \\ \frac{n}{2} + \frac{1}{2} & \text{if } n \text{ is odd} \end{cases}.$$

THEOREM 10.73. [**153**] *For a positive integer n, $\operatorname{ept}(K_{1,n}) = \Theta(\log n)$.*

THEOREM 10.74. [**81**] *For a positive integer n, $\operatorname{ept}(K_n) = \Theta(\log\log(n))$.*

The expected propagation time for random graphs depends on the edge probability of the random graph and is determined in [**124**]. The effect of graph operations on expected propagation time is discussed in [**81**]. Additional parameters related to probabilistic zero forcing are also discussed in [**153**].

10.8. Topics Not Covered

We have not covered propagation time for minor monotone floors of zero forcing parameters. Minor monotone floors involve forcing by 'hopping' and thus do not have unique final colorings. This necessitates a more technical definition of propagation time, as given in [**77**], where information on propagation time for minor monotone floors of zero forcing parameters can be found.

CHAPTER 11

Throttling

Throttling addresses the question of minimizing the sum (or in some cases, the product) of the resources used to accomplish a task (number of filled vertices, PMUs, or cops) and the time needed to complete that task (propagation time or capture time). Butler and Young [**74**] introduced the study of this question for standard zero forcing, Carlson et al. [**78**] studied throttling for PSD forcing, Breen et al. [**57**] studied throttling for Cops and Robbers, and Carlson [**77**] introduced the universal definition used here and studied throttling for standard zero forcing and minor monotone floors. As with propagation time discussed in Chapter 10, in this chapter we omit minor monotone floors, allowing simpler universal definitions of propagation time, but we also consider related parameters such Cops and Robbers and power domination. We also discuss product throttling of some parameters, which minimizes the product of the resources used to accomplish a task and the time needed to complete that task.

Throttling is a newer field than forcing and is a very active research area. Although the first throttling paper [**74**] appeared in 2013, there was not much additional activity until 2017, when the question of PSD throttling was raised at the AIM zero forcing workshop [**131**]. This led to the paper [**78**], which in turn sparked additional work, including [**46, 57, 77, 79, 92**].

11.1. Universal Definitions for Forcing Throttling

Suppose that X is a process that defines an X-set and for each X-set B defines a propagation time $\mathrm{pt}_X(G;B)$; X could be one of Z, Z_+, Z_-, γ_P, or c (with $\mathrm{pt}_c((G;B)$ interpreted as $\mathrm{capt}(G;B)$). The *X-throttling number* of B in G is

$$\mathrm{th}_X(G;B) = |B| + \mathrm{pt}_X(G;B)$$

with the convention that $\mathrm{pt}_X(G;B) = \infty$ if B is not an X-set. The *X-throttling number* of G is

$$\mathrm{th}_X(G) = \min_{B\subseteq V(G)}\{\mathrm{th}_X(G;B)\}.$$

An X-set B is an *optimal X-throttling set* for G if $\mathrm{th}_X(G;B) = \mathrm{th}_X(G)$.

In Figure 11.1 we illustrate the Z_+-propagation process and sets $B^{[i]}$ of filled vertices for a tree T_1 and optimal initial set B. Notice that the set B that realizes the X-throttling number is not necessarily a minimum X-forcing set (but can be). Examples where a larger set B is used and where a minimum set B is used are shown in Figures 11.1 and 11.2, respectively.

Define

$$\mathrm{th}_X(G,k) = k + \mathrm{pt}_X(G,k),$$

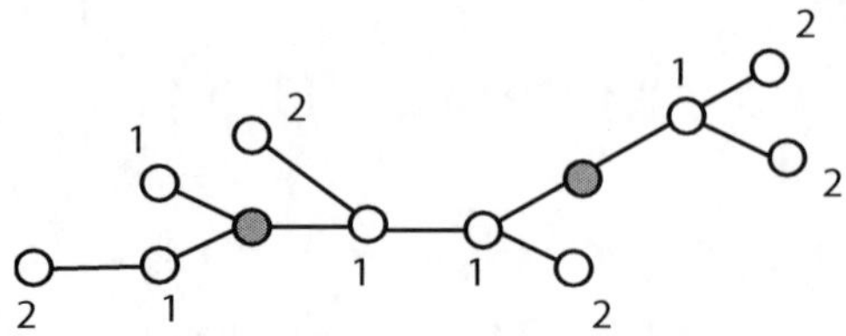

FIGURE 11.1. PSD throttling on the tree T_1, showing the time each vertex is filled using a PSD zero forcing set B of 2 vertices as shown: $\mathrm{pt}_+(T_1; B) = 2$ and $\mathrm{th}_+(T_1) = 2 + 2 = 4$. In this case the throttling number is not realized by a minimum PSD zero forcing set, since $\mathrm{Z}_+(T_1) = 1$ and $\mathrm{pt}_+(T_1) = 4$.

the minimum throttling that can be achieved with k vertices. With this definition,

$$\mathrm{th}_X(G) = \min_{X(G) \leq k \leq n} \mathrm{th}_X(G, k)$$

for a graph G of order n.

Here we list various bounds that apply to two or more of the throttling numbers discussed in subsequent sections (each section lists which of these remarks apply there).

OBSERVATION 11.1. *Let G be a graph of order n and let X be one of a color change rule or power domination or capture process. Then $\mathrm{th}_X(G) \leq n$ because $\mathrm{th}_X(G, n) = n$. If $X(G) = n$, then $\mathrm{th}_X(G) = n$ and a minimum X-forcing set is optimal. If $X(G) < n$, then $X(G) + 1 \leq \mathrm{th}_X(G)$ because $|B| < n$ implies $\mathrm{pt}_X(G; B) \geq 1$.*

PROPOSITION 11.2. *Let G be a graph of order n and let X be one of a color change rule or the power domination process. Then:*

(1) $k + \mathrm{pt}_X(G, k) \leq n$ for $k = X(G), \ldots, n - 1$.
(2) If $X(G) < n$, then $\mathrm{th}_X(G) = \min_{X(G) \leq k \leq X(G) + \mathrm{pt}_X(G) - 1} \mathrm{th}_X(G, k)$, i.e., we need consider only $k \leq X(G) + \mathrm{pt}_X(G) - 1$.

PROOF. (1): This is immediate because at least one vertex must be filled at each time step.

(2): Note $X(G) \leq X(G) + \mathrm{pt}_X(G) - 1$ since $X(G) < n$. If $X(G) + \mathrm{pt}_X(G) = n$, then $\mathrm{th}_X(G, X(G) + \mathrm{pt}_X(G) - 1) \leq n = \mathrm{th}_X(G, X(G) + \mathrm{pt}_X(G))$. So assume $X(G) + \mathrm{pt}_X(G) < n$. Then $\mathrm{pt}_X(G, X(G) + \mathrm{pt}_X(G)) \geq 1$ and $\mathrm{th}_X(G, X(G) + \mathrm{pt}_X(G)) \geq X(G) + \mathrm{pt}_X(G) + 1 > \mathrm{th}_X(G)$, so a set of size $X(G) + \mathrm{pt}_X(G)$ is not optimal. □

PROPOSITION 11.3. *Let G be a graph of order n and let X be one of a color change rule or the power domination or capture process. Then $\mathrm{th}_X(G) = X(G) + \mathrm{pt}_X(G)$ if $\mathrm{pt}_X(G) \leq 2$.*

PROOF. This is immediate if $\mathrm{pt}_X(G) \leq 1$, so suppose $\mathrm{pt}_X(G) = 2$. When Proposition 11.2 applies, $\mathrm{th}_X(G, X(G)) = X(G) + 2 \leq n$. Then for B with $X(G) < |B| < n$, we have $\mathrm{pt}_X(G; B) \geq 1$ and thus $\mathrm{th}_X(G; B) \geq X(G) + 2$. To complete the proof, it suffices to show that $\mathrm{capt}(G) = 2$ implies $\mathrm{th}_c(G) \leq n$. This is immediate if $c(G) \le n-2$. Furthermore, $c(G)=n$ implies $G=nK_1$ and $\mathrm{capt}(G)=0$ and $c(G)=n-1$ implies $G=K_2 \sqcup (n-2)K_1$ and $\mathrm{capt}(G)=1$. □

REMARK 11.4. Let X be one of a color change rule or power domination or capture process. Suppose G is a graph of order n and G' is an induced subgraph of G of order n'. Then

$$\mathrm{th}_X(G) \le n - n' + \mathrm{th}_X(G'),$$

by choosing $B = B' \cup (V(G)\setminus V(G'))$ where $\mathrm{th}_X(G';B') = \mathrm{th}_X(G')$. If we propagate only within G', the propagation time is $\mathrm{pt}_X(G';B')$, so $\mathrm{pt}_X(G;B) \le \mathrm{pt}_X(G';B')$ and $\mathrm{th}_X(G;B) \le n - n' + \mathrm{th}_X(G')$.

REMARK 11.5. Let G be a graph of order n and let X be a color change rule that requires adjacency and does not allow unfilled vertex forcing (this includes the power domination capture processes). Then from Remark 10.3, $\min_{X(G)\le k\le n}(k + \mathrm{rad}_k(G)) \le \mathrm{th}_X(G)$.

REMARK 11.6. Suppose $G = \sqcup_{i=1}^r G_i$ where the G_i are the connected components of G, and let X be a color change rule that requires adjacency (this includes the power domination process). It is immediate that $\mathrm{th}_X(G) \le \sum_{i=1}^r \mathrm{th}_X(G_i)$, but this is not tight if at least two components have edges. If $\mathrm{th}_X(G_i) = \mathrm{th}_X(G_i;k_i)$ and $\mathrm{pt}_X(G,k_i) = p_i$ for $i = 1,\dots,r$, then

$$\mathrm{th}_X(G) \le \sum_{i=1}^r k_i + \max_{i=1}^r p_i.$$

This bound can be tight, but it can also be a strict inequality, because when taking a disjoint union, it can be advantageous to use fewer vertices (since these sum) and a longer propagation time (because this takes the maximum rather than summing) than is optimal within a single component. See Remark 11.12 for more discussion. The next bound is weaker but is easier to use.

$$\mathrm{th}_X(G) \le \sum_{i=1}^r X(G_i) + \max_{i=1}^r \mathrm{pt}_X(G_i).$$

Given the difficulty of computing a throttling number from its parts, some results focus on connected graphs.

Computation. In [**61**] Brimkov, Carlson, Hicks, Patel, and Smith established a powerful theorem for showing that throttling a propagation process is NP-complete.

THEOREM 11.7. [**61**] *Let X be a color change rule or the power domination process. Suppose*

(1) For any graph G, there exist constants b, c such that for any X-forcing set, $\mathrm{pt}_X(G;S) < b = O(|V(G)|^c)$, and $\mathrm{pt}_X(G;S)$ and b can be computed in $O(|V(G)|^c)$ time.
(2) For any graphs G_1 and G_2, the X-forcing sets of $G_1 \sqcup G_2$ are of the form $\{S_1 \sqcup S_2\}$ where S_i is an X-forcing set of G_i, $i = 1,2$.
(3) For any graphs G_1 and G_2, and for any X-forcing sets S_i of $G_i, i = 1,2$, $\mathrm{pt}_X(G_1 \sqcup G_2; S_1 \sqcup S_2) = \max\{\mathrm{pt}_X(G_1;S_1), \mathrm{pt}_X(G_2;S_2)\}$.
(4) The problem of determining whether $X(G) < k$ is NP-Complete.

Then the problem of determining whether $\text{th}_X(G) < k$ *is NP-Complete.*

Applications of this theorem to particular X-throttling numbers are listed in the relevant sections.

Graph Operations. The effect of some graph operations on throttling number is the same for some of the zero forcing parameters and power domination, so the proofs is presented together.

PROPOSITION 11.8. *Let* X *be one of the graph parameters* Z, Z_+, Z_-, *or* γ_P, *with* th_X *one of* $\text{th}, \text{th}_+, \text{th}_-$, *or* th_{γ_P}. *Let* G *be a graph,* $v \in V(G)$, *and* $e = \{u, w\} \in E(G)$. *Then:*

(1) $\text{th}_X(G) - 1 \leq \text{th}_X(G - v)$.
(2) $\text{th}_X(G) - 1 \leq \text{th}_X(G/e)$.
(3) $\text{th}_X(G_e) \leq \text{th}_X(G) + 1$.

PROOF. $\text{th}_X(G) - 1 \leq \text{th}_X(G - v)$: This follows from Remark 11.4.

$\text{th}_X(G) - 1 \leq \text{th}_X(G/e)$: Let v_e denote the new vertex formed by identifying u and w in the contraction of e. Choose a set $B \subseteq V(G/e)$ such that $\text{th}_X(G/e) = \text{th}_X(G/e; B)$ and construct a propagating set of X-forces $\mathcal{F}$. If $v_e \in B$, then let $B' = B \setminus \{v_e\} \cup \{u, w\}$. If $v_e \notin B$, let $x \to v_e \in \mathcal{F}$; without loss of generality, assume x is a neighbor of w in G and define $B' = B \cup \{u\}$. In either case, $\text{pt}_X(G; B') \leq \text{pt}_X(G/e; B)$. Thus $\text{th}_X(G) \leq \text{th}_X(G/e) + 1$.

$\text{th}_X(G_e) \leq \text{th}_X(G) + 1$: Since $G = (G_e)/\{u, v_e\}$ (where v_e is the new vertex adjacent to u and w in G_e), $\text{th}_X(G_e) - 1 \leq \text{th}_X(G)$. □

PROPOSITION 11.9. *Let* X *be one of of the graph parameters* Z, Z_+, *or* γ_P, *with* th_X *one of* th, th_+, *or* th_{γ_P}. *Let* G *be a graph and* $e = \{u, w\} \in E(G)$. *Then:*

(1) $\text{th}_X(G) - 1 \leq \text{th}_X(G - e) \leq \text{th}_X(G) + 1$.
(2) $\text{th}_X(G) \leq \text{th}_X(G_e)$.

PROOF. $\text{th}_X(G) \leq \text{th}_X(G - e) + 1$: Choose a set $B \subseteq V(G - e)$ such that $\text{th}_X(G - e) = \text{th}_X(G - e; B)$ and construct a propagating set of X-forces $\mathcal{F}$; without loss of generality, w is filled before or at the same time as u. If w is filled before u, then $B' = B \cup \{u\}$; otherwise let $B' = B$. In either case, $\text{pt}_X(G; B') \leq \text{pt}_X(G - e; B)$. Thus $\text{th}_X(G) \leq \text{th}_X(G; B') \leq \text{th}_X(G - e) + 1$.

$\text{th}_X(G - e) \leq \text{th}_X(G) + 1$: Choose a set $B \subseteq V(G)$ such that $\text{th}_X(G) = \text{th}_X(G; B)$ and construct a propagating set of X-forces $\mathcal{F}$. If a force $w \to u$ (or $u \to w$) appears in $\mathcal{F}$, then $B' = B \cup \{u\}$ (or $B' = B \cup \{w\}$); otherwise let $B' = B$. In any case, $\text{pt}_X(G - e; B') \leq \text{pt}_X(G; B)$ and $|B'| \leq |B| + 1$. Thus $\text{th}_X(G - e) \leq \text{th}_X(G - e; B') \leq \text{th}_X(G) + 1$.

$\text{th}_X(G) \leq \text{th}_X(G_e)$: Let v_e denote the new vertex adjacent to u and w in edge subdivision. Choose a set $B \subseteq V(G_e)$ such that $\text{th}_X(G_e) = \text{th}_X(G_e; B)$ and construct a propagating set of X-forces $\mathcal{F}$.

Suppose first that v_e performs a force; without loss of generality $v_e \to w \in \mathcal{F}$. If $v_e \in B$, then let $B' = B \setminus \{v_e\} \cup \{w\}$. If $v_e \notin B$, then $u \to v_e \in \mathcal{F}$; define $B' = B$. In all cases, $\text{pt}_X(G; B') \leq \text{pt}_X(G_e; B)$ and $\text{th}_X(G) \leq \text{th}_X(G_e)$.

Finally, consider the case where v_e does not perform a force; without loss of generality, w is filled before or at the same time as u. If $v_e \in B$, define $B' = B \setminus \{v_e\} \cup \{u\}$. If $v_e \notin B$, define $B' = B$. In all cases, $\text{pt}_X(G; B') \leq \text{pt}_X(G_e; B)$ and $\text{th}_X(G) \leq \text{th}_X(G_e)$. □

As with zero forcing, a universal vertex plays a special role. Due to the similarity across variants of zero forcing, we state and prove a version for several variants, which is similar to the proofs for zero forcing but accounts for propagation time.

The next result is immediate from Propositions 9.16, 9.48, 9.88, and 10.4.

PROPOSITION 11.10. *Let X be one of the Z-color rule, the Z_+-color rule, or the Z_--color rule. Let H be a graph with no isolated vertices and let $G = H \vee K_1$ where $V(K_1) = \{u\}$. Then $\operatorname{th}_X(G) = \operatorname{th}_X(H) + 1$, there is an optimal X-throttling set $B \subseteq V(G)$ such that $u \in B$.*

PROPOSITION 11.11. *Let X be one of of the graph parameters Z, Z_+, or γ_P, with th_X one of $\operatorname{th}, \operatorname{th}_+$, or $\operatorname{th}_{\gamma_P}$. Let G and G' be graphs of order n and n', respectively. For $k \geq X(G)$,*

$$\operatorname{th}_X(G \,\square\, G') \leq n'k + \operatorname{pt}_X(G, k).$$

PROOF. Choose a set $S = \{x_1, \ldots, x_k\} \subseteq V(G)$ such that $\operatorname{pt}_X(G; S) = \operatorname{pt}_X(G, k)$. Define $B = \{(x_i, v') : v' \in V(G'), i = 1, \ldots, k\}$. Then B is a forcing set for $G \,\square\, G'$, and $\operatorname{pt}_X(G \,\square\, G'; B) \leq \operatorname{pt}_X(G; S)$. Thus $\operatorname{th}_X(G \,\square\, G'; B) \leq n'k + \operatorname{pt}_X(G, k)$. □

11.2. Z-throttling

Butler and Young introduced throttling in [**74**]. The *standard throttling number* or *Z-throttling number* of B in G is $\operatorname{th}(G; B) = |B| + \operatorname{pt}(G; B)$. The *standard k-throttling number* is $\operatorname{th}(G, k) = k + \operatorname{pt}(G, k)$. The *throttling number*, *standard throttling number*, or *Z-throttling number* of a graph G of order n is

$$\operatorname{th}(G) = \min_{B \subseteq V(G)} \operatorname{th}(G; B) = \min_{Z(G) \leq k \leq n} \operatorname{th}(G, k).$$

In the next figure we illustrate an optimal Z-throttling set for the tree T_1 shown in Figure 10.1 and other examples. That no lower value is possible is verified in [**184**, Example TH.2-1].

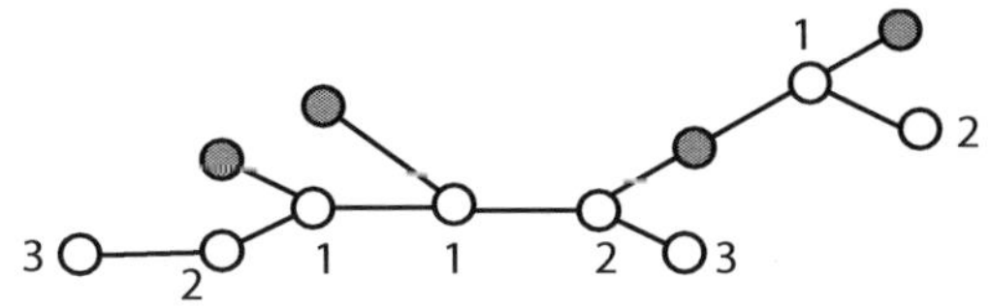

FIGURE 11.2. Throttling on a tree T_1, showing the time each vertex is filled. For T_1, the standard throttling number is realized by a minimum standard zero forcing set: $Z(T_1) = 4, \operatorname{pt}(T_1) = 3$, and $\operatorname{th}(T_1) = 7$.

REMARK 11.12. For a graph G of order n, $Z(G) + 1 \leq \operatorname{th}(G) \leq n$ (Observation 11.1). Propositions 11.2 and 11.3 and Remarks 11.4, 11.5, and 11.6 also apply to Z-throttling. The bound $\operatorname{th}_X(G) \leq \sum_{i=1}^r k_i + \max_{i=1}^r p_i$ in Remark 11.6 can be tight; for example $\operatorname{th}(P_n \sqcup P_2) = \operatorname{th}(P_n) + 1$ for $n \geq 2$. But it can also be a strict inequality. For example, $\operatorname{th}(3P_4) = 6$ using one vertex in each path and propagation time three, whereas $\operatorname{th}(P_4) = \operatorname{th}(P_4, 2) = 3$ and $\operatorname{pt}_Z(P_4, 2) = 1$ so $\sum_{i=1}^3 k_i + \max_{i=1}^3 p_i = 6 + 1 = 7$.

REMARK 11.13. It is well known that any vertex in a connected graph of order at least two can be excluded from a minimum standard zero forcing set (see Theorem 9.10). However, this is not true for an optimal standard throttling set, as can be seen by considering the wheel W_7 (and for some arbitrarily large n). Recall $\mathrm{Z}(W_n) = 3$ and let B_1 consist of the universal vertex and 2 consecutive cycle vertices. Then $\mathrm{pt}(W_7; B_1) = 2$, so $\mathrm{th}(W_7) = 5$ by Proposition 11.3. If the universal vertex u is not in a standard zero forcing set B_2 of size three, then B_2 consists of three consecutive vertices, $\mathrm{pt}(W_7; B_2) = 3$, and $\mathrm{th}(W_7; B_2) = 6$. This is illustrated in Example TH.2-2 in [**184**]. A nonempty intersection of optimal standard throttling sets is related to the fact that it may not be possible to exclude a vertex an efficient zero forcing set – see Proposition 10.8 and the discussion of wheels immediately following.

Butler and Young obtained a fundamental lower bound on throttling number, which appears in the next theorem, and showed this bound is realized by a path.

THEOREM 11.14. [**74**] *Let G be a graph of order n. Then* $\mathrm{th}(G) \geq \lceil 2\sqrt{n} - 1 \rceil$ *and this bound is tight.*

PROOF. For a zero forcing set B, let $b = |B|$ and $t = \mathrm{pt}_+(G; B)$. Recall that $\frac{n-b}{b} \leq t$ by Remark 10.7. Thus we want to minimize the value of $b + \left(\frac{n}{b} - 1\right)$. If we allow b and t to be nonnegative real numbers, the minimum value of $b + t$ is $2\sqrt{n} - 1$, achieved by using $b = \sqrt{n}$ and $t = \sqrt{n} - 1$. Since $2\sqrt{n} - 1$ is the minimum value of $b + t$ for $0 \leq b, t \in \mathbb{R}$ and $\frac{n-b}{b} \leq t$, adding the constraint that b and t are integers with $t \geq 0$ and $b \geq 1$ shows $b + t \geq \lceil 2\sqrt{n} - 1 \rceil$. □

For a positive integer k, Carlson [**77**] gave a characterization of graphs G having $\mathrm{th}(G) \leq k$.

THEOREM 11.15. [**77**] *Given a positive integer k,* $\mathrm{th}(G) \leq k$ *if and only if there exist integers $a \geq 1$ and $b \geq 0$ such that $a + b = k$ and G can be obtained from $K_a \,\square\, P_b$ by deleting edges in a K_a and contracting edges in a P_b.*

Computation. It is known that computing the throttling number number is NP-complete [**61**].

The function `th(G)` is available in [**184**] for computing the standard throttling number of a specific graph (that is not too large) by minimizing $k + \mathrm{pt}(G, k)$ over $\mathrm{Z}(G) \leq k \leq \min\{\mathrm{Z}(G) + \mathrm{pt}(G) - 1, n\}$ (see Proposition 11.2). The function `thk(G)` returns both $\mathrm{th}(G)$ and the lowest k such that $\mathrm{th}(G) = \mathrm{th}(G, k)$. The use of these functions is illustrated in Section 11.2 of [**184**].

Graph Operations. In the proof of the next result, we omit proofs of bounds established in Propositions 11.8 and 11.9 (but examples showing tightness are presented).

PROPOSITION 11.16. *Let G be a graph, $v \in V(G)$, and $e = \{u, w\} \in E(G)$.*

(1) $\mathrm{th}(G) - 1 \leq \mathrm{th}(G - v) \leq \mathrm{th}(G) + 1$. *If v is a twin, then* $\mathrm{th}(G - v) \leq \mathrm{th}(G)$.
(2) $\mathrm{th}(G) - 1 \leq \mathrm{th}(G - e) \leq \mathrm{th}(G) + 1$.
(3) $\mathrm{th}(G) - 1 \leq \mathrm{th}(G/e) \leq \mathrm{th}(G) + 1$.
(4) $\mathrm{th}(G) \leq \mathrm{th}(G_e) \leq \mathrm{th}(G) + 1$.

All bounds are tight.

PROOF. $\text{th}(G) - 1 \leq \text{th}(G - v)$ is tight: Deleting a leaf from the star $K_{1,n-1}$ with $n \geq 4$ shows that the lower bound for vertex deletion is tight, including for twins (see Theorem 11.23(6), the proof of which does not use this result).

$\text{th}(G - v) \leq \text{th}(G) + 1$ and $\text{th}(G - v) \leq \text{th}(G)$ when v is a twin: Choose a set $B \subseteq V(G)$ such that $\text{th}(G) = \text{th}(G; B)$ and construct a propagating set of Z-forces $\mathcal{F}$. Note that v is involved in at most one force. If a force $v \to u$ appears in $\mathcal{F}$ for some vertex u, then let $B' = B \cup \{u\}$; otherwise let $B' = B$. In either case, $\text{pt}(G - v; B') \leq \text{pt}(G; B)$ and $|B'| \leq |B| + 1$. Thus $\text{th}(G - v) \leq \text{th}(G - v; B') \leq \text{th}(G) + 1$.

Recall that $Z(K_3 \square P_3) = 3$ (Theorem 9.5(13)). It is straightforward to verify that $\text{pt}(K_3 \square P_3) = 2$ (see Example TH.2-3 in [**184**]), and thus $\text{th}(K_3 \square P_3) = 5$ by Proposition 11.3. Let v be a vertex in the middle copy of K_3 (see Figure 11.3). Then $Z(K_3 \square P_3 - v) = 2$. However, $\text{pt}(K_3 \square P_3 - v, 2) = 4$, $\text{pt}(K_3 \square P_3 - v, 3) = 3$, and $\text{pt}(K_3 \square P_3 - v, 4) = 2$, so $\text{th}(K_3 \square P_3 - v) = 6 = \text{th}(K_3 \square P_3) + 1$. Thus the upper bound for vertex deletion is tight.

FIGURE 11.3. A graph G and vertex v such that $\text{th}(G) = 5$ and $\text{th}(G - v) = 6$.

Now assume v has a twin w. At least one of v or w must be in B (Proposition 9.15), so $B' = B \setminus \{v\} \cup \{w\}$ is a standard zero forcing set for $G - v$, $|B'| \leq |B|$, and $\text{pt}(G - v; B') \leq \text{pt}(G; B)$. Thus $\text{th}(G - v) \leq \text{th}(G)$. It is easy to verify that $\text{th}(F_2) = 4$ for the friendship graph F_2. For a vertex v such that $\deg(v) = 2$, $\text{th}(F_2 - v) = 4 = \text{th}(F_2)$ (see Example TH.2-4 in [**184**]). So the upper bound for vertex deletion of a twin is tight.

$\text{th}(G) - 1 \leq \text{th}(G - e)$ is tight: Let G be the graph constructed from C_4 by adding an edge e between two vertices not adjacent in the cycle. Then $\text{th}(G) = 4$ and $\text{th}(G - e) = \text{th}(C_4) = 3 = \text{th}(G) - 1$ (see Example TH.2-5 in [**184**]). Thus the lower bound for edge deletion is tight.

$\text{th}(G - e) \leq \text{th}(G) + 1$ is tight: Let e be an edge of F_3 between two degree-2 vertices. Then $\text{th}(F_3) = 5$ and $\text{th}(F_3 - e) = 6$ (see Example TH.2-6 in [**184**]), so the upper bound for edge deletion is tight.

$\text{th}(G) \leq \text{th}(G/e) + 1$ and $\text{th}(G_e) \leq \text{th}(G) + 1$ are tight: Choose n such that $\text{th}(P_n) = \text{th}(P_{n-1}) + 1$. Since $P_{n-1} = P_n/e$ and $P_n = (P_{n-1})_e$, the bounds are tight.

$\text{th}(G/e) \leq \text{th}(G) + 1$: Choose a set $B \subseteq V(G)$ such that $\text{th}(G) = \text{th}(G; B)$ and construct a propagating set of Z-forces $\mathcal{F}$. If at least one of u or w is in B, then define $B' = B \setminus \{u, w\} \cup \{v_e\}$; otherwise $B' = B$. Thus $|B'| \leq |B|$. If neither u nor w performs a force, then $B'' = B'$. Now suppose at least one of u, w performs a force. Assume without loss of generality that $u \to y$ before (or at the same time) w performs a force. Define $B'' = B' \cup \{y\}$ (note only one force can be performed by u). Then $|B''| \leq |B| + 1$ and $\text{pt}(G/e; B'') \leq \text{pt}(G; B)$, so $\text{th}(G/e) \leq \text{th}(G) + 1$. Let $G = K_3 \square P_3$ and let e be an edge within the middle copy of K_3. Then $\text{th}(G) = 5$

and $\mathrm{th}(G/e) = 6$ [**77**] (see also Example TH.2-7 in [**184**]), so the upper bound for edge contraction is tight.

$\mathrm{th}(G) \leq \mathrm{th}(G_e)$ is tight: Choosing n such that $\mathrm{th}(P_n) = \mathrm{th}(P_{n-1})$ shows the bound is tight. □

THEOREM 11.17. *Let H be a graph and let G be constructed from H by adding a universal vertex u. Then $\mathrm{th}(G) = \mathrm{th}(H)+1$ and there is an optimal throttling set $B \subseteq V(G)$ such that $u \in B$ where u is the universal vertex added in constructing G from H.*

PROOF. All cases are covered by Proposition 11.10 except showing that $\mathrm{th}_X(H) \leq \mathrm{th}_X(G) - 1$ when H has one or more isolated vertices. Let $B \subseteq V(G)$ such that $\mathrm{th}(G) = \mathrm{th}(G; B)$ and let $\mathcal{F}$ be a propagating set of Z-forces of B.

If $u \notin B$, then the argument the proof of in Proposition 11.10 shows we can swap to have $u \in B$ unless an isolated vertex $x \in B$ and $x \to u$. In this case, $B^{(1)} = \{u\}$. Let $B' = B \cup \{u\}$, so $|B'| = |B| + 1$, $\mathrm{pt}(G; B') = \mathrm{pt}(G; B) - 1$, and $\mathrm{th}(G; B') = \mathrm{th}(G; B)$. Thus we assume $u \in B$.

As in the proof of Proposition 11.10, we can conclude $\mathrm{th}(H) \leq \mathrm{th}(G) - 1$ if u does not force or $u \to w$ and $v \in N_H(w)$. So suppose $u \to w \in \mathcal{F}$ and w is an isolated vertex in H. Necessarily $u \to w$ is last force performed, because every other vertex except w must be filled in order for u to force. Then $B'' = B \setminus \{u\} \cup \{w\}$ is a zero forcing set for H with propagating set of forces $\mathcal{F} \setminus \{u \to w\}$ and $|B''| = |B|$. Since $B^{(\mathrm{pt}(G))} = \{w\}$, $\mathrm{pt}(H; B'') = \mathrm{pt}(G; B) - 1$ and $\mathrm{th}(H) \leq \mathrm{th}(H; B'') = \mathrm{th}(G; B) - 1 = \mathrm{th}(G) - 1$. □

Theorem 11.17 allows the extension of small examples that show bounds are tight (presented in the proof of Proposition 11.16) to infinite families of examples.

COROLLARY 11.18. *Let G and G' be graphs of order n and n', respectively. For $k \geq \mathrm{Z}(G)$,*

$$\mathrm{th}(G \,\square\, G') \leq n'k + \mathrm{pt}(G, k)$$

and this bound is tight. Furthermore, $\mathrm{th}(P_{n'} \,\square\, G') = 2n' - 1$.

PROOF. The bound is immediate from Proposition 11.11, and the tightness follows from the second statement. Since $\mathrm{Z}(P_{n'}) = 1$ and $\mathrm{pt}(P_{n'}, 1) = n - 1$, $\mathrm{th}(P_{n'} \,\square\, G') \leq 2n' - 1$. Since the order of $P_{n'} \,\square\, G' = (n')^2$, Theorem 11.14 implies $\mathrm{th}(P_{n'} \,\square\, G') \geq 2n' - 1$. □

Extreme Values. The lower bound in Theorem 11.14 means that for each fixed positive integer k, the number of graphs having $\mathrm{th}(G) = k$ is finite.

REMARK 11.19. All the graphs having $\mathrm{th}(G) \leq 3$ are listed below.

(1) $\mathrm{th}(G) = 1$ if and only if $|V(G)| = 1$.
(2) $\mathrm{th}(G) = 2$ if and only if $|V(G)| = 2$.
(3) $\mathrm{th}(G) = 3$ if and only if $|V(G)| = 3$ or G is one of $2K_2$, P_4, or C_4.

If $\mathrm{th}(G) = 4$, then $4 \leq |V(G)| \leq 6$.

Carlson and Kritschgau [**79**] established that $\mathrm{th}(G) \geq n - k$ has a finite forbidden induced subgraph characterization and gave such a characterization of $\mathrm{th}(G) = n$ for connected graphs.

THEOREM 11.20. **[79]** *Let k be a fixed positive integer. The set of graphs G of order at least k such that* $\operatorname{th}(G) \geq |V(G)| - k$ *is characterized by a finite set of forbidden induced subgraphs.*

THEOREM 11.21. **[79]** *For a connected graph G of order n,* $\operatorname{th}(G) = n$ *if and only if G does not have P_4, C_4, or bowtie as an induced subgraph.*

The next result can be obtained from Theorem 11.21, but we we present a somewhat different proof.

THEOREM 11.22. *Let G be a graph of order n. The following are equivalent:*

(1) $\operatorname{th}(G) = n$.
(2) G is a threshold graph.
(3) G does not have P_4, C_4, or $2K_2$ as an induced subgraph.

PROOF. The equivalence of (2) and (3) is well known.

(1) $\Rightarrow$ (3): We prove the contrapositive. Let G be a graph of order n that has an induced subgraph H that is one of P_4, C_4, or $2K_2$. Note that $\operatorname{th}(H) = 3$, so $\operatorname{th}(G) \leq n - 4 + 3 = n - 1$ by Remark 11.4.

(2) $\Rightarrow$ (1): Now assume G is a threshold graph of order n. Recall that a threshold graph can be constructed from a single vertex by adding a sequence of isolated vertices and/or universal vertices (i.e., $H \vee K_1$). We prove $\operatorname{th}(G) = n$ by induction on n. When $n = 1$ the result is immediate. Assume $\operatorname{th}(H) = n$ for all threshold graphs of order n and let G be a threshold graph of order $n + 1$. Then there is a threshold graph H such that $G = H \sqcup K_1$ or $G = H \vee K_1$. If $G = H \sqcup K_1$, then $\operatorname{th}(G) = \operatorname{th}(H) + 1$ since the isolated vertex must be added to a standard zero forcing set for H to obtain a forcing set for G and does not affect propagation time. If $G = H \vee K_1$, then $\operatorname{th}(G) = \operatorname{th}(H) + 1$ by Theorem 11.17. □

Families of Graphs. In the next theorem we list the throttling numbers of several families of graphs.

THEOREM 11.23. *Let G be a graph of order n.*

(1) **[74]** *For* $n \geq 1$, $\operatorname{th}(P_n) = \lceil 2\sqrt{n} - 1 \rceil$.

(2) **[78]** *For* $n \geq 3$, $\operatorname{th}(C_n) = \begin{cases} \lceil 2\sqrt{n} - 1 \rceil & \text{unless } n = (2k+1)^2 \\ 2\sqrt{n} & \text{if } n = (2k+1)^2 \end{cases}$.

(3) For $n \geq 4$, $\operatorname{th}(W_n) = \begin{cases} \lceil 2\sqrt{n-1} \rceil & \text{unless } n = (2k+1)^2 + 1 \\ 2\sqrt{n-1} + 1 & \text{if } n = (2k+1)^2 + 1 \end{cases}$.

(4) For $n \geq 1$, $\operatorname{th}(K_n) = n$.

(5) For $n \geq 2$, $\operatorname{th}(K_n - e) = n$.

(6) $\operatorname{th}(K_{1,n-1}) = n$ *for* $n \geq 2$, *and* $\operatorname{th}(K_{\ell,n-\ell}) = n - 1$ *for* $n \geq 4$ *and* $2 \leq \ell \leq n - 2$.

(7) For $d \geq 1$, $\operatorname{th}(Q_d) = 2^{d-1} + 1$.

(8) For $k \geq 1$, $\operatorname{th}(F_k) = k + 2$.

(9) For $s \geq 1$, $\operatorname{th}(\mathrm{Half}_s) = 2s$.

(10) For $k \geq 1$, $\operatorname{th}(B_k) = k + 2$.

(11) $\operatorname{th}(\overline{K_n}) = n$.

(12) For $n \geq 4$, $\operatorname{th}(\overline{C_n}) = \begin{cases} n - 1 & \text{unless } n = 6 \\ 4 = n - 2 & \text{if } n = 6 \end{cases}$.

(13) *For* $n \geq 3$, $\operatorname{th}(\overline{P_n}) = n - 1$.

PROOF. Sources are provided for (1) and (2). (3) follows from (2) and Theorem 11.17. (4) and (5) follow from Theorem 11.22. (6), (7), (8), (11), (12), and (13) follow from Proposition 11.3 and their standard zero forcing numbers and propagation times. (9) follows from Theorem 11.22 since a half-graph is a threshold graph. For (10), note that $Z(B_k) = k$, $\operatorname{pt}(B_k) = \operatorname{pt}(B_k, k) = 3$, and $\operatorname{pt}(B_k, k+1) = 1$ using all the vertices in one copy of $K_{1,k}$ (recall $B_k = K_{1,k} \,\square\, P_2$). □

Recall that the throttling number of a Cartesian product of a path with a graph of the same order was determined in Corollary 11.18.

Nordhaus-Gaddum Problems. We apply previous results to determine bounds for the sum of the throttling numbers of a graph and its complement.

THEOREM 11.24. *Let G be a graph of order n. Then*

$$n + o(n) \leq \operatorname{th}(G) + \operatorname{th}(\overline{G}) \leq 2n$$

(even if each of G and $\overline{G}$ is required to have an edge) and the bounds are tight.

PROOF. Let G be a graph of order n. By Observation 11.1, $\operatorname{th}(G)+\operatorname{th}(\overline{G}) \leq 2n$ and this bound is tight when G is a threshold graph by Theorem 11.22 because the complement of a threshold graph is a threshold graph.

By Theorem 9.30, $Z(G) + Z(\overline{G}) \geq n - 2$. Since $\operatorname{pt}(G; B) \geq 1$ for any set $B \neq V(G)$, this implies $n \leq \operatorname{th}(G)+\operatorname{th}(\overline{G})$. Note that $\operatorname{th}(P_n)+\operatorname{th}(\overline{P_n}) = n+\lceil 2\sqrt{n} - 1\rceil - 1$ by Theorem 11.23, so the lower bound is $n + o(n)$. □

11.3. Z_+-throttling

Positive semidefinite throttling was introduced in [**78**]. The *PSD throttling number* or Z_+*-throttling number* of B in G is $\operatorname{th}_+(G; B) = |B| + \operatorname{pt}_+(G; B)$. The *PSD k-throttling number* is $\operatorname{th}_+(G, k) = k + \operatorname{pt}_+(G, k)$. The *PSD throttling number* or Z_+*-throttling number* of a graph G of order n is

$$\operatorname{th}_+(G) = \min_{B \subseteq V(G)} \operatorname{th}_+(G; B) = \min_{Z(G) \leq k \leq n} \operatorname{th}_+(G, k).$$

Figure 11.1 illustrates an optimal PSD throttling set for the tree T_1 shown in Figure 10.7 and other examples. The value $\operatorname{th}_+(T_1) = 4$ is verified in [**184**, Example TH.3-1].

OBSERVATION 11.25. *Observation 11.1, Propositions 11.2 and 11.3, and Remarks 11.4, 11.5, and 11.6 apply to Z_+-throttling.*

OBSERVATION 11.26. *Since any standard zero forcing set of a graph G is a PSD zero forcing set and the same set of forces is possible,*

$$\operatorname{pt}_+(G; B) \leq \operatorname{pt}(G; B), \ \operatorname{th}_+(G; B) \leq \operatorname{th}(G; B), \ \textit{and} \ \operatorname{th}_+(G) \leq \operatorname{th}(G).$$

The next theorem lists bounds for $\operatorname{th}_+(G)$.

THEOREM 11.27. [**78**] *Let G be a graph of order n.*

(1) *If $\Delta(G) = 2$, then* $\operatorname{th}_+(G) \geq \left\lceil \sqrt{2n} - \frac{1}{2} \right\rceil$.
(2) *If $\Delta(G) \geq 3$, then*

$$\operatorname{th}_+(G) \geq \left\lceil Z_+(G) + \log_{(\Delta(G)-1)} \left(\frac{(\Delta(G) - 2)n + 2\,Z_+(G)}{\Delta(G)\, Z_+(G)} \right) \right\rceil.$$

(3) $\text{th}_+(G) \le n - \alpha(G) + 1$, *where* $\alpha(G)$ *denotes the independence number of* G.

Each of these bounds is tight.

REMARK 11.28. Let T be a tree of order n. Since $\text{pt}_+(T; B) = \text{ecc}(B)$, $\text{th}_+(T) = \min_k(k + \text{rad}_k(T))$. Determination of a tight asymptotic upper bound on $\text{th}_+(T)$ is an active area of research. Since $\text{th}_+(P_n) = \lceil \sqrt{2n} - \frac{1}{2} \rceil$ [**78**], $c \ge \sqrt{2}$ in any inequality of the form $\text{th}_+(T) \le c\sqrt{n} + o(n)$. There are spiders with PSD throttling number greater than that of the path: It was shown in [**57**] that $\text{th}_+(S(4,3,2)) = 5 > 4 = \text{th}_+(P_{10})$ and this was generalized in [**250**] to $\text{th}_+(S(t, t-1, \dots, 3, 2) = t + 1 = \text{th}_+(P_{t(t+1)/2}) + 1$. A family of spiders $S_{n,a,c}$ of order n with $\text{th}_+(S_{n,a,c}) > 1.4502\sqrt{n}$ was constructed in [**152**]. It is known that the PSD throttling number is the same as the cop throttling number $\text{th}_c(T)$ for trees (see Section 11.6). It was shown in [**57**] that $\text{th}_c(T) \le 2\lfloor\sqrt{n}\rfloor$. This bound was improved in [**152**] to $\text{th}_c(T) \le \frac{\sqrt{14}}{2}\sqrt{n} + O(1) \approx 1.871\sqrt{n} + O(1)$.

For a positive integer t, Carlson and Kritschgau gave a characterization of graphs G having $\text{th}_+(G) \le t$. Since even stating the characterization is lengthy, we refer the reader to [**79**].

Computation. It is known that computing the PSD throttling number number is NP-complete [**61**].

The function `thp(G)` is available in in [**184**] for computing the PSD throttling number $\text{th}_+(G)$ of a specific graph G (that is not too large) by minimizing $k + \text{pt}_+(G, k)$ over $Z_+(G) \le k \le \min\{Z_+(G) + \text{pt}_+(G) - 1, n\}$ (see Proposition 11.2). The function `thpk(G)` returns both $\text{th}_+(G)$ and the lowest k such that $\text{th}_+(G) = \text{th}_+(G, k)$. The use of these functions is illustrated in Section 11.3 of [**184**].

Extreme Values. The next two theorems characterize graphs with extreme values of $\text{th}_+(G)$.

THEOREM 11.29. [**78**] *Let* G *be a graph of order* n.

(1) $\text{th}_+(G) = 1$ *if and only if* $n = 1$.
(2) $\text{th}_+(G) = 2$ *if and only if* $G = K_{1,n-1}$ *or* $G = 2K_1$.
(3) For a graph G, $\text{th}_+(G) = 3$ *if and only if at least one of the following is true:*
   (a) G *is disconnected and exactly one of the following holds:*
      (i) G *is* $3K_1$, *or*
      (ii) G *has two components, each component is a copy of* $K_{1,n-1}$ *or* K_1, *and at least one component has order greater than one.*
   (b) G *is a tree with diameter three or four, or*
   (c) G *is connected, and* G *has a cycle or is a tree with* $\text{diam}\, G = 5$, *and there exist* $v, u \in V(G)$ *such that:*
      (i) $N(u) \cup N(v) = V(G)$,
      (ii) $\deg(w) \le 2$ *for all* $w \notin \{v, u\}$, *and*
      (iii) if $w_1, w_2 \in N(u)$ *or* $w_1, w_2 \in N(v)$, *then* w_1 *is not adjacent to* w_2.

COROLLARY 11.30. *Let* G *be a connected graph of order* n. *Then*

(1) $\text{th}_+(G) = 1$ *if and only if* $n = 1$.
(2) $\text{th}_+(G) = 2$ *if and only if* $G = K_{1,n-1}$.

(3) $\text{th}_+(G) = 3$ *if and only if*
   (a) *G is a tree with diameter three or four, or*
   (b) *G has a cycle or G is a tree with* $\operatorname{diam} G = 5$, *and there exist $v, u \in V(G)$ such that:*
      (i) $N(u) \cup N(v) = V(G)$,
      (ii) $\deg(w) \le 2$ *for all $w \notin \{v, u\}$, and*
      (iii) *if $w_1, w_2 \in N(u)$ or $w_1, w_2 \in N(v)$, then w_1 is not adjacent to w_2.*

THEOREM 11.31. [**78**] *Let G be a connected graph of order n.*
(1) $\text{th}_+(G) = n$ *if and only if* $G = K_n$.
(2) $\text{th}_+(G) = n - 1$ *if and only if* $\alpha(G) = 2$ *and G does not have an induced C_5, house, or double diamond subgraph (see Figure 11.4).*

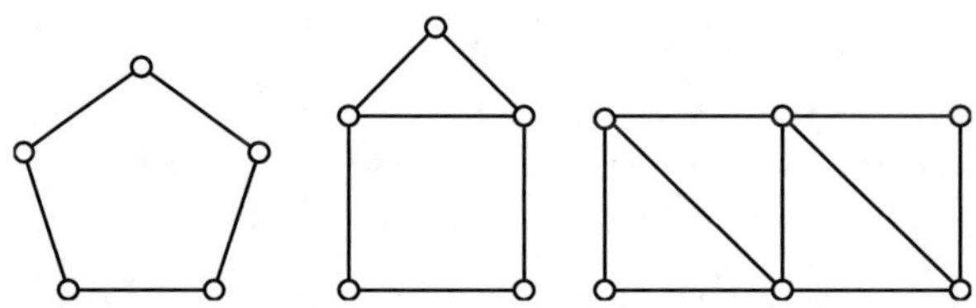

FIGURE 11.4. Graphs forbidden as induced subgraphs for $\text{th}_+(G) \ge n - 1$.

COROLLARY 11.32. *Let G be a graph of order n. Then*
(1) $\text{th}_+(G) = n$ *if and only if* $G = K_{n-r} \sqcup rK_1$.
(2) $\text{th}_+(G) = n - 1$ *if and only G has exactly one component H with an edge,* $\alpha(H) = 2$ *and H does not have an induced C_5, house, or double diamond subgraph.*

THEOREM 11.33. [**79**] *Let k be a fixed positive integer. The set of connected graphs G of order at least k such that* $\text{th}_+(G) \ge |V(G)| - k$ *is characterized by a finite set of forbidden subgraphs.*

Graph Operations. In the proof of the next result, proofs of bounds established in Propositions 11.8 and 11.9 are omitted (but examples showing tightness are presented).

PROPOSITION 11.34. *Let G be a graph.*
(1) *For any vertex v of G,* $\text{th}_+(G) - 1 \le \text{th}_+(G - v)$. *There is no useful lower bound for* $\text{th}_+(G) - \text{th}_+(G - v)$ *if the removal of v disconnects the graph.*
(2) *For any edge e of G,* $\text{th}_+(G) - 1 \le \text{th}_+(G - e) \le \text{th}_+(G) + 1$.
(3) *For any edge e of G,* $\text{th}_+(G) - 1 \le \text{th}_+(G/e)$. *There is no useful lower bound for* $\text{th}_+(G) - \text{th}_+(G/e)$.
(4) *For any edge e of G,* $\text{th}_+(G) \le \text{th}_+(G_e) \le \text{th}_+(G) + 1$.

All bounds are tight.

PROOF. $\text{th}_+(G) - 1 \le \text{th}_+(G - v)$ is tight: Deleting an end vertex from the path P_n where n is chosen such that $\text{th}_+(P_n) = \text{th}_+(P_{n-1}) + 1$ shows this bound for vertex deletion is tight (e.g., $\text{th}_+(P_4) = 3$ and $\text{th}_+(P_3) = 2$).

$\text{th}_+(G) - \text{th}_+(G-v)$ has no useful lower bound, as can be seen from the star $K_{1,n-1}$ with v the vertex of degree $n-1$.

$\text{th}_+(G) - 1 \le \text{th}_+(G-e)$ is tight: Let $F = P_4 \vee K_1$ and let e be an edge whose endpoints have degrees three and four; F and $F-e$ are shown in Figure 11.5. Then $\text{th}_+(F) = 4$ and $\text{th}_+(F-e) = 3$. Computations illustrating this example are presented in Example TH.3-3 in [**184**].

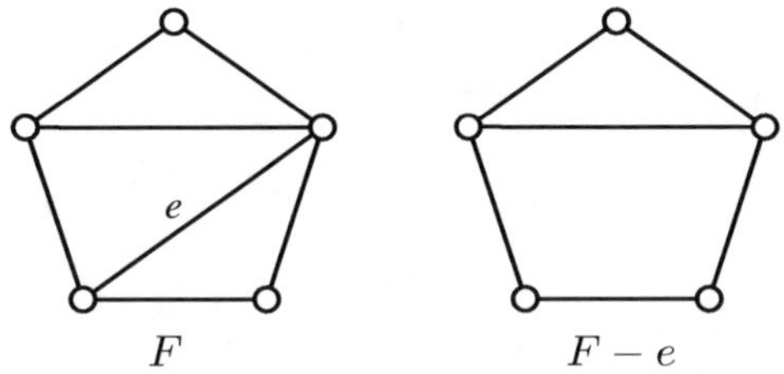

FIGURE 11.5. A graph F such that $\text{th}_+(F-e) = \text{th}_+(F) - 1$.

$\text{th}_+(G-e) \le \text{th}_+(G) + 1$ is tight, as can be seen by deleting an edge from the star $K_{1,n-1}$ with $n \ge 3$.

$\text{th}_+(G) - 1 \le \text{th}_+(G/e)$ is tight follows by contracting an edge of P_4.

$\text{th}_+(G) - \text{th}_+(G/e)$ has no useful lower bound: Recall that $B_k = K_{1,k} \square P_2$ and $B_k/e = F_k$ (see Example 9.53). Since $Z_+(B_k) = 2$ and $\text{pt}_+(B_k) = 1$, $\text{th}_+(B_k) = 3$, whereas $Z_+(F_k) = k+1$, $\text{pt}_+(F_k) = 1$, and $\text{th}_+(F_k) = k+2$.

$\text{th}_+(G) \le \text{th}_+(G_e)$ and $\text{th}_+(G_e) \le \text{th}_+(G) + 1$ are tight can be seen by subdividing or contracting an edge of P_4 shows this bound is tight. □

The next result was established in Proposition 11.10.

COROLLARY 11.35. *Let H be a graph in which each connected component has order at least two, and let G be constructed from H by adding a universal vertex u. Then* $\text{th}_+(G) = \text{th}_+(H) + 1$.

Corollary 11.35 is not true without the assumption that H is has no isolated vertices (recall that $K_{1,n} = K_1 \vee \overline{K_n}$). Corollary 11.35 allows the extension of small examples that show bounds are tight (presented in the proof of Proposition 11.34) to infinite families of examples.

COROLLARY 11.36. *Let G and G' be graphs of order n and n', respectively. For* $k \ge Z_+(G)$,

$$\text{th}_+(G \square G') \le n'k + \text{pt}_+(G,k)$$

and the bound is tight. Let T be a tree of order $n \ge 2$ and let G' be a graph of order $n' \ge 2$. Then $\text{th}_+(T \square G') \le n' + \text{rad}(T)$. *Furthermore,* $\text{th}_+(K_{1,n-1} \square K_r) = r+1$.

PROOF. The first inequality is immediate from Proposition 11.11. The last statement shows the bound is tight. Since $Z_+(T) = 1$ and $\text{pt}_+(T,1) = \text{rad}(T)$, $\text{th}_+(T \square K_{n'}) \le r + \text{rad}(T)$. Since $Z_+(K_{1,n-1} \square K_r) = r$ (Example 9.56) and $\text{pt}_+(K_{1,n-1} \square K_r) = 1$ (Proposition 10.29(3)), $\text{th}_+(K_{1,n-1} \square K_r) = r+1$. □

In general, PSD throttling number is not minor monotone. However, there is a relationship for connected minors of trees.

PROPOSITION 11.37. [**250**] *If T is a tree and T' is a connected minor of T, then* $\mathrm{th}_+(T') \le \mathrm{th}_+(T)$.

Families of Graphs. In the next theorem we list the PSD throttling numbers of several families of graphs.

THEOREM 11.38.

(1) [**78**] *For* $n \ge 1$, $\mathrm{th}_+(P_n) = \left\lceil \sqrt{2n} - \frac{1}{2} \right\rceil$.
(2) [**78**] *For* $n \ge 4$, $\mathrm{th}_+(C_n) = \left\lceil \sqrt{2n} - \frac{1}{2} \right\rceil$.
(3) For $n \ge 5$, $\mathrm{th}_+(W_n) = \left\lceil \sqrt{2(n-1)} - \frac{1}{2} \right\rceil + 1$.
(4) $\mathrm{th}_+(K_n) = n$.
(5) $\mathrm{th}_+(K_n - e) = n - 1$.
(6) For $n \ge 2$, $\mathrm{th}_+(K_{\ell,n-\ell}) = \ell + 1$ *for* $1 \le \ell \le \left\lfloor \frac{n}{2} \right\rfloor$.
(7) For $d \ge 1$, $\mathrm{th}_+(Q_d) = 2^{d-1} + 1$.
(8) For $k \ge 1$, $\mathrm{th}_+(F_k) = k + 2$.
(9) For $n \ge 5$, $\mathrm{th}_+(\overline{C_n}) = n - 2$.
(10) For $n \ge 5$, $\mathrm{th}_+(\overline{P_n}) = n - 2$.

PROOF. (4), (6), (7), (8), (9), and (10) follow from Proposition 11.3 and their PSD zero forcing numbers and propagation times. (3) follows from (2) and Theorem 11.35. (5) follows from Theorem 11.31. □

Nordhaus-Gaddum Problems. We apply previous results to determine bounds for the sum of the PSD throttling numbers of a graph and its complement.

PROPOSITION 11.39. *Let G be a graph of order $n \ge 2$. Then*

$$n + o(n) \le \mathrm{th}_+(G) + \mathrm{th}_+(\overline{G}) \le 2n.$$

If each of G and $\overline{G}$ is required to have an edge, then $\mathrm{th}_+(G) + \mathrm{th}_+(\overline{G}) \le 2n - 1$. *All the bounds are tight.*

PROOF. Let G be a graph of order n. By Observation 11.1, $\mathrm{th}_+(G) + \mathrm{th}_+(\overline{G}) \le 2n$, and $\mathrm{th}_+(K_n) = \mathrm{th}_+(\overline{K_n}) = n$. Suppose G and $\overline{G}$ are each required to have an edge. If at least one of $G, \overline{G}$ has PSD throttling number less than n, then $\mathrm{th}_+(G) + \mathrm{th}_+(\overline{G}) \le 2n - 1$. Suppose $\mathrm{th}_+(G) = n$. Then $G = K_{n-r} \sqcup rK_1$ with $1 \le r \le n-1$, and $\overline{G} = (n-r)K_1 \vee K_r$. Since $1 \le r \le n-1$, $\overline{G}$ is connected. This implies $\mathrm{th}_+(\overline{G}) \le n - 1$ since $\overline{G} \ne K_n$. For $r = 2$, $G = K_2 \sqcup (n-2)K_1 = \overline{K_n - e}$, $\overline{G} = K_n - e$, $\mathrm{th}_+(\overline{K_n - e}) = n$, and $\mathrm{th}_+(K_n - e) = n - 1$.

By Theorem 9.60, $\mathrm{Z}_+(G) + \mathrm{Z}_+(\overline{G}) \ge n - 2$. Since $\mathrm{pt}_+(G; B) \ge 1$ for any set $B \ne V(G)$, this implies $n \le \mathrm{th}_+(G) + \mathrm{th}_+(\overline{G})$. Note that $\mathrm{th}_+(P_n) + \mathrm{th}_+(\overline{P_n}) = n + \left\lceil \sqrt{2n} - \frac{1}{2} \right\rceil - 2$ by Theorem 11.38, so the lower bound is $n + o(n)$. □

11.4. Z_--throttling

Skew throttling was introduced in [**92**]. The *skew throttling number* (or *Z_--throttling number*) of B in G is $\mathrm{th}_-(G; B) = |B| + \mathrm{pt}_-(G; B)$. The *skew k-throttling number* is $\mathrm{th}_-(G, k) = k + \mathrm{pt}_-(G, k)$. The *skew throttling number* (or *Z_--throttling number*) of a graph G of order n is

$$\mathrm{th}_-(G) = \min_{B \subseteq V(G)} \mathrm{th}_-(G; B) = \min_{\mathrm{Z}_-(G) \le k \le n} \mathrm{th}_-(G, k).$$

In the next figure we illustrate an optimal skew throttling set for the tree T_1 shown in Figure 10.9 and other examples. That no lower value is possible is verified in [**184**, Example TH.4-1].

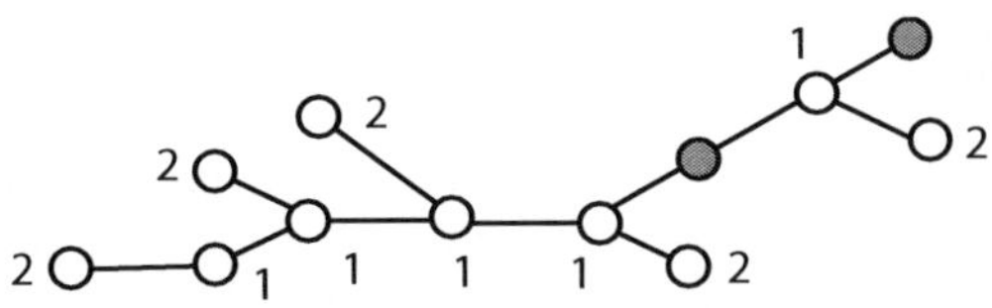

FIGURE 11.6. Skew throttling on the tree T_1, showing the time each vertex is filled. The skew throttling number is realized by a minimum skew forcing set, since $\mathrm{Z}_-(T_1) = 2, \mathrm{pt}_-(T_1) = 2$, and $\mathrm{th}_-(T_1) = 4$.

OBSERVATION 11.40. *Observation 11.1, Propositions 11.2 and 11.3, and Remarks 11.4, and 11.6 apply to* Z$_-$*-throttling. However, Remark 11.5 does not (and the* 0*-radius is not defined).*

OBSERVATION 11.41. *Any standard zero forcing set is a skew zero forcing set, so* $\mathrm{pt}_-(G; B) \leq \mathrm{pt}(G; B)$, $\mathrm{th}_-(G; B) \leq \mathrm{th}(G; B)$, *and* $\mathrm{th}_-(G) \leq \mathrm{th}(G)$.

OBSERVATION 11.42. *If a graph G of order n has an edge uv, then* $\mathrm{th}_-(G) \leq n - 1$, *by considering* $B = V(G) \setminus \{u, v\}$.

OBSERVATION 11.43. *For any graph with an edge,* $\delta(G) \leq \mathrm{th}_-(G)$ *because* $\delta(G) - 1 \leq \mathrm{Z}_-(G)$ *(Remark 9.63) and* $\mathrm{Z}_-(G) + 1 \leq \mathrm{th}_-(G)$ *(Observation 11.1).*

Computation. The function `ths(G)` is available in in [**184**] for computing the skew throttling number $\mathrm{th}_-(G)$ of a specific graph G (that is not too large) by minimizing $k + \mathrm{pt}_-(G, k)$ over $\mathrm{Z}_-(G) \leq k \leq \min\{\mathrm{Z}_-(G) + \mathrm{pt}_-(G) - 1, n\}$ (see Remark 11.1). The function `thsk(G)` returns both $\mathrm{th}_-(G)$ and the lowest k such that $\mathrm{th}_-(G) = \mathrm{th}_-(G, k)$. The use of these functions is illustrated in Section 11.4 of [**184**].

Extreme Values. The next theorem characterizes graphs with very low and very high skew throttling numbers. For $s, t \geq 0$, define $H(s, t)$ by $V(H(s, t)) = \{b\} \sqcup \{x_i, y_i : i = 1, \ldots, s\} \sqcup \{z_i, w_i : i = 1, \ldots, t\}$ and $E(H(s, t)) = \{bx_i, x_iy_i : i = 1, \ldots, s\} \cup \{bz_i, bw_i, z_iw_i : i = 1, \ldots, t\}$. A cograph is a graph that does not contain P_4 as an induced subgraph.

THEOREM 11.44. [**92**] *Let G be a graph of order n.*

(1) $\mathrm{th}_-(G) = 1$ *if and only if* $G = K_1$ *or* $G = rK_2$ *for* $r \geq 1$.
(2) A graph G has $\mathrm{th}_-(G) = 2$ *if and only if G is one of* $2K_1$, $H(s, t) \sqcup rK_2$ *with* $r + s + t \geq 1$, *or* $\left(\widetilde{G} \circ K_1\right) \sqcup rK_2$ *where each component of* $\widetilde{G}$ *has an edge.*
(3) $\mathrm{th}_-(G) = n$ *if and only if* $G = nK_1$.
(4) $\mathrm{th}_-(G) = n - 1$ *if and only if G is a cograph, does not have an induced* $2K_2$, *and has at least one edge.*

COROLLARY 11.45. *Let G be a connected graph of order n.*

(1) $\mathrm{th}_-(G) = 1$ *if and only if* $G = K_1$.

(2) A graph G has $\text{th}_-(G) = 2$ *if and only if G is $H(s,t)$ with $s+t \geq 1$ or* $\left(\widetilde{G} \circ K_1\right)$ *where $\widetilde{G}$ is connected.*

(3) $\text{th}_-(G) = n$ *if and only if* $n = 1$.

(4) $\text{th}_-(G) = n - 1$ *if and only if G is a connected cograph of order at least two that does not have an induced $2K_2$.*

Graph Operations. Many graph operation bounds for standard throttling have analogs for skew throttling, and again there is a connection to the forcing number bounds (see for example, edge deletion). In the proof of the next result, proofs of bounds established in Proposition 11.8 are omitted (but examples showing tightness are presented).

PROPOSITION 11.46. *Let G be a graph, $v \in V(G)$, and $e = \{u, w\} \in E(G)$.*

(1) $\text{th}_-(G) - 1 \leq \text{th}_-(G - v) \leq \text{th}_-(G) + 1$.

(2) $\text{th}_-(G) - 2 \leq \text{th}_-(G - e) \leq \text{th}_-(G) + 2$.

(3) $\text{th}_-(G) - 1 \leq \text{th}_-(G/e) \leq \text{th}_-(G) + 1$.

(4) $\text{th}_-(G) - 1 \leq \text{th}_-(G_e) \leq \text{th}_-(G) + 1$.

All bounds are tight.

PROOF. $\text{th}_-(G) - 1 \leq \text{th}_-(G - v)$ is tight: Delete a vertex from the complete graph K_n with $n \geq 3$.

$\text{th}_-(G - v) \leq \text{th}_-(G) + 1$: The proof is the same as the proof in Proposition 11.16 that $\text{th}(G - v) \leq \text{th}(G) + 1$. For tightness, let $G = rK_2$ with $r \geq 2$, so $\text{th}_-(G - v) = 1 + 1 = \text{th}_-(G) + 1$.

$\text{th}_-(G) - 2 \leq \text{th}_-(G - e) \leq \text{th}_-(G) + 2$: Choose an optimal skew throttling set $B \subseteq V(G - e)$ (respectively, $B \subseteq V(G)$). Define $B' = B \cup \{u, w\}$. Then $\text{pt}_-(G; B') \leq \text{pt}_-(G - e; B)$ (respectively, $\text{pt}_-(G - e; B') \leq \text{pt}_-(G; B)$) and $|B'| \leq |B| + 2$. Thus $\text{th}_-(G) \leq \text{th}_-(G; B') \leq \text{th}_-(G - e) + 2$ (respectively, $\text{th}_-(G - e) \leq \text{th}_-(G - e; B') \leq \text{th}_-(G) + 2$). For tightness: Let $G = C_4 \sqcup P_4$ and let e be an edge of the C_4, so $C_4 \sqcup P_4 - e = 2P_4$. Then $\text{th}_-(G - e) = 2 = \text{th}_-(G) - 2$; this is illustrated in Example TH.4-2 in [**184**]. Let $G = rK_2$ with $r \geq 2$, so $\text{th}_-(G - e) = 2 + 1 = \text{th}_-(G) + 2$.

$\text{th}_-(G) - 1 \leq \text{th}_-(G/e)$ and $\text{th}_-(G_e) \leq \text{th}_-(G) + 1$ are tight: Choose n such that $\text{th}_-(P_n) = \text{th}_-(P_{n-1}) + 1$ and note that $P_{n-1} = P_n/e$ and $P_n = (P_{n-1})_e$.

Let v_e denote the new vertex formed by identifying u and w in contraction or the new vertex adjacent to u and w in edge subdivision.

$\text{th}_-(G/e) \leq \text{th}_-(G) + 1$: Choose a set $B \subseteq V(G)$ such that $\text{th}_-(G) = \text{th}_-(G; B)$ and construct a propagating set of Z_--forces $\mathcal{F}$. If at least one of u or w is in B, then define $B' = B \setminus \{u, w\} \cup \{v_e\}$; otherwise $B' = B$. Thus $|B'| \leq |B|$. If neither u nor w performs a force, then let $B'' = B'$. So suppose at least one of u, w performs a force. Assume without loss of generality that $u \to y$ before (or at the same time) w performs a force. Define $B'' = B' \cup \{y\}$ (note only one force can be performed by u). Then $|B''| \leq |B| + 1$ and $\text{pt}_-(G/e; B'') \leq \text{pt}_-(G; B)$, so $\text{th}_-(G/e) \leq \text{th}_-(G) + 1$. Example TH.4-3 in [**184**] shows this bound is tight since $\text{th}_-(G/e) = \text{th}_-(G) + 1$.

$\text{th}_-(G) - 1 \leq \text{th}_-(G_e)$: Choose a set $B \subseteq V(G_e)$ such that $\text{th}_-(G_e) = \text{th}_-(G_e; B)$ and construct a propagating set of Z_--forces $\mathcal{F}$. If $v_e \in B$, then let $B' = B \setminus \{v_e\} \cup \{u, w\}$; use the same forces (except remove $v_e \to u$ or $v_e \to w$ if it is in $\mathcal{F}$). Suppose now that $v_e \notin B$. If v_e does not perform a force, then let $B' = B$. So assume $v_e \to w \in \mathcal{F}$. At the time this force happens, u is already filled but u may not be able to force w, so define $B' = B \cup \{w\}$. In all cases, $\mathrm{pt}_-(G; B') \leq \mathrm{pt}_-(G_e; B)$ and $\mathrm{th}_-(G) \leq \mathrm{th}_-(G-e)+1$. Example TH.4-3 in [**184**] presents a graph G such that $\mathrm{th}_-(G_e) = \mathrm{th}_-(G) - 1$. □

The next result was established in Proposition 11.10.

COROLLARY 11.47. *Let H be a graph in which each connected component has order at least two, and let G be constructed from H by adding a universal vertex u. Then* $\mathrm{th}_-(G) = \mathrm{th}_-(H) + 1$.

Corollary 11.47 is not true for a graph with isolated vertices. Let u be the universal vertex of $K_{1,n-1}$, so $K_{1,n-1} - u = (n-1)K_1$, and $\mathrm{th}_-(K_{1,n-1}) = n-1 = \mathrm{th}_-(\overline{K_{n-1}})$.

It is not the case that $\mathrm{th}_-(G \Box G') \leq n'k + \mathrm{pt}_+(G,k)$ for $k \geq \mathrm{Z}_-(G)$. Recall that $\mathrm{Z}_-(P_n \Box P_n) = n$ (see Proposition 9.94) but $\mathrm{Z}_-(P_{2k}) = 0$. Thus $\mathrm{th}_-(P_{2k} \Box P_{2k}) \geq 2k$ but $(2k)(0) + \mathrm{pt}_+(P_{2k}, 0) = k$.

Families of Graphs. In the next theorem we list the skew throttling numbers of several families of graphs.

THEOREM 11.48. *Let G be a graph of order n.*

(1) [**92**] *For $n \geq 3$,* $\mathrm{th}_-(P_n) = \left\lceil \sqrt{2(n+1)} - \frac{3}{2} \right\rceil$.
(2) [**92**] *For $n \geq 4$,* $\mathrm{th}_-(C_n) = \left\lceil \sqrt{2n} - \frac{1}{2} \right\rceil$.
(3) For $n \geq 5$, $\mathrm{th}_-(W_n) = \left\lceil \sqrt{2(n-1)} + \frac{1}{2} \right\rceil$.
(4) [**92**] $\mathrm{th}_-(K_n) = n - 1$.
(5) $\mathrm{th}_-(K_n - e) = n - 1$.
(6) [**92**] $K_{n_1,n_2,\dots,n_s}$ *with $s \geq 2$ and $n = n_1 + n_2 + \cdots + n_s$ has* $\mathrm{th}_-(K_{n_1,n_2,\dots,n_s}) = n - 1$.
(7) [**92**] $\mathrm{th}_-(Q_k) = 2^{k-1} + 1$.
(8) $\mathrm{th}_-(F_k) = 2$.
(9) $\mathrm{th}_-(B_k) = 3$.
(10) $\mathrm{th}_-(\overline{P_n}) = n - 2$ *for $n \geq 4$.*
(11) $\mathrm{th}_-(\overline{C_n}) = n - 2$ *for $n \geq 5$.*

PROOF. (5), (8), (9), (10), and (11) follow from Proposition 11.3 and their skew zero forcing numbers and propagation times. (3) follows from (2) and Theorem 11.47. □

Additional results, including skew throttling numbers of balanced spiders and a lower bound in terms of diameter for graphs with no leaves, can be found in [**92**].

Nordhaus-Gaddum Problems. We apply previous results to determine upper bounds for the sum of the skew throttling numbers of a graph and its complement.

PROPOSITION 11.49. *Let G be a graph of order $n \geq 2$. Then* $\mathrm{th}_-(G) + \mathrm{th}_-(\overline{G}) \leq 2n - 1$ *and the bound is tight. If each of G and $\overline{G}$ is required to have an edge, then* $\mathrm{th}_-(G) + \mathrm{th}_-(\overline{G}) \leq 2n - 2$ *and the bound is tight.*

PROOF. Let G be a graph of order n. By Theorem 11.44, $\text{th}_-(G) = n$ if and only if $G = nK_1$, which together with Observation 11.1 implies $\text{th}_-(G) \leq n-1$ if G has an edge. Since $\text{th}_+(K_n) = n-1$ and $\text{th}_-(\overline{K_n}) = n$, equality without the assumption that each has an edge is realized by complete graph. Now suppose each of G and $\overline{G}$ has an edge. Then $\text{th}_-(G) \leq n-1$ and $\text{th}_-(\overline{G}) \leq n-1$, so $\text{th}_-(G) + \text{th}_-(\overline{G}) \leq 2n-2$. Let G be a threshold graph. Then is $\overline{G}$ is also a threshold graph, and $\text{th}_-(G) = \text{th}_-(\overline{G}) = n-1$. □

11.5. Throttling for Power Domination

The *power domination throttling number* or γ_P*-throttling number* of S in G is $\text{th}_{\gamma_P}(G;S) = |S| + \text{pt}_{\gamma_P}(G;S)$. Define $\text{th}_{\gamma_P}(G,k) = k + \text{pt}_{\gamma_P}(G,k)$. The *power domination throttling number* or γ_P*-throttling number* of a graph G of order n is

$$\text{th}_{\gamma_P}(G) = \min_{S \subseteq V(G)} \text{th}_{\gamma_P}(G;S) = \min_{\gamma_P(G) \leq k \leq n} \text{th}_{\gamma_P}(G,k).$$

The study of power domination throttling was introduced in [**61**].

In the next figure we illustrate an optimal power domination throttling set for the tree T_1 shown in Figure 10.13 and other examples. That no lower value is possible is verified in [**184**, Example TH.5-1].

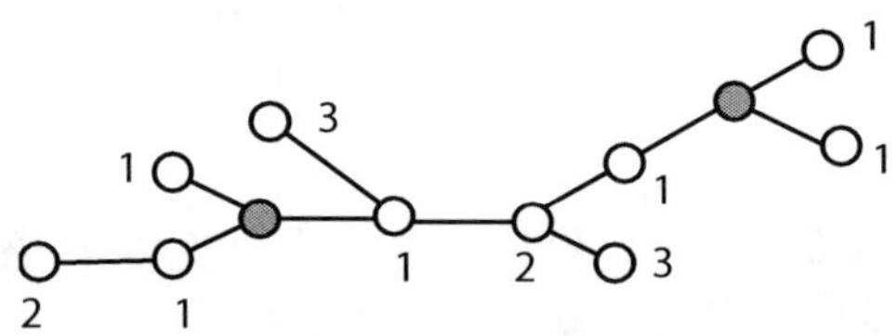

FIGURE 11.7. Power domination throttling on the tree T_1, showing the time each vertex is filled. The power domination throttling number is realized by a minimum power domination forcing set: $\gamma_P(T_1) = 2, \text{pt}_{\gamma_P}(T_1) = 3, \text{th}_{\gamma_P}(T_1) = 5$.

OBSERVATION 11.50. *Observation 11.1, Propositions 11.2 and 11.3, and Remarks 11.4, 11.5, and 11.6 apply to power domination throttling.*

OBSERVATION 11.51. *For any graph G, $\text{th}_{\gamma_P}(G) \leq \gamma(G)+1$ since $\text{pt}_{\gamma_P}(G,\gamma(G)) = 1$. Thus $\text{th}_{\gamma_P}(G) = \text{th}_{\gamma_P}(G,k)$ for some k with $\gamma_P(G) \leq k \leq \gamma(G)$.*

OBSERVATION 11.52. *Any standard zero forcing set is a power dominating set, so $\text{pt}_{\gamma_P}(G;B) \leq \text{pt}(G;B)$, $\text{th}_{\gamma_P}(G;B) \leq \text{th}(G;B)$, and $\text{th}_{\gamma_P}(G) \leq \text{th}(G)$.*

The next theorem gives a lower bound for $\text{th}_{\gamma_P}(G)$ analogous to the Butler-Young lower bound for throttling number in Theorem 11.14.

THEOREM 11.53. [**61**] *If G is a graph of order n that has an edge and Δ denotes its maximum degree, then $\text{th}_{\gamma_P}(G) \geq \left\lceil 2\sqrt{\frac{n}{\Delta}} - \frac{1}{\Delta} \right\rceil$; this bound is tight.*

Computation. It is shown in [**61**] that computing the power domination throttling number number is NP-complete.

The function `thpd(G)` is available in in [**184**] for computing the power domination throttling number $\text{th}_{\gamma_P}(G)$ of a specific graph G (that is not too large) by minimizing $k+\text{pt}_{\gamma_P}(G,k)$ over $\gamma_P(G) \leq k \leq \gamma(G)$ (see Observation 11.51). The function `thpdk(G)` displays both $\text{th}_{\gamma_P}(G)$ and the lowest k such that $\text{th}_{\gamma_P}(G) = \text{th}_{\gamma_P}(G,k)$. The use of these functions is illustrated in Section 11.5 of [**184**].

Extreme Values. The next two theorems characterize graphs with extreme values of $\text{th}_{\gamma_P}(G)$ without assuming the graph is connected.

THEOREM 11.54. [**61**] *Let G be a graph of order n.*

(1) $\text{th}_{\gamma_P}(G) = 1$ *if and only if* $n = 1$.
(2) $\text{th}_{\gamma_P}(G) = 2$ *if and only* $n \geq 2$ *and G has a universal vertex or* $G = 2K_1$.

THEOREM 11.55. [**61**] *Let G be a graph of order n.*

(1) $\text{th}_{\gamma_P}(G) = n$ *if and only if* $G = nK_1$.
(2) $\text{th}_{\gamma_P}(G) = n-1$ *if and only if G is of one of the forms* $P_3 \sqcup (n-3)K_1$, $C_3 \sqcup (n-3)K_1$, $P_4 \sqcup (n-4)K_1$, $C_4 \sqcup (n-4)K_1$, *or* $K_2 \sqcup K_2 \sqcup (n-3)K_1$.

For standard, PSD, and skew zero forcing, we are able obtain results on extreme values of the throttling number for connected graphs from the general result, or vice versa. Theorem 11.54 immediately establishes the analogous result on low values for connected graphs.

COROLLARY 11.56. *Let G be a connected graph of order n.*

(1) $\text{th}_{\gamma_P}(G) = 1$ *if and only if* $n = 1$.
(2) $\text{th}_{\gamma_P}(G) = 2$ *if and only* $n \geq 2$ *and G has a universal vertex.*

However high values of the power domination number are much lower when the graph is required to be connected. Recall that $\gamma_P(G) \leq \gamma(G)$ and thus $\text{th}_{\gamma_P}(G) \leq \gamma(G) + 1$ for all graphs G. Since $\gamma(G) \leq \left\lfloor \frac{n}{2} \right\rfloor$ for a graph of order n having no isolated vertices [**174**, Theorem 2.1],

$$\text{th}_{\gamma_P}(G) \leq \left\lfloor \frac{n}{2} \right\rfloor + 1$$

for a connected graph G of order $n \geq 2$. Examples with $\text{th}_{\gamma_P}(G) = \left\lfloor \frac{n}{3} \right\rfloor + 1$ follow immediately from examples with $\gamma_P(G) = \left\lfloor \frac{n}{3} \right\rfloor$. We can do very slightly better, as illustrated in the next example.

EXAMPLE 11.57. Let H be a connected graph of order $k \geq 2$ and let G' be obtained from H by adding two leaves to each vertex in H (so $G' \in \mathfrak{T}(H)$; see Theorem 9.155). Create G by adding an additional leaf one of the leaves of G'; the graph G created from C_3 is shown in Figure 11.8. Then $\text{th}_{\gamma_P}(G) = k+2 = \left\lfloor \frac{n}{3} \right\rfloor + 2$ because $\gamma_P(G) = k$, $\text{pt}_{\gamma_P}(G) = 2$, and $\gamma(G) = k+1$. The computation $\text{th}_{\gamma_P}(G) = 5$ for G created from C_3 is Example TH.5-2 in [**184**].

Families of Graphs. In the next theorem we list the power domination throttling numbers of several families of graphs; these values are listed explicitly in [**61**] or are immediate from Theorem 11.54. In Proposition 11.59 we establish the value of the power domination throttling number for several additional families.

THEOREM 11.58. [**61**]

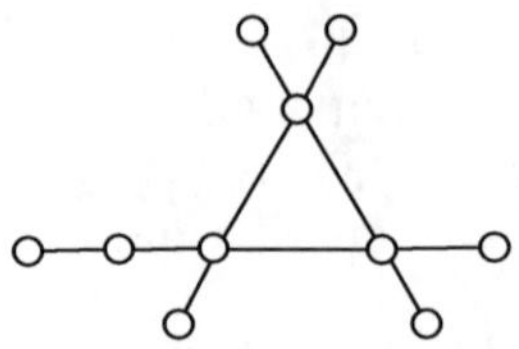

FIGURE 11.8. A graph G with $\text{th}_{\gamma_P}(G) = \lfloor \frac{n}{3} \rfloor + 2$.

(1) *For* $n \geq 1$, $\text{th}_{\gamma_P}(P_n) = \lceil \sqrt{2n} - \frac{1}{2} \rceil$.
(2) *For* $n \geq 4$, $\text{th}_{\gamma_P}(C_n) = \lceil \sqrt{2n} - \frac{1}{2} \rceil$.
(3) *For* $n \geq 4$, $\text{th}_{\gamma_P}(W_n) = 2$.
(4) $\text{th}_{\gamma_P}(K_n) = 2$.
(5) *For* $n \geq 3$, $\text{th}_{\gamma_P}(K_n - e) = 2$.
(6) *For* $n \geq 2$, $\text{th}_{\gamma_P}(K_{1,n-1}) = 2$, *and* $\text{th}_{\gamma_P}(K_{\ell,n-\ell}) = \ell + 1$ *for* $2 \leq \ell \leq \lfloor \frac{n}{2} \rfloor$.
(7) *For* $k \geq 1$, $\text{th}_{\gamma_P}(F_k) = 2$.
(8) *For* $k \geq 1$, $\text{th}_{\gamma_P}(B_k) = 2$.

PROPOSITION 11.59.
(1) *For* $n \geq 3$, $\text{th}_{\gamma_P}(\overline{C_n}) = 3$
(2) *For* $n \geq 3$, $\text{th}_{\gamma_P}(\overline{P_n}) = 3$.
(3) *For* $k \geq 1$, $\text{th}_{\gamma_P}(Q_{2^k}) = 2^{2^k-k-1} + 2$.

PROOF. Let $G = \overline{C_n}$ or $G = \overline{P_n}$. The cases $n = 3, 4$ are straightforward to verify, so assume $n \geq 5$. Then G does not have a universal vertex, so $\text{th}_{\gamma_P}(G) \geq 3$. Any one vertex is a power dominating set but not a dominating set, and the power propagation time is two.

By Theorem 9.147, $\gamma_P(Q_{2^k}) = 2^{2^k-k-1}$. Since power domination is achieved by dominating a copy of Q_{2^k-1}, $\text{pt}_{\gamma_P}(Q_{2^k}) = 2$. Thus $\text{th}_{\gamma_P}(Q_{2^k}) = 2^{2^k-k-1} + 2$ by Proposition 11.3. □

Graph Operations. In the proof of the next result, proofs of bounds established in Proposition 11.8 and 11.9 are omitted; examples showing tightness are presented in all cases.

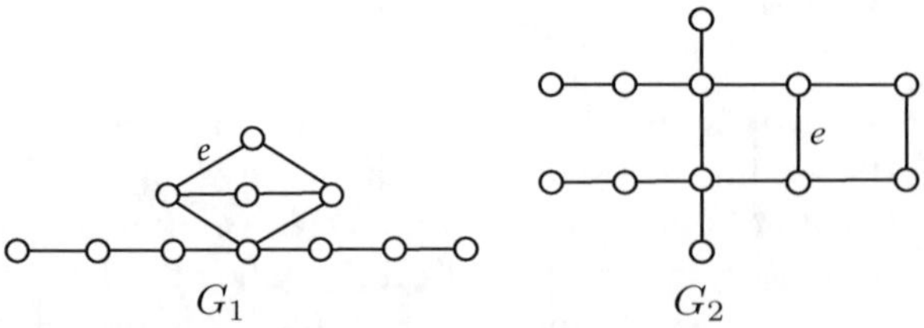

FIGURE 11.9. Graphs G_1 and G_2 with $\text{th}_{\gamma_P}(G_1 - e) = \text{th}_{\gamma_P}(G_1) - 1$ and $\text{th}_{\gamma_P}(G_2/e) = \text{th}_{\gamma_P}(G_2) + 1$.

THEOREM 11.60. *Let* G *be a graph.*
(1) *For any vertex* v *of* G, $\text{th}_{\gamma_P}(G) - 1 \leq \text{th}_{\gamma_P}(G - v)$. *There is no useful lower bound for* $\text{th}_{\gamma_P}(G) - \text{th}_{\gamma_P}(G - v)$.

(2) For any edge e of G, $\mathrm{th}_{\gamma_P}(G) - 1 \le \mathrm{th}_{\gamma_P}(G - e) \le \mathrm{th}_{\gamma_P}(G) + 1$.
(3) For any edge e of G, $\mathrm{th}_{\gamma_P}(G) - 1 \le \mathrm{th}_{\gamma_P}(G/e) \le \mathrm{th}_{\gamma_P}(G) + 1$.
(4) For any edge e of G, $\mathrm{th}_{\gamma_P}(G) \le \mathrm{th}_{\gamma_P}(G_e) \le \mathrm{th}_{\gamma_P}(G) + 1$.

All bounds are tight.

PROOF. $\mathrm{th}_{\gamma_P}(G)-1 \le \mathrm{th}_{\gamma_P}(G-v)$ is tight: Delete an end vertex from the path P_n where n is chosen such that $\mathrm{th}_{\gamma_P}(P_{n-1}) = \mathrm{th}_{\gamma_P}(P_n) - 1$. (e.g., $\mathrm{th}_{\gamma_P}(P_4) = 3$ and $\mathrm{th}_{\gamma_P}(P_3) = 2$).

No useful lower bound for $\mathrm{th}_{\gamma_P}(G)-\mathrm{th}_{\gamma_P}(G-v)$: Consider $H \in \mathfrak{T}$ (see Theorem 9.155), $G = H \vee K_1$, and $H = G - v$.

$\mathrm{th}_{\gamma_P}(G) \le \mathrm{th}_{\gamma_P}(G - e) + 1$ is tight: Let G_1 be the graph shown on the left in Figure 11.9. Then $\mathrm{th}_{\gamma_P}(G_1) = 5$ and $\mathrm{th}_{\gamma_P}(G_1 - e) = 4$. Computations illustrating this example are presented in Example TH.5-3 in [**184**].

$\mathrm{th}_{\gamma_P}(G - e) \le \mathrm{th}_{\gamma_P}(G) + 1$ is tight: Let $F = P_4 \vee K_1$ and let e be an edge whose endpoints have degrees three and four; F and $F-e$ are shown in Figure 11.5. Then $\mathrm{th}_{\gamma_P}(F) = 2$ and $\mathrm{th}_{\gamma_P}(F - e) = 3$. Computations illustrating this example are presented in Example TH.5-4 in [**184**].

$\mathrm{th}_{\gamma_P}(G) \le \mathrm{th}_{\gamma_P}(G/e) + 1$ and $\mathrm{th}_{\gamma_P}(G_e) \le \mathrm{th}_{\gamma_P}(G) + 1$ are tight: Contract an edge of P_4 (or subdivide an edge of P_3).

$\mathrm{th}_{\gamma_P}(G/e) \le \mathrm{th}_{\gamma_P}(G) + 1$: Let $e = \{u, w\}$ and let v_e denote the new vertex resulting from the contraction of e. Choose a set $S \subseteq V(G)$ such that $\mathrm{th}_{\gamma_P}(G) = \mathrm{th}_{\gamma_P}(G; S)$ and construct a propagating set of power domination forces $\mathcal{F}$. If at least one of u or w is in S, then define $S' = S\setminus\{u, w\}\cup\{v_e\}$; otherwise $S' = S$. Thus $|S'| \le |S|$. Observe that $N[S] \subseteq N[S']$. If neither u nor w performs a force after the domination step, then $S'' = S'$. Now suppose at least one of u, w performs a force after the domination step and $v_e \notin S'$. Define $S'' = S' \cup \{v_e\}$. Then $|S''| \le |S| + 1$ and $\mathrm{pt}_{\gamma_P}(G/e; S'') \le \mathrm{pt}_{\gamma_P}(G; S)$, so $\mathrm{th}_{\gamma_P}(G/e) \le \mathrm{th}_{\gamma_P}(G) + 1$. Let G_2 be the graph shown on the right in Figure 11.9. Then $\mathrm{th}_{\gamma_P}(G_2) = 4$ and $\mathrm{th}_{\gamma_P}(G_2/e) = 5$. Computations illustrating this example are presented in Example TH.5-5 in [**184**].

$\mathrm{th}_{\gamma_P}(G) \le \mathrm{th}_{\gamma_P}(G_e)$ is tight: Subdivide an edge of P_4. □

11.6. Throttling for Cops and Robbers

Throttling has been studied for the game of Cops and Robbers and results for cop throttling have implications for PSD throttling. The *cop throttling number* of a graph G of order n was introduced in [**57**] and defined to be $\mathrm{th}_c(G) = \min_{c(G)\le k\le n}\{k+\mathrm{capt}_k(G)\}$ (if $k < c(G)$, then the k-capture time is infinite). Since the capture time is one for a dominating set that is not all vertices, a commonly used equivalent definition is

$$\mathrm{th}_c(G) = \min_{c(G)\le k\le \gamma(G)} \{k + \mathrm{capt}_k(G)\}.$$

The *cop throttling number* of S in G is $\mathrm{th}_c(G; S) = |S|+\mathrm{capt}(G; S)$, if S is a capture set and infinity otherwise, so $\mathrm{th}_c(G) = \min_{S\subseteq V(G)} \mathrm{th}_c(G)$ is an another equivalent definition.

In the next figure we illustrate an optimal cop throttling set for the tree T_1 shown in Figure 10.15 and other examples.

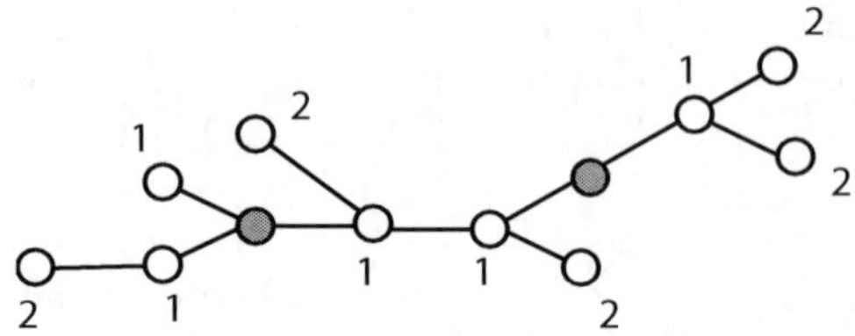

FIGURE 11.10. Cop throttling on the tree T_1, showing the time the cops would catch the robber if the robber were on that vertex using the set S of 2 vertices shown in gray: $\text{capt}(T_1) = 2$ and $\text{th}_c(T_1) = 2 + 2 = 4$. The cop throttling number of T_1 is not realized by with the minimum number of cops, since $c(T_1) = 1$ and $\text{capt}(T_1) = 4$ (as shown in Figure 10.15).

REMARK 11.61. Observation 11.1, Proposition 11.3, and Remarks 11.4, 11.5, and 11.6 apply to cop throttling. Proposition 11.2 does not apply to cop throttling, because $\text{capt}(G) > n$ is possible when when $c(G) \geq 2$ (see Theorem 10.66).

Since a dominating set is a capture set with capture time equal to one, the next bound is immediate, and examples showing that this bound is tight are presented in [**57**].

OBSERVATION 11.62. *For every graph G, $\text{th}_c(G) \leq \gamma(G) + 1$. If G has no isolated vertices, then $\text{th}_c(G) \leq \frac{n}{2} + 1$ where n is the order of G.*

The next result relates PSD throttling and cop throttling. Although we have included parts of this result in relevant earlier sections, for historical reasons we reproduce the entire result here.

THEOREM 11.63. [**57**] *Suppose $S \subseteq V(G)$ is a PSD zero forcing set. Then S is a capture set, $\text{capt}(G; S) \leq \text{pt}_+(G; S)$, $\text{th}_c(G; S) \leq \text{th}_+(G; S)$, $c(G) \leq \text{Z}_+(G)$, and $\text{th}_c(G) \leq \text{th}_+(G)$. If T is a tree, then $\text{th}_c(T) = \text{th}_+(T)$.*

Extreme Values. The next theorem characterizes graphs with very low values of $\text{th}_c(G)$.

THEOREM 11.64. [**57**] *Let G be a graph of order n.*

(1) $\text{th}_c(G) = 1$ *if and only if* $G = K_1$.
(2) $\text{th}_c(G) = 2$ *if and only if* $n \geq 2$ *and G satisfies one of the following conditions:*
  (a) $\gamma(G) = 1$.
  (b) $G = 2K_1$.
(3) $\text{th}_c(G) = 3$ *if and only if* $n \geq 3$ *and G satisfies one of the following conditions:*
  (a) $\gamma(G) > 2$ *and there exists* $z \in V(G)$ *such that*
    (i) for all $v \in V(G)$, $\text{dist}(z, v) \leq 2$, *and*
    (ii) for all $w \in G - N[z]$, *there is a vertex* $u \in N[z]$ *such that* $N[w] \subset N[u]$.
  (b) $\gamma(G) = 2$.
  (c) $G = 3K_1$.

COROLLARY 11.65. *Let G be a connected graph of order n. Then*

(1) $\text{th}_c(G) = 1$ *if and only if* $n = 1$.
(2) $\text{th}_c(G) = 2$ *if and only if* $\gamma(G) = 1$.
(3) $\text{th}_c(G) = 3$ *if and only if* G *satisfies one of the following conditions:*
  (a) $\gamma(G) > 2$ *and there exists* $z \in V(G)$ *such that*
    (i) for all $v \in V(G)$, $\text{dist}(z, v) \leq 2$, *and*
    (ii) for all $w \in G - N[z]$, *there is a vertex* $u \in N[z]$ *such that* $N[w] \subset N[u]$.
  (b) G *is connected and* $\gamma(G) = 2$.

It was shown in [**57**] that $\text{th}_c(G) = O(\sqrt{n})$ (where n is the order of G) for trees, unicyclic graphs, incidence graphs of finite projective planes $G(\mathbb{P}_q)$ (see Example 9.162), grids, and hypercubes. A square root bound for chordal graphs was established in [**46**] and improved by Geneson:

THEOREM 11.66. [**152**] *Let G be a chordal graph of order n. Then*

$$\text{th}_c(G) \leq \frac{\sqrt{14}}{2}\sqrt{n} + O(1) \approx 1.871\sqrt{n} + O(1).$$

Since $\text{th}_c(T) = \text{th}_+(T)$ for tree T, Remark 11.28 concerning bounds on $\text{th}_+(T)$ also applies to $\text{th}_c(T)$, and provides more background on the asymptotic upper bound for $\text{th}_c(G)$ for chordal graphs. It was asked in [**57**] whether whether $\text{th}_c(G) = O(\sqrt{n})$. A positive answer to this question would have implied Meyniel's Conjecture, since $c(G) \leq \text{th}_c(G)$. However it is not true in general. Figure 11.11 illustrates the method used to establish Theorem 11.67.

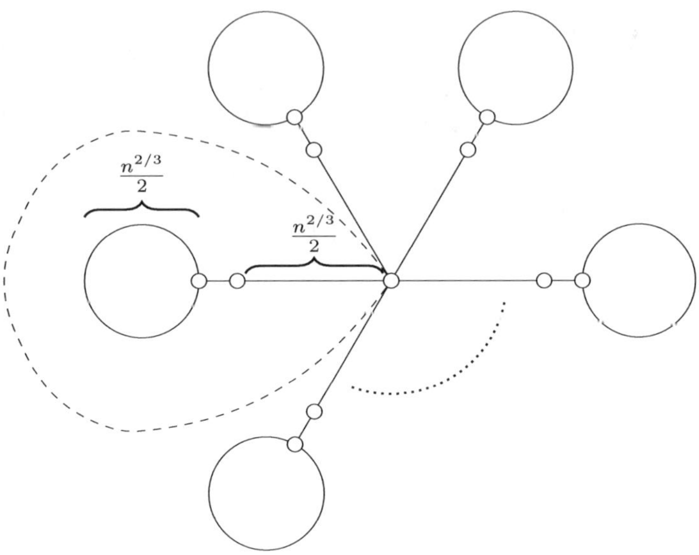

FIGURE 11.11. Construction of the connected graph H_n of order n with $\text{th}_c(H_n) = \Omega(n^{2/3})$. Each of the circles is a graph with cop number $\Omega\left(\sqrt{n^{2/3}}\right) = \Omega(n^{1/3})$ by using the existence of connected graphs on s vertices with cop number at least $\sqrt{\frac{s}{8}}$ for all $s \geq 72$ [**48**, Theorem 3.8].

THEOREM 11.67. [**46**] *There exist connected graphs of arbitrarily large orders n with cop throttling number $\Omega(n^{2/3})$.*

A (slightly) sublinear bound for cop throttling number is given in [**46**].

PROBLEM 11.68. *Find an (asymptotically) tight upper bound for cop throttling number.*

Families of Graphs. In the next theorem we list the cop throttling numbers of several families of graphs. In some cases the result is listed explicitly by Breen et al. whereas in others it is immediate from results in that paper (including Theorem 11.64).

THEOREM 11.69. [**57**]

(1) *For* $n \geq 1$, $\operatorname{th}_c(P_n) = \lceil \sqrt{2n} - \frac{1}{2} \rceil$.
(2) *For* $n \geq 4$, $\operatorname{th}_c(C_n) = \lceil \sqrt{2n} - \frac{1}{2} \rceil$.
(3) $\operatorname{th}_c(K_n) = 2$.
(4) $\operatorname{th}_c(K_n - e) = 2$.
(5) $\operatorname{th}_c(K_{1,n-1}) = 2$ *and* $\operatorname{th}_+(K_{\ell,n-\ell}) = \ell + 1$ *for* $2 \leq \ell \leq n-2$.
(6) *For* $k \geq 1$, $\operatorname{th}_c(F_k) = 2$.
(7) *For* $n \geq 5$, $\operatorname{th}_c(\overline{C_n}) = 3$.
(8) *For* $n \geq 5$, $\operatorname{th}_c(\overline{P_n}) = 3$.

The second part of the next result will be used in the discussion of the effect of edge deletion.

PROPOSITION 11.70.

(1) $\operatorname{th}_c(G(\mathbb{P}_q)) = q + 3$ *where* $G(\mathbb{P}_q)$ *is the incidence graph of the projective plane* $\mathbb{P}_q$ *(see Example 9.162).*
(2) *Let* G *be constructed from two disjoint copies of* $G(\mathbb{P}_q)$ *(the incidence graph of the projective plane* $\mathbb{P}_q$*) connected by an edge* $e = uv$ *where* u *is in one copy of* $\mathbb{P}_q$ *and* v *is in the other. Then* $\operatorname{th}_c(G) \leq q + 5$.

PROOF. (1): $\operatorname{th}_c(G(\mathbb{P}_q)) = q+3 = (q+1)+2 = c(G(\mathbb{P}_q)) + \operatorname{capt}(G(\mathbb{P}_q))$ follows Proposition 11.3 (see also Example 9.162 and Proposition 10.70).

(2): Use $q + 2$ cops and place them all on u initially. If the robber is placed in the copy of $G(\mathbb{P}_q)$ that contains u, the robber can be caught in 2 rounds using $q+1$ cops and leaving one of u. If the robber is placed in the copy of $G(\mathbb{P}_q)$ that contains v, the first round is used to move all cops to v and the robber can be caught in 2 rounds using $q + 1$ cops and leaving one on v. So $q + 2$ cops can catch the robber in at most 3 rounds and $\operatorname{th}_c(G) \leq q + 5$. □

Graph Operations. Some (but not all) of the bounds on the effects of graph operations that work for PSD throttling also work for cop throttling.

PROPOSITION 11.71. *Let* G *be a graph.*

(1) *For any vertex* v *of* G, $\operatorname{th}_c(G) - \operatorname{th}_c(G - v) \leq 1$. *There is no useful lower bound for* $\operatorname{th}_c(G) - \operatorname{th}_c(G - v)$.
(2) *For any edge* e *of* G, $\operatorname{th}_c(G) - \operatorname{th}_c(G - e) \leq 1$. *There is no useful lower bound for* $\operatorname{th}_c(G) - \operatorname{th}_c(G - e)$.

All bounds are tight.

PROOF. $\text{th}_c(G) - \text{th}_c(G - v) \leq 1$: Choose an optimal set for $G - v$ and place an additional cop on v. Deleting an end vertex from the path P_n where n is chosen such that $\text{th}_c(P_{n-1}) = \text{th}_c(P_n) - 1$ shows this bound is tight (e.g., $\text{th}_c(P_4) = 3$ and $\text{th}_c(P_3) = 2$).

No useful lower bound on $\text{th}_c(G) - \text{th}_c(G - v)$: Let $G = K_1 \vee H$ where H a graph that has a large cop throttling number (e.g., $\text{th}_c(P_n) \approx \sqrt{2n}$).

$\text{th}_c(G) - \text{th}_c(G - e) \leq 1$: Let $e = uw$ and define $S' = S \cup \{u\}$ (note that another cop is added to u even if u already has one or more cops). Then $\text{capt}(G; S') \leq \text{capt}(G; S)$ by leaving one cop on u and applying the capture process for S.

No useful lower bound on $\text{th}_c(G) - \text{th}_c(G - e)$: Let G be the graph in Proposition 11.70(2).; Then $\text{th}_c(G) \leq q + 5$ and $G - e = 2G(\mathbb{P}_q)$, so $\text{th}_c(G - e) \geq c(G - e) + 2 = 2q + 4$. □

11.7. Product Throttling

All the throttling discussed so far involves minimizing the sum of the number of vertices and the propagation time. Product throttling minimizes the product of the number of vertices and the propagation time. To make product throttling nontrivial, the case of a zero product (where each vertex is filled and the propagation time is zero) must be excluded. This can be done by requiring that the cost of positioning resources (e.g., cops) to fill vertices be considered (e.g., by adding one to the propagation time before multiplying by the number of vertices used); this is called *product throttling with initial cost* or *initial cost product throttling*. The study of product throttling was initiated in [**46**] for Cops and Robbers with the initial cost definition (the terminology 'initial cost' was introduced in [**15**]): The *product cop throttling number with initial cost* is

$$\text{th}_c^\times(G) = \min_{c(G) \leq k \leq n} \{k(1 + \text{capt}_k(G))\}.$$

Equivalently, $\text{th}_c^\times(G) = \min_{S \subseteq V(G)} \text{th}_c^\times(G; S)$ where $\text{th}_c^\times(G; S) = |S|(1+\text{capt}(G; S))$.

Alternatively, the initial set consisting of all vertices filled must be excluded (or a requirement that at least one time step be performed must be added). In the study of product throttling for power domination in [**14**], at least the domination step was required, and the product throttling number was defined as the product of the number of PMUs and the power propagation time. Requiring at least one time step and excluding the zero round case are effectively the same for connected graphs of order at least two, but it is more convenient to exclude the zero round case by requiring that the number of vertices used is less than the order (this avoids introducing a new definition of propagation time). We refer to this as *product throttling with no initial cost* or *no initial cost product throttling*. For a connected graph G of order $n \geq 2$, the *product power throttling number with no initial cost* is

$$\text{th}_{\text{pd}}^*(G) = \min_{\gamma_P(G) \leq k < n} k \, \text{pt}_{\gamma_P}(G, k);$$

equivalently, $\text{th}_{\text{pd}}^*(G) = \min_{S \subsetneq V(G)} \text{th}_{\text{pd}}^*(G; S)$ where $\text{th}_{\text{pd}}^*(G; S) = |S| \, \text{pt}_{\gamma_P}(G; S)$ for $S \neq V(G)$. (Note that what [**15**] and we denote by $\text{th}_{\text{pd}}^*(G)$ is denoted by $\text{th}_{\text{pd}}^\times(G)$ in [**14**].)

A universal approach to product throttling is presented in [**15**], including a comparison of the two types of product throttling and studies of product throttling for standard zero forcing, PSD forcing, power domination, and Cops and Robbers (skew forcing presents difficulties since arbitrarily large connected graphs can have skew forcing number zero). Using the two definitions given for product cop throttling and product power throttling as models, there are two universal definitions for product throttling:

$$\text{(with initial cost)}\quad \mathrm{th}_X^\times(G) = \min_{X(G)\le k\le n} k(1+\mathrm{pt}_X(G,k))$$

and

$$\text{(with no initial cost)}\quad \mathrm{th}_X^*(G) = \min_{X(G)\le k< n} k\,\mathrm{pt}_X(G,k),$$

where the definition of $\mathrm{th}_X^*(G)$ applies only to connected graphs of order at least two. The notation $\mathrm{th}_X^\times(G,k) = k(1+\mathrm{pt}_X(G,k))$ and $\mathrm{th}_X^*(G,k) = k\,\mathrm{pt}_X(G,k)$ is also used. As noted in [**15**] (where these definitions are introduced), the initial cost product throttling definition could be enhanced by considering a variable initial cost, i.e., $\mathrm{th}_X^\times(G) = \min_{X(G)\le k\le n} k(\omega+\mathrm{pt}_X(G,k))$. The minimum number of vertices that can be used to achieve propagation time p is denoted by $k_X(G,p) = \min\{k : \mathrm{pt}_X(G,k) = p\}$; the case of $k_X(G,1)$ is of particular interest.

In the remainder of this section we first review results for standard zero forcing, which is trivial for initial cost and not really throttling for no initial cost (the lack of interesting standard zero forcing product throttling problems may have delayed the study of product throttling). Then elementary universal results are presented. Finally, we summarize additional results specific to the individual parameters Cops and Robbers, power domination, and PSD zero forcing (including consequences of the universal results).

Product throttling for standard zero forcing.

REMARK 11.72. For any graph G of order n, $\mathrm{th}^\times(G) = n$ (achieved by filling all vertices) because

$$|B|(1+\mathrm{pt}(G;B)) \ge |B|\left(1+\frac{n-|B|}{|B|}\right) = n$$

by Equation (10.1).

THEOREM 11.73. [**15**] *For any graph G order $n\ge 2$, $\mathrm{th}^*(G)$ is the least k such that $\mathrm{pt}(G,k)=1$, i.e., $\mathrm{th}^*(G) = k(G,1)$. Furthermore, $k(G,1)\ge \frac{n}{2}$.*

Graphs attaining the extreme values $\mathrm{th}^*(G) = \frac{n}{2}$ and $\mathrm{th}^*(G) = n-1$ are characterized in [**15**].

Elementary results for universal product throttling. The universal definitions and results listed here may possibly apply to additional parameters.

OBSERVATION 11.74. [**15**] *Let X be one of PSD zero forcing, power domination, or Cops and Robbers, and let G be a graph of order n. Then:*

(1) If G is connected, $n\ge 2$, and $X(G)\le k\le n$, then $\mathrm{th}_X^*(G) < \mathrm{th}_X^\times(G)$ because $\mathrm{th}_X^\times(G,k) = k + \mathrm{th}_X^*(G,k)$.

(2) $\mathrm{th}_X^\times(G) \ge \min_{X(G)\le k\le n} k(1+\mathrm{rad}_k(G))$ and $\mathrm{th}_X^*(G) \ge \min_{X(G)\le k<n} k\,\mathrm{rad}_k(G)$ (G must be connected and $n\ge 2$ for $\mathrm{th}_X^*(G)$ to be defined).

OBSERVATION 11.75. [**15**] *Let X be one of PSD zero forcing, power domination, or Cops and Robbers, and let G be a graph of order n. Then:*

(1) $X(G) \leq \text{th}_X^\times(G) \leq n$. *If* G *is connected and* $n \geq 2$, *then* $X(G) + 1 \leq \text{th}_X^\times(G)$.
(2) $\text{th}_X^\times(G) \leq X(G)(1 + \text{pt}_X(G))$.
(3) $\text{th}_X(G) \leq \text{th}_X^\times(G)$. *Furthermore,* $\text{th}_X^\times(G) = \text{th}_X(G)$ *if and only if* $\text{th}_X(G) = \text{th}_X(G, 1)$ *or* $\text{th}_X(G) = n$.
(4) $\text{th}_X^\times(G) \leq 2k_X(G, 1)$.
(5) $X(G) \geq \frac{n}{2}$ *implies* $\text{th}_X^\times(G) = n$.

OBSERVATION 11.76. [**15**] *Let* X *be one of PSD zero forcing, power domination, or Cops and Robbers, and let* G *be a connected graph of order* $n \geq 2$. *Then:*
(1) $X(G) \leq \text{th}_X^*(G) \leq k_X(G, 1) \leq n - 1$.
(2) $\text{th}_X^*(G) \leq X(G)\,\text{pt}_X(G)$ *and* $\text{th}_X^*(G) = X(G)$ *if and only if* $\text{pt}_X(G) = 1$.
(3) *Suppose* G' *is a connected graph of order* $n' \geq 2$ *that is an induced subgraph of* G *such that* $\text{th}_X^*(G') = k_X(G', 1)$. *Then* $\text{th}_X^*(G) \leq n - n' + \text{th}_X^*(G')$.

PROPOSITION 11.77. [**15**] *Let* X *be one of PSD zero forcing, power domination, or Cops and Robbers, and let* G *be a connected graph of order* $n \geq 2$. *Then* $\text{th}_X^*(G) \geq \text{th}_X(G) - 1$. *If* $\text{th}_X(G) = \text{th}_X(G, 1)$ *or* $\text{th}_X(G) = \text{th}_X(G, k_X(G, 1))$, *then* $\text{th}_X^*(G) = \text{th}_X(G) - 1$.

Low values of product throttling numbers can be determined by factoring the values. With suitable adjustment of what it means to have $X(G) = k$ or $\text{pt}_X(G) = p$ for $k, p = 1, 2, 3$, the next result produces characterizations of graphs with $\text{th}_c^\times(G) = \ell$, $\text{th}_c^*(G) = \ell$, $\text{th}_{\text{pd}}^\times(G) = \ell$, $\text{th}_{\text{pd}}^*(G) = \ell$, $\text{th}_+^\times(G) = \ell$, and $\text{th}_+^*(G) = \ell$ for low ℓ.

REMARK 11.78. [**15**] Let X be one of PSD zero forcing, power domination, or Cops and Robbers, and let G be a connected graph of order $n \geq 2$. Then:
(a) No graph of order two or more has $\text{th}_X^\times(G) = 1$.
(b) $\text{th}_X^*(G) = 1$ if and only if $X(G) = 1$ and $\text{pt}_X(G) = 1$.
(c) $\text{th}_X^\times(G) = 2$ if and only if $X(G) = 1$ and $\text{pt}_X(G) = 1$.
(d) $\text{th}_X^*(G) = 2$ if and only if G satisfies exactly one of the following conditions:
  (i) $X(G) \leq 2$, $\text{pt}_X(G, 2) = 1$, and $\text{pt}_X(G, 1) > 2$.
  (ii) $X(G) = 1$ and $\text{pt}_X(G, 1) = 2$.
(e) $\text{th}_X^\times(G) = 3$ if and only if G satisfies exactly one of the following conditions:
  (i) $n = 3$ and $X(G) > 1$.
  (ii) $X(G) = 1$ and $\text{pt}_X(G, 1) = 2$.
(f) $\text{th}_X^*(G) = 3$ if and only if G satisfies exactly one of the following conditions:
  (i) $X(G) \leq 3$, $\text{pt}_X(G, 3) = 1$, $\text{pt}_X(G, 2) > 1$, and $\text{pt}_X(G, 1) > 3$.
  (ii) $X(G) = 1$, $\text{pt}_X(G, 1) = 3$, and $\text{pt}_X(G, 2) > 1$.
(g) $\text{th}_X^\times(G) = 4$ if and only if G satisfies exactly one of the following conditions:
  (i) $n = 4$, $X(G) > 1$, and $\text{pt}_X(G, 2) > 1$.
  (ii) $X(G) \leq 2$, $\text{pt}_X(G, 2) = 1$, and $\text{pt}_X(G, 1) > 3$.
  (iii) $X(G) = 1$ and $\text{pt}_X(G, 1) = 3$.

Next we consider the specific parameters Cops and Robbers, power domination, and PDS zero forcing. Recall that $\text{pt}_{\gamma_P}(G, \gamma(G)) = 1$ and $\text{capt}_{\gamma(G)}(G) = 1$. For a connected graph G of order $n \geq 2$, $\gamma(G) \leq \frac{n}{2}$ [**174**, Theorem 2.4].

Product throttling for Cops and Robbers. Product throttling for Cops and Robbers with initial cost was the first form of product throttling studied [**46**]. Observation 11.75 applies to $\text{th}_c^\times(G)$. In particular, $\text{th}_c^\times(G) \leq c(G)(1 + \text{capt}(G))$,

$\mathrm{th}_c^\times(G) \le 2\gamma(G)$, and $c(G) \le \mathrm{th}_c^\times(G) \le n$ where $n = |V(G)|$. Examples of graphs with $\mathrm{th}_c^\times(G) < c(G)(1 + \mathrm{capt}(G))$ and $\mathrm{th}_c^\times(G) < 2\gamma(G)$ are also provided in [**46**]. Low values of $\mathrm{th}_c^\times(G)$ are characterized in [**46**] (and are easy to derive using Remark 11.78). The next result is more substantial.

THEOREM 11.79. [**46**] *If H is a chordal graph, then* $\mathrm{capt}_k(H) = \mathrm{rad}_k(H)$ *and*

$$\mathrm{th}_c^\times(H) = 1 + \mathrm{rad}(H) = c(H) + \mathrm{capt}(H).$$

The next result is immediate from Theorem 11.79 and the characterization of low values. The value of $\mathrm{th}_c^\times(P_n)$ is given in [**46**].

COROLLARY 11.80.

(1) $\mathrm{th}_c^\times(P_n) = 1 + \lceil \frac{n-1}{2} \rceil$.
(2) $\mathrm{th}_c^\times(K_n) = 2$.
(3) $\mathrm{th}_c^\times(K_{1,n-1}) = 2$ *for* $n \ge 2$ *and* $\mathrm{th}_c^\times(K_{r,n-r}) = 4$ *for* $2 \le r \le n-2$.

Product throttling for Cops and Robbers with no initial cost was introduced in [**15**]. Observation 11.76 applies to no initial cost product throttling for Cops and Robbers. In particular, $c(G) \le \mathrm{th}_c^*(G) \le \gamma(G) \le \frac{n}{2}$ and $\mathrm{th}_c^*(G) \le c(G)\,\mathrm{capt}(G)$ for a connected graph G of order $n \ge 2$. Low values of $\mathrm{th}_c^*(G)$ are characterized in [**15**] (using Remark 11.78). It was shown in [**14**] that $\mathrm{th}_{\mathrm{pd}}^*(G) = \gamma(G)$ for unit interval graphs, and this was extended to $\mathrm{th}_c^*(G)$.

THEOREM 11.81. [**15**] *If G is a connected unit interval graph of order at least two, then* $\mathrm{th}_c^*(G) = \gamma(G)$.

Additional results concerning $\mathrm{th}_c^\times(G)$ (respectively, $\mathrm{th}_c^*(G)$) can be found in [**46**] (respectively, [**15**]).

Product throttling for power domination. Product throttling with no initial cost was introduced for power domination in [**14**]. Observation 11.76 applies to $\mathrm{th}_{\mathrm{pd}}^*(G)$. In particular, $\mathrm{th}_{\mathrm{pd}}^*(G) \le \gamma(G) \le \frac{n}{2}$ and $\mathrm{th}_{\mathrm{pd}}^*(G) \le \gamma_P(G)\,\mathrm{pt}_{\gamma_P}(G)$ for a connected graph G of order $n \ge 2$. Conditions on the graph that ensure $\mathrm{th}_{\mathrm{pd}}^*(G) = \gamma(G)$ were presented and the next theorem lists several families of graphs having $\mathrm{th}_{\mathrm{pd}}^*(G) = \gamma(G)$. Examples of graphs with $\mathrm{th}_{\mathrm{pd}}^*(G) < \gamma(G)$ and $\mathrm{th}_{\mathrm{pd}}^*(G) < \gamma_P(G)\,\mathrm{pt}_{\gamma_P}(G)$ are also provided in [**14**].

THEOREM 11.82. [**14**] *For the following graphs G,* $\mathrm{th}_{\mathrm{pd}}^*(G) = \gamma(G)$:

(1) For $n \ge 2$, P_n, C_n, K_n, and $K_{r,n-p}$ for $1 \le r \le n-1$.
(2) A connected unit interval graph of order at least two.
(3) For $2 \le n \le m$, $K_n \Box K_m$.
(4) For $n, m \ge 2$, $J_n \Box J_m$ where $J_r = P_r$ or $J_r = C_r$ (where $C_2 = P_2$).

Low values of $\mathrm{th}_{\mathrm{pd}}^*(G)$ are characterized in [**14**] (and are easy to derive using Remark 11.78). The next result addresses the highest value.

THEOREM 11.83. [**14**] *Let G be a connected graph of order $n \ge 2$. Then* $\mathrm{th}_{\mathrm{pd}}^*(G) = \frac{n}{2}$ *if and only if* $G = (H \circ K_1) \circ K_1$ *for some connected graph H,* $G = C_4 \circ K_1$*, or* $G = C_4$.

Product throttling for power domination with initial cost was introduced in [**15**]. Observation 11.75 applies to $\mathrm{th}_{\mathrm{pd}}^\times(G)$. In particular, $\mathrm{th}_{\mathrm{pd}}^\times(G) \le \gamma_P(G)(1 + \mathrm{pt}_{\gamma_P}(G))$, $\mathrm{th}_{\mathrm{pd}}^\times(G) \le 2\gamma(G)$, and $\mathrm{th}_{\mathrm{pd}}^\times(G) \le |V(G)|$.

THEOREM 11.84. [**15**] [1]

(1) $\text{th}_{\text{pd}}^{\times}(P_n) = \text{th}_{\text{pd}}^{\times}(C_n) = 1 + \lceil \frac{n-1}{2} \rceil$.
(2) $\text{th}_{\text{pd}}^{\times}(K_n) = 2$.
(3) $\text{th}_{\text{pd}}^{\times}(K_{1,n-1}) = 2$ *for* $n \geq 2$, $\text{th}_{\text{pd}}^{\times}(K_{2,n-2}) = 3$ *for* $n \geq 4$, *and* $\text{th}_{\text{pd}}^{\times}(K_{p,q}) = 4$ *for* $p, q \geq 3$.
(4) *For* $m \geq 2$, $\text{th}_{\text{pd}}^{\times}(P_2 \square P_m) = m$ *if* $m \equiv 0 \mod 3$ *and* $\text{th}_{\text{pd}}^{\times}(P_2 \square P_m) = m+1$ *if* $m \not\equiv 0 \mod 3$.

Low values of $\text{th}_{\text{pd}}^{\times}(G)$ are characterized in [**14**] (and are easy to derive using Remark 11.78). A family of graphs G_n of order n with $\text{th}_{\text{pd}}^{\times}(G_n) = \frac{6}{7}n$ is presented in [**15**]. Additional results concerning $\text{th}_{\text{pd}}^{*}(G)$ (respectively, $\text{th}_{\text{pd}}^{\times}(G)$) can be found in [**14**] and [**15**] (respectively, [**15**]).

Product throttling for PSD zero forcing. Both types of product throttling for PSD zero forcing were introduced in [**15**]. Observation 11.75 applies to $\text{th}_{+}^{\times}(G)$, and in particular, $\text{Z}_+(G) \geq \frac{n}{2}$ implies $\text{th}_{+}^{\times}(G) = n$, as illustrated in the next example. Low values of $\text{th}_{+}^{\times}(G)$ are characterized in [**15**] (using Remark 11.78) and illustrated in the next example.

EXAMPLE 11.85. [**15**] For $n, d \geq 1$, $\text{th}_{+}^{\times}(K_n) = n$ and $\text{th}_{+}^{\times}(Q_d) = 2^d$. For $n \geq 2$, $K_{1,n-1} = 2$.

The next result follows from Theorems 10.68 and 11.79 since a tree is chordal.

COROLLARY 11.86. [**15**] *Let* T *be a tree. Then* $\text{th}_{+}^{\times}(T, k) = \text{th}_{c}^{\times}(T, k)$ *and* $\text{th}_{+}^{\times}(T) = \text{th}_{c}^{\times}(T) = 1 + \text{rad}\, T$. *In particular,* $\text{th}_{+}^{\times}(P_n) = 1 + \lceil \frac{n-1}{2} \rceil$.

Observation 11.76 applies to $\text{th}_{+}^{*}(G)$. In particular, $\text{Z}_+(G) \leq \text{th}_{+}^{*}(G) < n - 1$ and $\text{th}_{\text{pd}}^{*}(G) \leq \text{Z}_+(G)\,\text{pt}_+(G)$. Low values of $\text{th}_{+}^{*}(G)$ are characterized in [**15**] (using Remark 11.78) and illustrated in the next example.

EXAMPLE 11.87. [**15**] For $n \geq 2$, $\text{th}_{+}^{*}(P_n) = \gamma(P_n) = \lceil \frac{n}{3} \rceil$. For $n \geq 4$, $\text{th}_{+}^{*}(C_n) = \gamma(C_n) = \lceil \frac{n}{3} \rceil$. For $n \geq 2$, $\text{th}_{+}^{*}(K_n) = \text{Z}_+(K_n) = n - 1$. For $1 \leq p \leq q$, $\text{th}_{+}^{*}(K_{p,q}) = \text{Z}_+(K_{p,q}) = p$. For $d \geq 1$, $\text{th}_{+}^{*}(Q_d) = \text{Z}_+(Q_d) = 2^{d-1}$.

PROPOSITION 11.88. [**15**] *For a graph* G *of order* $n \geq 2$, $\text{th}_{+}^{*}(G) \leq n - \alpha(G)$.

COROLLARY 11.89. [**15**] *For a connected graph* G *of order* $n \geq 2$, $\text{th}_{+}^{*}(G) = n - 1$ *if and only if* $G = K_n$.

THEOREM 11.90. [**15**] *Let* G *be a connected graph of order* $n \geq 2$. *Then* $\text{th}_{+}^{*}(G) = n - 2$ *if and only if* $\alpha(G) = 2$ *and* G *does not have an induced* C_5 *or house subgraph (see Figure 11.12).*

Additional results concerning $\text{th}_{+}^{\times}(G)$ and $\text{th}_{+}^{*}(G)$ can be found in [**15**].

---

[1] Remark 7.6 in [**15**] incorrectly states that power domination on trees behaves like Cops and Robbers (it does not) and that if T is a tree, then $\text{th}_{\text{pd}}^{\times}(T) = 1 + \text{rad}\, T$ (this is incorrect, as can be seen by considering a double star $DS(p, q)$ with $p, q \geq 2$.

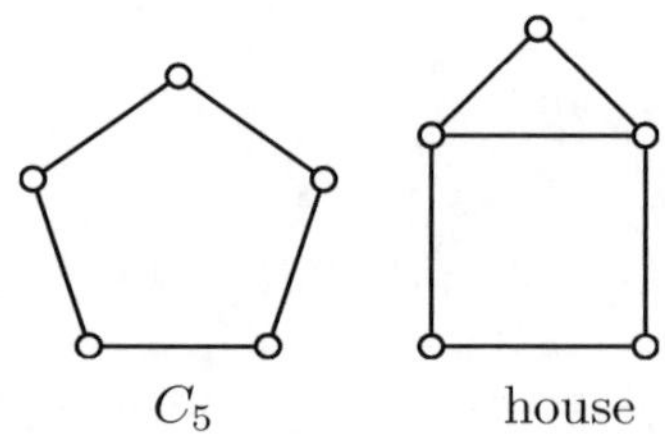

FIGURE 11.12. Graphs forbidden as induced subgraphs for $\text{th}_+^*(G) = n - 2$.

11.8. Topics Not Covered

We have not covered throttling for minor monotone floors of zero forcing parameters, which is studied in [**77**] and [**79**], since they need a more complicated definition of propagation time (recall that minor monotone floors of zero forcing parameters were also omitted from Chapter 10, Propagation Time and Capture Time).

Weighted throttling involves taking a linear combination of number of filled vertices/PMUs/cops and propagation time/capture time. This subject has been mentioned in some of the papers cited. In most cases the methods for weighted throttling are the same as for throttling and the results are analogous, but the values of throttling numbers differ depending on the weight. Information on weighted throttling can be found in [**74**] for standard throttling and in [**78**] for PSD throttling.

There has not been much work on throttling numbers of random graphs, but in many cases the values can be inferred from information about parameter values for random graphs (see Section 9.11).

APPENDIX A

Graph Terminology and Notation

In this appendix we review standard graph terminology and notation. Graph terminology is presented (and linear algebra terminology is not) because terminology is much more standard in linear algebra than in graph theory. A very elementary example is the ambiguity of the term *graph*, where there is a difference of opinion even among highly regarded reference works as to whether *graph* means a simple graph (no loops or multiple edges allowed) or a general graph (loops and multiple edges permitted). Hence we provide graph terminology here, whereas standard references such as [**178, 196, 271**] can be consulted for for linear algebra terminology.

Let $G = (V(G), E(G))$ be a ***graph***. The finite sets $\boldsymbol{V(G)}$ and $\boldsymbol{E(G)}$ are the ***vertex set*** and the ***edge set*** of G, respectively, where an edge is a two-element set of vertices; the edge $\{v, u\}$ is also denoted by uv. By our definition, a graph is necessarily ***simple***, i.e., does not contain multiple edges or loops, is undirected, and finite.

The ***order*** of G is the number of vertices $|V(G)|$ and the ***size*** of G is the number of edges $|E(G)|$.

In a graph, two vertices v and u are ***adjacent*** if uv is an edge, denoted by $\boldsymbol{v \sim u}$. The notation $\boldsymbol{v \not\sim u}$ means $v \neq u$ and $uv \notin E(G)$. The edge uv is ***incident*** to v and u (and vice versa).

A graph H is a ***subgraph*** of a graph G if $V(H) \subseteq V(G)$ and $E(H) \subseteq E(G)$; this is denoted by $\boldsymbol{H \leq G}$. Equivalently, G is a ***supergraph*** of H and $\boldsymbol{G \geq H}$.

For $U \subseteq V(G)$, the ***subgraph of G induced by U***, denoted by $\boldsymbol{G[U]}$, is the subgraph of G with $V(G[U]) = U$ and $E(G[U]) = \{uv : uv \in E(G) \text{ and } u, v \in E(G)\}$; $G[U]$ is also called an ***induced subgraph*** of G.

A subgraph H of G is a ***spanning subgraph*** if $V(H) = V(G)$.

The ***degree*** of a vertex v, denoted by $\mathbf{deg}_G(\boldsymbol{v})$, is the number of vertices that are adjacent to v; when the context is clear, the subscript is omitted.

The ***maximum degree*** of G is $\boldsymbol{\Delta(G)} = \max_{v \in V(G)} \deg_G(v)$.

The ***minimum degree*** of G is $\boldsymbol{\delta(G)} = \min_{v \in V(G)} \deg_G(v)$.

A graph is ***regular*** (or ***k-regular***) if every vertex has degree k for some fixed nonnegative integer k.

A ***high-degree vertex*** of a graph is a vertex of degree at least 3.

The (open) ***neighborhood*** of a vertex v is $\boldsymbol{N_G(v)} = \{u \in V(G) : v \sim u\}$.

The ***closed neighborhood*** is $\boldsymbol{N_G[v]} = N_G(v) \cup \{v\}$.

A vertex of degree one is called a ***leaf***.

A vertex that is adjacent to every other vertex is called a ***universal vertex***.

The ***independence number*** of G, denoted by $\boldsymbol{\alpha(G)}$, is the maximum cardinality of a set of vertices such that no two of which are adjacent.

Two graphs G and G' are ***isomorphic***, denoted by $\boldsymbol{G \cong G'}$, if there is a bijection $\varphi : V(G) \to V(G')$ such that $g \sim h$ if and only if $\varphi(g) \sim \varphi(h)$.

Several types of graph families are used in other definitions. These are listed here, before introducing additional families later in this appendix (see also Table A.1).

A ***complete graph*** K_n is a graph of order n that contains all possible edges. A ***clique*** of G is a subgraph of G that is a complete graph.

A ***path*** P_n is a graph with $V(P_n) = \{v_1, \ldots, v_n\}$ and $E(P_n) = \{\{v_i, v_{i+1}\} : i = 1, \ldots, n-1\}$.

A ***cycle*** C_n is a graph obtained from a path P_n just described by adding the edge $\{v_1, v_n\}$.

A ***walk*** on G is a sequence of vertices $v_0, v_1, \ldots, v_k$ such that $v_{i-1} \sim v_i$ for $i = 1, \ldots, k$; here k is the ***length*** of the walk, which is equal to the number of edges in the walk. In addition to referring to the graph P_n, the term ***path*** is also applied to a walk that does not repeat any vertex. The ***length*** of a path or cycle is the number of edges in the graph.

The ***distance*** between two vertices v and u, denoted by $\mathbf{dist}_G\boldsymbol{(v,u)}$, is the length of the shortest walk connecting them (necessarily a path).

The ***diameter*** of G, denoted by $\mathbf{diam}\boldsymbol{(G)}$, is the maximum distance among every possible pair of vertices in G.

The ***girth*** of G, denoted by $\boldsymbol{g(G)}$, is the minimum of the lengths of the cycles in G.

A graph is ***connected*** if for any two distinct vertices v and u there is a walk from v to u; otherwise G is ***disconnected***. A ***connected component*** of a graph is a maximal connected subgraph.

The ***connectivity*** or ***vertex connectivity***, denoted by $\boldsymbol{\kappa(G)}$, of a connected graph $G \neq K_n$ is the minimum number of vertices whose deletion disconnects G, and $\kappa(K_n) = n - 1$; if G is disconnected, then $\kappa(G) = 0$.

A ***path cover*** of G is a collection of induced paths in G with the property that every vertex of G is in exactly one path.

The ***path cover number of*** G, denoted by $\mathbf{P}\boldsymbol{(G)}$, is the minimum number of paths in a path cover of G.

A set of subgraphs of G, each of which is a clique and such that every edge of G is contained in at least one of these cliques, is called a ***clique cover*** of G (or an ***edge clique cover***).

The ***clique cover number*** of G, denoted by $\mathbf{cc}\boldsymbol{(G)}$, is the smallest number of cliques in a clique cover of G.

In a graph G, vertices v and w that have the same set of neighbors (except possibly v and w) are called ***twins***. If $N_G(v) = N_G(w)$, then v and w are ***independent twins***, whereas if $N_G[v] = N_G[w]$, then v and w are ***adjacent twins***.

A ***matching*** of G, is a set of independent edges in G, and the ***matching number*** of G, denoted by $\mathbf{match}\boldsymbol{(G)}$, is the maximum size of a matching.

Various graph operations are defined on one or two graphs. Let $G = (V, E)$ and $G' = (V', E')$ be graphs.

The ***complement*** of G, denoted by $\overline{G}$, is the graph $(V, \overline{E})$ where $uv \in \overline{E}$ if and only if $uv \notin E$ for $u, v \in V$ and $u \neq v$.

If $U \subsetneq V$, then $\boldsymbol{G - U}$ is the induced subgraph obtained from G by removing every vertex $u \in U$ and all edges that are incident to u; $G - \{u\}$ can be written as $\boldsymbol{G - u}$.

If $e \in E$, then $\boldsymbol{G - e}$ is the graph obtained from G by deleting the edge e.

If $e = uv$ is an edge of G, then $\boldsymbol{G/e}$ is the graph obtained from G by identifying u and v and suppressing any loops or multiple edges that arise in this process. We say that the edge e has been ***contracted*** and call G/e an ***contraction*** of G.

We say G' is a ***minor*** of G if G' can be obtained from G by a sequence of edge deletions, isolated vertex deletions, and edge contractions (in any order); this is denoted by $\boldsymbol{G' \preceq G}$.

The ***subdivision*** of edge $e = uv$ in G, denoted by $\boldsymbol{G_e}$, is the graph obtained from G by adding a new vertex w, adding two edges uw and vw and removing the edge e.

The ***complete edge subdivision*** of G, denoted by $\hat{\overline{G}}$, is the graph obtained from G by performing an edge subdivision once on each edge of G.

If $E \neq \emptyset$, the ***line graph*** of a graph G, denoted by $\boldsymbol{L(G)}$, is the graph having vertex set E, with two vertices in $L(G)$ adjacent if and only if the corresponding edges share an endpoint in G.

The ***r-th power*** of G, denoted by $\boldsymbol{G^r}$, is the graph on vertex set $V(G)$ where two vertices v, u are adjacent if $\mathrm{dist}_G(v, u) \leq r$.

The ***r-th strong power*** of G, denoted by $\boldsymbol{G^{(r)}}$ is the graph on vertex set $V(G)$ where two vertices v, u are adjacent if $\mathrm{dist}_G(v, u) = r$.

A vertex v of a connected graph G is called a ***cut-vertex*** if $G - v$ is disconnected.

A ***block*** of G is a maximal induced subgraph that contains no cut-vertex for itself; that is, $G[W]$ is a block if $G[W]$ has no cut-vertex and $G[W']$ has a cut-vertex whenever $W \subsetneq W'$. For any graph, its edges can be partitioned into the edges of each block, and two adjacent blocks share a cut-vertex.

The ***union*** of G and G' is the graph $\boldsymbol{G \cup G'} = (V \cup V', E \cup E')$. If $V \cap V' = \emptyset$, then the union is ***disjoint*** and can be denoted by $\boldsymbol{G \sqcup G'}$.

If $V \cap V' \neq \emptyset$, then the ***intersection*** of G and G' is the graph $\boldsymbol{G \cap G'} = (V \cap V', E \cap E')$.

Suppose G_i are graphs of order at least two for $i = 1, \ldots, h$, there is a vertex v such that for all $i \neq j$, $G_i \cap G_j = \{v\}$, and $G = \bigcup_{i=1}^{h} G_i$. If $h \geq 2$, then clearly v is a cut-vertex of G, and any graph with a cut-vertex can be expressed as such a decomposition with $h \geq 2$. When all $G_i - v$ are required to be connected, $G = \bigcup_{i=1}^{h} G_i$ is called the ***cut-vertex decomposition*** at v.

If G and G' are disjoint, the ***join*** of G and G', denoted by $\boldsymbol{G \vee G'}$, is the graph obtained from $G \sqcup G'$ by adding all possible edges between V and V'.

If G and G' are disjoint, the ***Cartesian product*** of G and G', denoted by $\boldsymbol{G \,\Box\, G'}$, is a graph with $V(G \,\Box\, G') = V \times V'$ such that (v, v') and (u, u') are adjacent if and only if (1) $v = u$ and $v'u' \in E'$, or (2) $v' = u'$ and $vu \in E$.

If G and G' are disjoint, the ***tensor product*** (also called the ***direct product***) of G and G', denoted by $\boldsymbol{G \times G'}$, is a graph with $V(G \times G') = V \times V'$ such that (v, v') and (u, u') are adjacent if $vu \in E$ and $v'u' \in E'$.

If G and G' are disjoint, the ***strong product*** of G and G', denoted by $\boldsymbol{G \boxtimes G'}$, is a graph on $V \times V'$ such that (v, v') and (u, u') are adjacent if and only if (1)

$v = u$ and $v'u' \in E'$, or (2) $v' = u'$ and $vu \in E$, or (3) $vu \in E$ and $v'u' \in E'$. Observe that $G \boxtimes G' = G \square G' \cup G \times G'$.

If G and G' are disjoint, the ***lexicographic product*** of G and G', denoted by $\boldsymbol{G * G'}$, is a graph on $V \times V'$ such that two vertices (v, v') and (u, u') are adjacent in $G * H$ if either (1) $vu \in E(G)$, or (2) $v = u$ and $v'u' \in E(H)$.

If G and G' are disjoint and G has order n, the ***corona*** of G with G', denoted by $\boldsymbol{G \circ G'}$, is the graph obtained from the disjoint union of G and n copies of G' by joining the ith vertex of G and vertices of the ith copy of G' for each i.

Note that $G' \square G \cong G \square G'$, $G' \times G \cong G \times G'$, and $G' \boxtimes G \cong G \boxtimes G'$, whereas $G' * G$ need not be isomorphic to $G * G'$ and $G' \circ G$ need not be isomorphic to $G \circ G'$.

Next we define additional graph families.

An ***empty graph*** $\overline{K_n}$ is a graph of order n with no edges.

A ***complete bipartite graph*** $K_{p,q}$ is a graph with $V(K_{p,q}) = X \sqcup Y$ such that $|X| = p$ and $|Y| = q$ and $E(K_{p,q}) = \{xy : x \in X \text{ and } y \in Y\}$; $K_{1,n-1}$ is called a ***star***.

A ***complete multipartite graph*** $K_{n_1,\dots,n_r}$ is a graph whose vertices can be partitioned into r parts of sizes $n_1, \dots, n_r$ and the set of edges is exactly the set of all possible edges between distinct parts.

A ***hypercube*** $Q_d = K_2 \square \cdots \square K_2$ is the Cartesian product of d copies of K_2.

The graph $P_s \square P_t$ is called an $s \times t$ ***grid*** (or a ***grid graph***).

A ***wheel*** is a graph of the form $W_n = C_{n-1} \vee K_1$.

For $k \geq 3$ and $r \geq 1$, a ***generalized wheel*** $GW(k, r)$ is a graph of order $kr + 1$ constructed from $C_k \square P_r$ by adding a new vertex adjacent to every vertex in one end copy of C_k. Note that a generalized wheel $GW(k, 1)$ is a wheel W_{k+1}.

For $k \geq 2$, the kth-***supertriangle***, denoted by T_k, is an equilateral triangular grid with k vertices on each side.

The sth-***half-graph***, denoted by Half_s, is the graph constructed from a copy of K_s with vertices $\{a_1, \dots, a_s\}$ and a copy of $\overline{K_s}$ with vertices $\{b_1, \dots, b_s\}$ by adding all edges $\{a_i, b_j\}$ such that $i + j \leq s + 1$.

The ***diamond*** is the graph obtained from C_4 by adding an edge.

For $k \geq 2$, the kth-***necklace***, denoted by N_k, is the 3-regular graph obtained from k copies of the diamond by adding k edges connecting pairs of vertices of degree 2; N_k can also be constructed from C_{3k} by adding an adjacent twin of every third vertex.

For $s \geq 2$ and $n \geq s+2$, a ***broom graph*** $B_n(s)$ is a graph of order n constructed from a $K_{1,s}$ and a P_{n-s} by identifying an end point of the path with the center of the star; i.e., the union of a star and a path where the intersection is one vertex that is the center of the star and an endpoint of the path.

For $k \geq 2$, the kth-***book***, denoted by B_k, is the graph obtained from k copies of C_4 by identifying an edge from each C_4; equivalently, $B_k = K_{1,k} \square P_2$.

For $k \geq 2$, the kth-***friendship graph***, denoted by F_k, is the graph with $2k + 1$ vertices and $3k$ edges constructed by joining one vertex to k disjoint copies of K_2; equivalently, F_k is a union of k copies of K_3 that share one common vertex.

The (k, p)-***lollipop***, denoted by $L_{k,p}$, is the graph constructed by connecting a K_k to a P_p by a bridge from an endpoint of the P_p to a vertex of the K_k.

A ***tree*** is a connected graph that does not contains a cycle as a subgraph.

A ***spider*** or ***generalized star*** is a tree that has at most one high-degree vertex; a path is a spider. Equivalently, a spider is a graph created from a star by (zero or more) edge subdivisions. A spider can be described in terms of lengths of its legs; e.g., $S(5,3,2)$ is a tree of order 11 with one vertex adjacent to three paths of orders 5, 3, and 2.

A ***double path*** is constructed by starting with two disjoint paths P_s and P_t where $s,t \geq 3$ and then adding one edge between any one degree-2 vertex of P_s and any one degree-2 vertex of P_t.

A ***double generalized star*** is obtained from two disjoint generalized stars, each of which has a high-degree vertex, by adding an edge between the two high-degree vertices.

For $p,q \geq 2$, a ***double star*** $DS(p,q)$ is a tree of order $p+q+2$ that has two adjacent vertices of degrees $p+1$ and $q+1$ and every other vertex is a leaf.

A ***circulant graph*** is defined by its order $n \geq 3$ and its connection set $S \subseteq \{1,2,\dots,\lfloor \frac{n}{2} \rfloor\}$ and is denoted by $\mathrm{Circ}(n,S)$. Let the vertices be $v_0,\dots,v_{n-1}$. Then v_i and v_j are adjacent if and only if $|i-j| \in S$ (with arithmetic performed modulo n). A ***consecutive circulant*** is a circulant graph of the form $\mathrm{Circ}(n,[k])$ with $k < \frac{n}{2}$ (for n even, $\mathrm{Circ}(n,[\frac{n}{2}]) = K_n$ and this graph is not usually viewed as a circulant).

For convenience, Table A.1 lists graph symbols and names. The "First use" is the section other than this one in which the graph first appears.

First use	Graph symbol	Name	Order
1.4	K_n	complete graph	n
1.4	$\overline{K_n}$	empty graph	n
1.3	P_n	path	n
2.2	C_n	cycle	n
1.4	$K_{1,n-1}$	star	n
2.1	$K_{n_1,\dots,n_r}$	complete multipartite graph	$n_1+\cdots+n_r$
9.1	Q_d	hypercube	2^d
9.1	W_n	wheel	n
9.9	$GW(k,r)$	generalized wheel	$kr+1$
9.1	T_k	supertriangle	$\frac{1}{2}k(k+1)$
9.1	Half_s	half-graph	$2s$
9.1	N_k	necklace	$4k$
9.1	$B_n(s)$	broom	n
9.3	B_k	book	$2k+2$
9.4	F_k	friendship graph	$2k+1$
8.2	$L_{k,p}$	lollipop	$k+p$
2.3	$S(\ell_1,\dots,\ell_r)$	spider or generalized star	$1+\ell_1+\cdots+\ell_r$
11.7	$DS(p,q)$	double star	$p+q+2$
2.4	$\mathrm{Circ}(n,S)$	circulant graph	n

TABLE A.1. Table of graph names and symbols.

A ***split graph*** is a graph G with the property that its vertices can be partitioned into two sets V_0 and V_1 such that V_0 is an independent set and V_1 induces a clique.

For any positive integer k, a ***k-tree*** is a graph that is obtained from K_{k+1} by repeatedly (possibly zero times) adding a new vertex and joining it to an existing k-clique. Thus, a 1-tree is a tree, and a tree of order two or more is a 1-tree.

A ***linear k-tree*** is constructed iteratively by starting with K_{k+1} and connecting each new vertex to the vertices of an existing K_k that includes a vertex of degree k.

A ***linear tree*** is a tree having a path that is incident to each vertex of degree 3 or more. Note that a linear tree is a 1-tree but not a linear 1-tree (which is a path).

For a graph G, the minimum k such that G is a subgraph of some k-tree is called the ***tree-width*** of G, denoted by $\operatorname{tw}(G)$.

A ***polygonal path*** is a "path" of cycles built up one cycle at a time by identifying an edge of a new cycle with an edge of the most recently added cycle that has a vertex of degree 2. A polygonal path has also been called an ***LSEAC*** graph, a ***2-connected partial linear 2-tree***, or a linear 2-tree by some authors (the last of these terms is incorrect, since a polygonal path need not be a 2-tree).

An ***interval graph*** is a graph that has a representation in which each vertex is assigned to a closed interval of $\mathbb{R}$ such that two vertices are adjacent exactly when the two corresponding intervals intersect.

A ***unit interval graph*** is an interval graph that has a representation in which the intervals are of unit length.

A ***cograph*** or a ***decomposable graph*** is a graph that can be constructed from isolated vertices by a sequence of joins or disjoint unions. Equivalently, a cograph is a graph that does not have have P_4 as an induced subgraph.

A ***threshold graph*** is a graph that can be constructed from a single vertex by a sequence of the following two operations: (1) Add a single isolated vertex. (2) Add a universal vertex. Note that (1) is a disjoint union and (2) is a join, so any threshold graph is a cograph. In fact, the family of threshold graphs is the intersection of cographs and split graphs, A graph G is a threshold graph if and only if G does not have any of P_4, C_4, or $2K_2$ as an induced subgraph.

A k-regular graph G of order n is ***strongly regular*** with parameters (n, k, a, c) if every pair of adjacent vertices has a common neighbors, and every pair of distinct nonadjacent vertices has c common neighbors.

A graph G is ***chordal*** if G has no induced cycle of length four or more. A connected chordal graph G can be built successively by adding cliques with vertex sets $X_1, \dots, X_k$ in such that for each $i = 2, \dots, k$, $X_i \cap (\cup_{j=1}^{i-1} X_j) \neq \emptyset$ and there exists an s with $1 \leq s \leq i-1$ such that $X_i \cap (\cup_{j=1}^{i-1} X_j) \subseteq X_s$. This implies $G[\cup_{j=1}^{i} X_j]$ is a connected chordal graph, $X_i \cap (\cup_{j=1}^{i-1} X_j)$ induces a clique, and $V(G) = \cup_{j=1}^{k} X_j$. The ordered sets $X_1, \dots, X_k$ are called a ***clique decomposition*** of G.

A graph is ***planar*** if it can be drawn on the plane such that its edges intersect only at their endpoints.

A graph is ***outerplanar*** if it can be drawn on the plane such that its edges intersect only at their endpoints and all vertices lie on the outer (infinite) face.

There are several matrices associated with a graph G with $V(G) = \{1, \dots, n\}$.

The ***adjacency matrix*** of G is the $n \times n$ matrix A_G whose i, j-entry is 1 if ij is an edge of G and is 0 otherwise.

The ***diagonal degree matrix*** of G is $D_G = [d_{ij}]$ where $d_{ii} = \deg(i)$ and $d_{ij} = 0$ for $i \neq j$.

The ***Laplacian matrix*** of G is $L_G = D_G - A_G$.

Observe that $A_G, L_G \in \mathcal{S}(G)$. It is immediate that $L_G\mathbb{1}_n = \mathbf{0}_n$ (where $\mathbb{1}_n$ is an all ones vector and $\mathbf{0}_n$ is an all zero vector) and well-known that L_G is positive semidefinite.

Asymptotic comparison is applied to graph parameters such as order and size in Part 4.

Let, f, g be real valued functions of the natural or real numbers, i.e., $f, g : \mathbb{N} \to \mathbb{R}$ or $f, g : \mathbb{R} \to \mathbb{R}$.

f is $\boldsymbol{O(g)}$ (***big-oh*** of g) if f does not grow faster than a constant multiple of g, i.e., there exist constants C, k such that $|f(x)| \le C|g(x)|$ for all $x \ge k$.

f is $\boldsymbol{o(g)}$ (***little-oh*** of g) if $\lim_{x\to\infty} \left|\frac{f(x)}{g(x)}\right| = 0$.

f is $\boldsymbol{\Omega(g)}$ (***big-omega*** of g) if f grows at least as fast as a constant multiple of g, i.e., there exist constants c, k such that $|f(x)| \ge c|g(x)|$ for all $x \ge k$.

f is $\boldsymbol{\Theta(g)}$ (***big-theta*** of g) if f is both $O(g)$ and $\Omega(g)$, i.e., there exist constants $c, C, k > 0$ such that $c|g(x)| \le |f(x)| \le C|g(x)|$ for all $x \ge k$.

Bibliography

[1] A. Aazami. Hardness results and approximation algorithms for some problems on graphs. Thesis (Ph.D.), University of Waterloo, 2008.
http://hdl.handle.net/10012/4147.

[2] A. Aazami. Domination in graphs with bounded propagation: algorithms, formulations and hardness results. *J. Combin. Optim.*, 19: 429–456, 2010.

[3] B. Ahmadi, F. Alinaghipour, M. Cavers, S. Fallat, K. Meagher, and S. Nasser. Minimum number of distinct eigenvalues of graphs. *Electron. J. Linear Algebra*, 26: 673-691, 2013.

[4] J. Ahn, C. Alar, B. Bjorkman, S. Butler, J. Carlson, A. Goodnight, H. Knox, C. Monroe, and M.C. Wigal. Ordered multiplicity inverse eigenvalue problem for graphs on six vertices. *Electron. J. Linear Algebra*, 37: 316–358, 2021.

[5] M. Aigner and M. Fromme. A game of Cops and Robbers. *Discrete Appl. Math.*, 8: 1–11, 1984.

[6] AIM Minimum Rank – Special Graphs Work Group (F. Barioli, W. Barrett, S. Butler, S.M. Cioaba, D. Cvetković, S.M. Fallat, C. Godsil, W. Haemers, L. Hogben, R. Mikkelson, S. Narayan, O. Pryporova, I. Sciriha, W. So, D. Stevanović, H. van der Holst, K. Vander Meulen, and A. Wangsness Wehe). Zero forcing sets and the minimum rank of graphs. *Linear Algebra Appl.*, 428: 1628–1648, 2008.

[7] AIM Minimum Rank Graph Catalog. L. Hogben, W. Barrett, J. Grout, H. van der Holst, K. Rasmussen, A. Smith, and D. Young, editors. Online catalog of minimum rank of families of graphs.
http://admin.aimath.org/resources/graph-invariants/minimumrankoffamilies/.

[8] J.S. Alameda, E. Curl, A. Grez, L. Hogben, O'N. Kingston, A. Schulte, D. Young, and M. Young. Families of graphs with maximum nullity equal to zero-forcing number. *Special Matrices*, 6: 56–67, 2018.

[9] American Institute of Mathematics Research Community, *Inverse eigenvalue problems for graphs and zero forcing.*
https://aimath.org/programs/researchcommunities/iepg/.

[10] American Institute of Mathematics SQuaRE, "Minimum rank of symmetric matrices described by a graph," held February 2008 and 2009 in Palo Alto, CA.

[11] American Institute of Mathematics Workshop, "Spectra of Families of Matrices described by Graphs, Digraphs, and Sign Patterns," held Oct. 23-27, 2006 in Palo Alto, CA.
http://aimath.org/pastworkshops/matrixspectrum.html.

[12] American Mathematical Society. Mathematics Research Community *Finding Needles in Haystacks: Approaches to Inverse Problems using Combinatorics and Linear Algebra.* Originally scheduled for June 2020, held June 6-11, 2021.
https://www.ams.org/programs/research-communities/2021MRC-Haystacks.

[13] D. Amos, Y. Caro, R. Davila, and R. Pepper. Upper bounds on the k-forcing number of a graph. *Discrete Appl. Math.*, 181: 1–10, 2015.

[14] S.E. Anderson, K.L. Collins, D. Ferrero, L. Hogben, C. Mayer, A.N. Trenk, S. Walker. Product throttling for power domination.
https://arxiv.org/abs/2010.16315.

[15] S.E. Anderson, K.L. Collins, D. Ferrero, L. Hogben, C. Mayer, A.N. Trenk, S. Walker. Product throttling. In *Research Trends in Graph Theory and Applications* (D. Ferrero, L. Hogben, S.R. Kingan, and G.L. Matthews, editors), Association for Women in Mathematics Series, Springer, New York, 2021, pp. 11–50.

[16] V.I. Arnold. On matrices depending on parameters. In *Vladimir I. Arnold - Collected Works, vol II, Hydrodynamics, Bifurcation Theory, and Algebraic Geometry 1965-1972* (A. Givental, B. Khesin, A. Varchenko, V. Vassiliev, O. Viro, editors). Springer-Verlag, Berlin, 2014 (originally published in 1971).

[17] R.F. Bailey and R. Craigen. Orthogonal Matrices with Zero Diagonal. *Electron. J. Linear Algebra*, 35: 307–318, 2019.

[18] F. Barioli, W. Barrett, S. Fallat, H.T. Hall, L. Hogben, B. Shader, P. van den Driessche, and H. van der Holst. Zero forcing parameters and minimum rank problems. *Linear Algebra Appl.*, 433: 401–411, 2010.

[19] F. Barioli, W. Barrett, S. Fallat, H.T. Hall, L. Hogben, B. Shader, P. van den Driessche, and H. van der Holst. Parameters related to tree-width, zero forcing, and maximum nullity of a graph. *J. Graph Theory*, 72: 146–177, 2013.

[20] F. Barioli, W. Barrett, S. Fallat, H.T. Hall, L. Hogben, and H. van der Holst. On the graph complement conjecture associated with minimum rank. *Linear Algebra Appl.*, 436: 4373–4391, 2012.

[21] F. Barioli and S.M. Fallat. On two conjectures regarding an inverse eigenvalue problem for acyclic symmetric matrices. *Electron. J. Linear Algebra*, 11: 41–50, 2004.

[22] F. Barioli and S. Fallat. On the minimum rank of the join and decomposable graphs. *Linear Algebra Appl.*, 421: 252–263, 2007.

[23] F. Barioli and S.M. Fallat. On the eigenvalues of generalized stars and double generalized stars. *Linear Multilinear Algebra*, 53: 269–291, 2005.

[24] F. Barioli, S.M. Fallat, H.T. Hall, D. Hershkowitz, L. Hogben, H. van der Holst, and B. Shader. On the minimum rank of not necessarily symmetric matrices: a preliminary study. *Electron. J. Linear Algebra*, 18: 126–145, 2009.

[25] F. Barioli, S.M. Fallat, and L. Hogben. Computation of minimal rank and path cover number for graphs. *Linear Algebra Appl.*, 392: 289–303, 2004.

[26] F. Barioli, S.M. Fallat, and L. Hogben. On the difference between the maximum multiplicity and path cover number for tree-like graphs. *Linear Algebra Appl.*, 409: 13–31, 2005.

[27] F. Barioli, S.M. Fallat, and L. Hogben. A variant on the graph parameters of Colin de Verdière: Implications to the minimum rank of graphs. *Electron. J. Linear Algebra*, 13: 387–404, 2005.

[28] F. Barioli, S. Fallat, L. Mitchell, and S. Narayan. Minimum semidefinite rank of outerplanar graphs and the tree cover number. *Electron. J. Linear Algebra*, 22: 10–21, 2011.

[29] W. Barrett, R. Bowcutt, M. Cutler, S. Gibelyou, and K. Owens. Minimum rank of edge subdivisions of graphs. *Electron. J. Linear Algebra*, 18: 530–563, 2009.

[30] W. Barrett, S. Butler, M. Catral, S. Fallat, H.T. Hall, L. Hogben, and M. Young. The maximum nullity of a complete subdivision graph is equal to its zero forcing number. *Electron. J. Linear Algebra*, 27: 458–477, 2014.

[31] W. Barrett, S. Butler, S.M. Fallat, H.T. Hall, L. Hogben, J.C.-H. Lin, B. Shader, and M. Young. The inverse eigenvalue problem of a graph: Multiplicities and minors. *J. Combin. Theory Ser. B*, 142: 276–306, 2020.

[32] W. Barrett, S. Fallat, H.T. Hall, and L. Hogben. Note on Nordhaus-Gaddum problems for Colin de Verdière type parameters. *Electron. J. Combin.*, 20: #P56, 2013.

[33] W. Barrett, S.M. Fallat, H.T. Hall, L. Hogben, J.C.-H. Lin, and B. Shader. Generalizations of the Strong Arnold Property and the minimum number of distinct eigenvalues of a graph. *Electron. J. Combin.*, 24: #P2.40, 2017.

[34] W.W. Barrett, H. van der Holst, and R. Loewy. Graphs whose minimal rank is two. *Electron. J. Linear Algebra*, 11: 258–280, 2004.

[35] W. Barrett, A. Lazenby, N. Malloy, C. Nelson, W. Sexton, R. Smith, J. Sinckovic, and T. Yang. The combinatorial inverse eigenvalue problems: complete graphs and small graphs with strict inequality. *Electron. J. Linear Algebra*, 26: 656–672, 2013.

[36] W. Barrett, C. Nelson, J. Sinckovic, and T. Yang. The combinatorial inverse eigenvalue problems II: all cases for small graphs. *Electron. J. Linear Algebra*, 27: 742–778, 2014.

[37] L.B. Beasley, R.A. Brualdi, B.L. Shader. Combinatorial orthogonality. In *Combinatorial and graph-theoretical problems in linear algebra* (R.A. Brualdi, S. Friedland, V. Klee, editors), IMA Vol. Math. Appl., 50, Springer, New York, 1993, 207–218.

[38] K.F. Benson, D. Ferrero, M. Flagg, V. Furst, L. Hogben, V. Vasilevska, and B. Wissman. Zero forcing and power domination for graph products. *Australas. J. Combin.*, 70: 221–235, 2018.

[39] K.F. Benson, D. Ferrero, M. Flagg, V. Furst, L. Hogben, V. Vasilevska. Note on Nordhaus-Gaddum problems for power domination. *Discrete Appl. Math.*, 251: 103–113, 2018.

[40] R. Bergen, S. Fallat, A. Gorr, F. Ihringer, K. Meagher, A. Purdyk, B. Yang, and G. Yu. Infection in Hypergraphs. *Discrete Appl. Math.*, 237: 43–56, 2018.

[41] A. Berman, S. Friedland, L. Hogben, U.G. Rothblum, and B. Shader. Minimum rank of matrices described by a graph or pattern over the rational, real and complex numbers. *Electron. J. Combin.*, 15: #P25, 2008.

[42] B. Bjorkman. Infectious power domination of hypergraphs. *Discrete Math.*, 343: 111724 (12 pp.), 2020.

[43] B. Bjorkman Morrison. Variations on the power domination problem: Hypergraphs and robustness. Dissertation (Ph.D.), Iowa State University, 2020.
`https://aimath.org/~hogben/Bjorkman_Dissertation_FINAL.pdf`.

[44] B. Bjorkman, L. Hogben, S. Ponce, C. Reinhart, and T. Tranel. Applications of analysis to the determination of the minimum number of distinct eigenvalues of a graph. *Pure Appl. Funct. Anal.*, 3: 537–563, 2018.

[45] D. Bokal, G. Fijavž, and B. Mohar. The minor crossing number. *SIAM Journal on Discrete Mathematics*, 20: 344–356, 2006.

[46] A. Bonato, J. Breen, B. Brimkov, J. Carlson, S. English, J. Geneson, L. Hogben, K.E. Perry, and C. Reinhart. Cop throttling number: Bounds, values, and variants. *J. Comb.*, 13: 79–103, 2022.

[47] A. Bonato, P. Golovach, G. Hahn, and J. Kratochvíl. The capture time of a graph. *Discrete Math.*, 309: 5588–5595, 2009.

[48] A. Bonato and R.J. Nowakowski. *The game of Cops and Robbers on graphs.* American Mathematical Society, Providence, 2011.

[49] A. Bonato, X. Pérez-Giménez, P. Prałat, and B. Reiniger. The Game of overprescribed Cops and Robbers played on graphs. *Graphs Combin.*, 57: 801–815, 2017.

[50] A. Bonato and P. Prałat. *Graph Searching Games and Probabilistic Methods.* CRC Press, Boca Raton, 2017.

[51] M. Booth, P. Hackney, B. Harris, C.R. Johnson, M. Lay, T.D. Lenker, L.H. Mitchell, S.K. Narayan, A. Pascoe, and B.D. Sutton. On the minimum semidefinite rank of a simple graph. *Linear Multilinear Algebra*, 59: 483–506, 2011.

[52] M. Booth, P. Hackney, B. Harris, C.R. Johnson, M. Lay, L.H. Mitchell, S.K. Narayan, A. Pascoe, K. Steinmetz, B.D. Sutton, and W. Wang. On the minimum rank among positive semidefinite matrices with a given graph. *SIAM J. Matrix Anal. Appl.*, 30: 731–740, 2008.

[53] C. Bozeman. Connections to zero forcing: Tree covers and power domination. Dissertation (Ph.D.), Iowa State University, 2018.

[54] C. Bozeman. On the power propagation time of a graph.
`https://arxiv.org/pdf/1608.01189.pdf`.

[55] C. Bozeman, B. Brimkov, C. Erickson, D. Ferrero, M. Flagg, and L. Hogben. Restricted power domination and zero forcing problems. *J. Combin. Optim.*, 37: 935–956, 2019.

[56] C. Bozeman, AV. Ellsworth, L. Hogben, J.C.-H. Lin, G. Maurer, K. Nowak, A. Rodriguez, and J. Strickland. Minimum rank of graphs with loops. *Electron. J. Linear Algebra*, 27: 907–934, 2014.

[57] J. Breen, B. Brimkov, J. Carlson, L. Hogben, K.E. Perry, and C. Reinhart. Throttling for the game of Cops and Robbers on graphs. *Discrete Math.*, 341: 2418–2430, 2018.

[58] B. Brešar, C. Bujtás, T. Gologranc, S. Klavžar, G. Košmrlj, B. Patkós, Z. Tuza, M. Vizer. Grundy dominating sequences and zero forcing sets. *Discrete Optim.*, 26: 66–77, 2017.

[59] W. G. Bridges and R.A Mena. Multiplicative cones–a family of three eigenvalue graphs. *Aequationes Math.*, 22: 208–214, 1981.

[60] B. Brimkov. MATLAB code for the paper "Throttling for the game of Cops and Robbers on graphs."
`http://www.caam.rice.edu/~bb19/throttling.html`.

[61] B. Brimkov, J. Carlson, I.V. Hicks, R. Patel, and L. Smith. Power domination throttling. *Theoret. Comput. Sci.*, 795: 142–153, 2019.

[62] B. Brimkov, C.C. Fast, and I.V. Hicks. Computational approaches for zero forcing and related problems. *European J. Oper. Res.*, 273: 889–903, 2019.

[63] B. Brimkov and I.V. Hicks. Complexity and computation of connected zero forcing. *Discrete Appl. Math.*, 229: 31–45, 2017.

[64] B. Brimkov and Z. Scherr. An exact algorithm for the minimum rank of a graph. `https://arxiv.org/abs/1912.00158`.

[65] R.A. Brualdi. From the editor-in-chief. Errata, additional references and other comments. *Linear Algebra Appl.*, 608: 407–409, 2021.

[66] R.A. Brualdi and H.J. Ryser. *Combinatorial Matrix Theory.* Cambridge University Press, Cambridge, 1991.

[67] R. Brualdi and B. Shader. *Matrices of Sign-solvable Linear Systems.* Cambridge University Press, Cambridge, 1995.

[68] D.J. Brueni and L.S. Heath. The PMU placement problem. *SIAM J. Discrete Math.*, 19: 744–761, 2005.

[69] D. Burgarth, D. D'Alessandro, L. Hogben, S. Severini, and M. Young. Zero forcing, linear and quantum controllability for systems evolving on networks. *IEEE Trans. Auto. Control*, 58: 2349–2354, 2013.

[70] D. Burgarth and V. Giovannetti. Full control by locally induced relaxation. *Phys. Rev. Lett.*, PRL 99, 100501, 2007.

[71] S. Butler, L. DeLoss, J. Grout, H.T. Hall, J. LaGrange, J.C.-H. Lin, T. McKay, J. Smith, G. Tims, and B. Wissman. Minimum Rank, Zero Forcing, Power Domination (*Sage* programs for calculating bounds on the minimum rank of a graph, and for computing zero forcing parameters and power domination number). `https://sage.math.iastate.edu/home/pub/125/`.

[72] S. Butler, C. Erickson, S. Fallat, H.T. Hall, B. Kroschel, J.C.-H. Lin, B. Shader, N. Warnberg, and B. Yang. Properties of a q-analogue of zero forcing. *Graphs Combin.*, 36: 1401–1419, 2020.

[73] S. Butler, J. Grout, and H.T. Hall. Using variants of zero forcing to bound the inertia set of a graph. *Electron. J. Linear Algebra*, 30 (2015): 1–18.

[74] S. Butler and M. Young. Throttling zero forcing propagation speed on graphs. *Australas. J. Combin.*, 57: 65–71, 2013.

[75] D. de Caen, E.R. van Dam, and E. Spence. A non-regular analogue of conference graphs. *J. Combin. Theory Ser. A*, 88: 194–204, 1999.

[76] T.R. Cameron. Personal communication. 2022.

[77] J. Carlson. Throttling for Zero Forcing and Variants. *Australas. J. Combin.*, 75: 96–112, 2019.

[78] J. Carlson, L. Hogben, J. Kritschgau, K. Lorenzen, M.S. Ross, S. Selken, and V. Valle Martinez. Throttling positive semidefinite zero forcing propagation time on graphs. *Discrete Appl. Math.*, 254: 33–46, 2019.

[79] J. Carlson and J. Kritschgau. Various characterizations of throttling numbers. *Discrete Appl. Math.*, 294: 85–97, 2021.

[80] M. Catral, A. Cepek, L. Hogben, M. Huynh, K. Lazebnik, T. Peters, and M. Young. Zero forcing number, maximum nullity, and path cover number of subdivided graphs. *Electron. J. Linear Algebra*, 23: 906–922, 2012.

[81] Y. Chan, E. Curl, J. Geneson, L. Hogben, K. Liu, I. Odegard, and M.S. Ross. Using Markov chains to determine expected propagation time for probabilistic zero forcing. *Electron. J. Linear Algebra*, 36: 318–333, 2020.

[82] G.J. Chang, P. Dorbec, M. Montassier, and A. Raspaud. Generalized power domination of graphs. *Discrete Appl. Math.*, 160: 1691–1698, 2012.

[83] G.J. Chang and N. Roussel. On the k-power domination of hypergraphs. *J. Combin. Optim.*, 30: 1095–1106, 2015.

[84] G.-S. Cheon, S.-G. Hwang, S.-H. Rim, B.L. Shader, S.-Z. Song. Sparse orthogonal matrices. *Linear Algebra Appl.*, 373: 211–222, 2003.

[85] K. Chilakamarri, N. Dean, C.X. Kang, and E. Yi. Iteration Index of a Zero Forcing Set in a Graph. *Bull. Inst. Combin. Appl.*, 64: 57–72, 2012.

[86] I. Cinzori, C.R. Johnson and H. Lang. An explicit upper bound on disparity for trees of a given diameter. To appear in *Linear Multilinear Algebra.*

[87] N.E. Clarke, G. MacGillivray. Characterizations of k-copwin graphs. *Discrete Math.*, 312 (2012) 1421–1425.

[88] Y. Colin de Verdière. Sur un nouvel invariant des graphes et un critère de planarité. *J. Combin. Theory Ser. B*, 50:11–21, 1990.

[89] Y. Colin de Verdière. On a new graph invariant and a criterion for planarity. In *Graph Structure Theory* (N. Robertson and P. Seymour, editors), Contemporary Mathematics 147, American Mathematical Society, Providence, RI, 1993, pp. 137–147.

[90] Y. Colin de Verdière. Multiplicities of eigenvalues and tree-width graphs. *J. Combin. Theory Ser. B*, 74: 121–146, 1998.

[91] R. Courant and D. Hilbert, *Methods of Mathematical Physics.* Wiley-Interscience, New York, 1953.

[92] E. Curl, J. Geneson, and L. Hogben. Skew throttling. *Australas. J. Combin.*, 78: 177–190, 2020.

[93] B.A. Curtis and B.L. Shader. Sign patterns of orthogonal matrices and the strong inner product property. *Linear Algebra Appl.*, 592: 228–259, 2020.

[94] J.L.R. D'Alembert. Recherches sur la courbe que forme une corde tendüe mise en vibration. *Histoire de l'Académie Royale des Sciences et des Belles Lettres de Berlin pour l' année MDCCXLVII*, 214–219, 1749.

[95] J.L.R. D'Alembert. Suite des recherches sur la courbe que forme une corde tendüe, mise en vibration. *Histoire de l'Académie Royale des Sciences et des Belles Lettres de Berlin pour l' année MDCCXLVII*, 220–249, 1749.

[96] J.L.R. D'Alembert. Addition au mémoire sur la courbe que forme une corde tendüe, mise en vibration. *Histoire de l'Académie Royale des Sciences et des Belles Lettres de Berlin pour l' année 1750*, 355–360, 1752.

[97] F.S. Dahlgren, Z. Gershkoff, L. Hogben, S. Motlaghian, and D. Young. The inverse eigenvalue problem for hollow matrices described by graphs. Preprint.

[98] E.R. van Dam. The combinatorics of Dom de Caen. *Des. Codes Cryptogr.*, 34: 137–148. 2005.

[99] E.R. van Dam. Nonregular graphs with three eigenvalues. *J. Combin. Theory Ser. B*, 73: 101–118, 1998.

[100] E.R. van Dam and J.H. Koolen. A new family of distance-regular graphs with unbounded diameter. *Invent. Math.*, 162: 189–193, 2005.

[101] E.R. van Dam, J.H. Koolen, and Z.-J. Xia. Graphs with many valencies and few eigenvalues. *Electron. J. Linear Algebra*, 28: 12–24, 2015.

[102] A. Daniilidis, J. Malick, and H. Sendov. Spectral (isotropic) manifolds and their dimension. *J. Anal. Math.*, 128: 369–397, 2016.

[103] R. Davila and M.A. Henning. Total forcing and zero forcing in claw-free cubic graphs. *Graphs Combin.*, 34: 1371–1384, 2018.

[104] R. Davila and M.A. Henning. On the total forcing number of a graph. *Discrete Appl. Math.*, 257. 115 127, 2019.

[105] R. Davila, M.A. Henning, C. Magnant, and R. Pepper. Bounds on the connected forcing number of a graph. *Graphs Combin.*, 34: 1159–1174, 2018.

[106] R. Davila, T. Kalinowski, and S. Stephen. A lower bound on the zero forcing number. *Discrete Appl. Math.*, 250: 363–367, 2018.

[107] R. Davila and F. Kenter. Bounds for the zero-forcing number of graphs with large girth. *Theory Appl. Graphs*, 2, no. 2, Art. 1, 8 pp., 2015.

[108] L. DeAlba, J. Grout, L. Hogben, R. Mikkelson, and K. Rasmussen. Universally optimal matrices and field independence of the minimum rank of a graph. *Electron. J. Linear Algebra*, 18: 403–419, 2009.

[109] L.M. DeAlba, T.L. Hardy, I.R. Hentzel, L. Hogben, and A. Wangsness. Minimum rank and maximum eigenvalue multiplicity of symmetric tree sign patterns. *Linear Algebra Appl.*, 418: 389–415, 2006.

[110] N. Dean, A. Ilic, I. Ramirez, J. Shen, and K. Tian. On the power dominating sets of hypercubes. *IEEE 14th International Conference on Computational Science and Engineering (CSE)*, 488–491, 2011.

[111] L. DeLoss, J. Grout, L. Hogben, T. McKay, J. Smith, and G. Tims. Techniques for determining the minimum rank of a small graph. *Linear Algebra Appl.*, 432: 2995–3001, 2010.

[112] R. Diestel. *Graph Theory*, 5th edition. Springer, Berlin, 2017.

[113] V. Dobrynin. On the rank of a matrix associated with a graph. *Discrete Math.*, 276: 169–175, 2004.

[114] A.L. Dontchev and R.T. Rockafeller. *Implicit Functions and Solution Mappings: A View form Variational Analysis*, 2nd edition. Springer Series in Operations Research and Financial Engineering, Springer, New York, 2014.

[115] M. Doob. Graphs with a small number of distinct eigenvalues. *Ann. N.Y. Acad. Sci.*, 175: 104–110, 1970.

[116] P. Dorbec, S. Varghese, A. Vijayakumar. Heredity for generalized power domination. *Discrete Math. Theor. Comput. Sci.*, 18: no. 3, Paper No. 5, 11 pp., 2016.

[117] A.C. Downing and A.S. Householder. Some inverse characteristic value problems, *J. Assoc. Comput. Mach.*, 3: 203–207, 1956.

[118] J.H. Drew, C.R. Johnson, D.D Olesky, and P. van den Driessche. Spectrally arbitrary patterns. *Linear Algebra Appl.*, 308: 121–137, 2000.

[119] A.L. Duarte. Construction of acyclic matrices from spectral data. *Linear Algebra Appl.*, 113:173–182, 1989.

[120] J. Earl. A Song of Symmetry and Strikeouts. Thesis (Ph.D.), University of Wyoming, 2020.

[121] C.J. Edholm, L. Hogben, M. Huynh, J. LaGrange, and D.D. Row. Vertex and edge spread of zero forcing number, maximum nullity, and n minimum rank of a graph. *Linear Algebra Appl.*, 436: 4352–4372, 2012.

[122] J. Ekstrand, C. Erickson, H.T. Hall, D. Hay, L. Hogben, R. Johnson, N. Kingsley, S. Osborne, T. Peters, J. Roat, A. Ross, D.D. Row, N. Warnberg, and M. Young. Positive semidefinite zero forcing. *Linear Algebra Appl.*, 439: 1862–1874, 2013.

[123] J. Ekstrand, C. Erickson, D. Hay, L. Hogben, and J. Roat. Note on positive semidefinite maximum nullity and positive semidefinite zero forcing number of partial 2-trees. *Electron. J. Linear Algebra*, 23: 79–97, 2012.

[124] S. English, C. MacRury, P. Prałatt. Probabilistic zero forcing on random graphs. *European J. Combin.*, 91, Paper No. 103207, 22 pp., 2021.

[125] L. Euler. Sur la vibration des cordes. *Histoire de l'Académie Royale des Sciences et des Belles Lettres de Berlin pour l' année MDCCXLVIII*, 69–85, 1750.

[126] S.M. Fallat, H.T. Hall, J.C.-H. Lin, and B.L. Shader. The bifurcation lemma for strong properties in the inverse eigenvalue problem of a graph. *Linear Algebra Appl.* 648: 70–87, 2022.

[127] S. Fallat and L. Hogben. The minimum rank of symmetric matrices described by a graph: A survey. *Linear Algebra Appl.*, 426: 558–582, 2007.

[128] S. Fallat and L. Hogben. Minimum Rank, Maximum Nullity, and Zero Forcing Number of Graphs. In *Handbook of Linear Algebra*, 2nd edition (L. Hogben, editor), CRC Press, Boca Raton, 2014, pp. 46-1–46-36.

[129] S. Fallat and C.R. Johnson. Olga, matrix theory, and the Taussky unification problem. *Linear Algebra Appl.*, 280: 39–49, 1998.

[130] S. Fallat and S.A. Mojallal. On the minimum number of distinct eigenvalues of a threshold graph. *Linear Algebra Appl.*, 642: 1–29, 1998.

[131] S. Fallat, S. Severini, and M. Young. AIM workshop: Zero forcing and its variants. American Institute of Mathematics, Jan. 30-Feb 3, 2017, `https://aimath.org/pastworkshops/zeroforcing.html`.

[132] W.E. Ferguson, Jr. The construction of Jacobi and periodic Jacobi matrices with prescribed spectra. *Math. Comp.*, 35: 1203–1220, 1980.

[133] R. Fernandes. The maximum multiplicity and the two largest multiplicities of eigenvalues in a Hermitian matrix whose graph is a tree. *Special Matrices*, 3, 2015.

[134] R. Fernandes and C.M. da Fonseca. The Inverse Eigenvalue problem for Hermitian matrices whose graphs are cycles. *Linear Multilinear Algebra*, 57: 673–682, 2009.

[135] D. Ferrero, M. Flagg, H.T. Hall, L. Hogben, J.C.-H. Lin, S.A. Meyer, S. Nasserasr, and B. Shader. Rigid linkages and partial zero forcing. *Electron. J. Combin.*, 26: #P2.43, 2019.

[136] D. Ferraro, L. Hogben, F.H.J. Kenter, and M. Young. Note on power propagation time and lower bounds for the power domination number. *J. Combin. Optim.*, 34: 736–641, 2017.

[137] M. Fiedler. Theory of Graphs and Its Applications. In *Proceedings of the Symposium held in Smolenice in June 1963.* Publishing House of the Czechoslovakia Academy of Sciences, Prague, 1964.

[138] M. Fiedler. A characterization of tridiagonal matrices. *Linear Algebra Appl.*, 2: 191–197, 1969.

[139] M. Fiedler. A property of eigenvectors of nonnegative symmetric matrices and its applications to graph theory. *Czech. Math. J.*, 25: 618–633, 1975.

[140] M. Fiedler. *Matrices and graphs in geometry.* Encyclopedia of Mathematics and Applications, Cambridge University Press, Cambridge, 2011.

[141] G. Fijavž and D. Wood. Graph minors and minimum degree. *Electron. J. Combin.* 17: #P151, 2010.

[142] F.V. Fomin, P.A. Golovach, J. Kratochvíl, N. Nisse. Pursuing a fast robber on a graph. *Theoret. Comput. Sci.*, 411: 1167–1181, 2010.

[143] C.M. da Fonseca. A lower bound for the number of distinct eigenvalues of some real symmetric matrices. *Electron. J. Linear Algebra*, 21: 3–11, 2010.

[144] P. Frankl. Cops and robbers in graphs with large girth and Cayley graphs. *Discrete Appl. Math.*, 17: 301–305, 1987.

[145] S. Friedland. Matrices with prescribed off-diagonal elements. *Israel J. Math.* 11: 184–189, 1972.

[146] P.M. Gadea, J. Muñoz Masqué, and I.V. Mykytyuk. *Analysis and Algebra on Differentiable Manifolds*, 2nd edition. Springer, Netherlands, 2013.

[147] F.R. Gantmacher and M.G. Krein. Oscillation Matrices and Kernels and Small Vibrations of Mechanical Systems. State Publishing House of Technical–Theoretical Literature, Moscow, Leningard, 1950.

[148] F. Gantmacher and M. Krein. *Oszillationslnatrizen, Oszillationskerne und kleine Schwingungen nzechanischer Systenze,* translated by Alfred Stohr, Akademie-Verlag, Berlin, 1960.

[149] C. Garnett and B. Shader. The nilpotent-centralizer method for spectrally arbitrary patterns. *Linear Algebra Appl.*, 43: 3836–3850, 2011.

[150] C.M. Garnett and B.L. Shader. A proof of the T_n conjecture: Centralizers, Jacobians and spectrally arbitrary sign patterns. *Linear Algebra Appl.*, 436: 4451–4458, 2012.

[151] T. Gavenčiak. Cop-win graphs with maximum capture-time. *Discrete Math.*, 310: 1557–1563, 2010.

[152] J. Geneson. Throttling numbers for adversaries on connected graphs. `https://arxiv.org/abs/1906.07178`.

[153] J. Geneson and L. Hogben. Expected propagation time for probabilistic zero forcing. To appear in *Australas. J. Combin.*. Preliminary version at `https://arxiv.org/abs/1812.10476`.

[154] J. Genin and J.S. Maybee. Mechanical vibration trees. *J. Math. Anal. Appl.* 45: 646–763, 1974.

[155] M. Gentner, L.D. Penso, D. Rautenbach, and U.S. Souza. Extremal values and bounds for the zero forcing number. *Discrete Appl. Math.* 214: 196–200, 2016.

[156] M. Gentner and D. Rautenbach. Some bounds on the zero forcing number of a graph. *Discrete Appl. Math.* 236: 203–213, 2018.

[157] J.P. Georges and D.W. Mauro. On generalized Petersen graphs labeled with a condition at distance two. *Discrete Math.*, 259: 311–318, 2002.

[158] G.M.L. Gladwell. *Inverse Problems in Vibration*, 2nd edition. Kluwer Academic Publishers, New York, 2005.

[159] C.D. Godsil and G.F. Royle. *Algebraic Graph Theory.* Springer-Verlag, New York, 2001.

[160] C.D. Godsil. Spectra of Trees. *Convexity and Graph Theory (Jerusalem, 1981), Ann. Discrete Math.*, 20: 151–159, 1984.

[161] C. Gordon, D.L. Webb, and S. Wolpert. One cannot hear the shape of a drum. *Bulletin AMS*, 27:134-138, 1992.

[162] L.J. Gray and D.G. Wilson. Construction of a Jacobi matrix from spectral data. *Linear Algebra Appl.*, 14: 131–134, 1976.

[163] C. Grood, J.A. Harmse, L. Hogben, T. Hunter, B. Jacob, A. Klimas, and S. McCathern, Minimum rank of zero-diagonal matrices described by a graph. *Electron. J. Linear Algebra,* 27: 458–477, 2014.

[164] I. Gutman. The energy of a graph. *Ber. Math.-Statist. Sekt. Forsch. Graz* 103: 1–22, 1978.

[165] I. Gutman. The energy of a graph: old and new results. In *Algebraic Combinatorics and Applications* (A. Betten, A. Kohner, R. Laue, and A.Wassermann, editors), Springer, Berlin, 2001, pp. 196–211.

[166] P. Hackney, B. Harris, M. Lay, L.H. Mitchell, S.K. Narayan, and A. Pascoe. Linearly independent vertices and minimum semidefinite rank. *Linear Algebra Appl.*, 431: 1105–1115, 2009.

[167] H.T. Hall. Proof of the δ-conjecture. Discrete Math Seminar, Iowa State University, October 18, 2011.

[168] H.T. Hall. Personal communication on the minimum number of distinct eigenvalues, 2020.

[169] H.T. Hall. Personal communication on q of a disjoint union, 2021.

[170] H.T. Hall, L. Hogben, R. Martin, and B. Shader. Expected values of parameters associated with the minimum rank of a graph. *Linear Algebra Appl.*, 433: 101–117, 2010.

[171] F. Harary. The determinant of the adjacency matrix of a graph. *SIAM Review*, 4: 202–210, 1962.

[172] F. Harary and M. Livingston. Independent Domination in Hypercubes. *Appl. Math. Lett.*, 6: 27–28, 1993.

[173] T.W. Haynes, S.M. Hedetniemi, S.T. Hedetniemi, and M.A. Henning. Domination in graphs applied to electric power networks. *SIAM J. Discrete Math.*, 15: 519–529, 2002.

[174] T.W. Haynes, S.M. Hedetniemi, and P.J. Slater. *Fundamentals of Domination in Graphs.* Marcel Dekker, Inc., New York, 1998.

[175] N.J. Higham. *Functions of Matrices: Theory and Computation.* Society for Industrial and Applied Mathematics, Philadelphia, 2008.

[176] H. Hochstadt. On some inverse problems in matrix theory. *Arch. Math.*, 2: 201–207, 1967.

[177] A. Hoffmann and L. Volkmann. On unique k-factors and unique $[1,k]$-factors in graphs. *Discrete Math.*, 278: 127–138, 2004.

[178] L. Hogben, editor. *Handbook of Linear Algebra*, 2nd edition. CRC Press, Boca Raton, 2014.

[179] L. Hogben. Orthogonal representations, minimum rank, and graph complements. *Linear Algebra Appl.*, 428: 2560–2568, 2008.

[180] L. Hogben. Minimum rank problems. *Linear Algebra Appl.*, 432: 1961–1974, 2010.

[181] L. Hogben. Nordhaus-Gaddum problems for Colin de Verdière type parameters, variants of tree-width, and related parameters. In *Recent Trends in Combinatorics* (A. Beveridge, J.R. Griggs, L. Hogben, G. Musiker, P. Tetali, editors), IMA Volumes in Mathematics and its Applications 159, Springer, New York, 2016, pp. 275–294.

[182] L. Hogben. Zero forcing and maximum nullity for hypergraphs. *Discrete Appl. Math.*, 282: 122–135, 2020.

[183] L. Hogben. *Sage* code for minimum rank/maximum nullity, variants of zero forcing, and power domination (MaxNull_ZeroForcing_Variants.sws). `https://sage.math.iastate.edu/home/pub/133/`. PDF available at `https://aimath.org/~hogben/MaxNullity_ZeroForcing_Variants--Sage.pdf`.

[184] L. Hogben. *Sage* code for variants of propagation time and throttling (PropagationTime_Throttling.sws). `https://sage.math.iastate.edu/home/pub/134/`. PDF available at `https://aimath.org/~hogben/PropagationTime_Throttling--Sage.pdf`.

[185] L. Hogben and H. van der Holst. Forbidden minors for the class of graphs G with $\xi(G) \leq 2$. *Linear Algebra Appl.*, 423: 42–52, 2007.

[186] L. Hogben, M. Hunnell, K. Liu, H. Schuerger, B. Small, and Y. Zhang. Upper Bounds for Positive Semidefinite Propagation Time. *Discrete Math.*, 345: 112967 (11 pp.), 2022.

[187] L. Hogben, M. Huynh, N. Kingsley, S. Meyer, S. Walker, and M. Young. Propagation time for zero forcing on a graph. *Discrete Appl. Math*, 160: 1994–2005, 2012.

[188] L. Hogben, K. Palmowski, D. Roberson, and M. Young. Fractional Zero Forcing via Three-color Forcing Games. *Discrete Appl. Math.*, 213: 114–129, 2016.

[189] H. van der Holst. A short proof of the planarity characterization of Colin de Verdière. *J. Combin. Theory Ser. B*, 65: 269–272, 1995.

[190] H. van der Holst. Topological and Spectral Graph Characterizations. Thesis (Ph.D.), Universiteit van Amsterdam, Amsterdam, 1996.

[191] H. van der Holst. On the "Largeur d'Arborescence." *J. Graph Theory*, 41: 24–52, 2002.

[192] H. van der Holst. Graphs whose positive semi-definite matrices have nullity at most two. *Linear Algebra Appl.*, 375: 1–11, 2003.

[193] H. van der Holst. Three-connected graphs whose maximum nullity is at most three. *Linear Algebra Appl.*, 429: 625–632, 2008.

[194] H. van der Holst. On the maximum positive semi-definite nullity and the cycle matroid of graphs. *Electron. J. Linear Algebra*, 18: 192–201, 2009.

[195] H. van der Holst, L. Lovász, and A. Schrijver. The Colin de Verdière graph parameter. In *Graph Theory and Combinatorial Biology* (L. Lovász, A. Gyárfás, G. Katona, A. Recski, and L. Székely, editors), János Bolyai Mathematical Society, Budapest, 1999, pp. 29–85.

[196] R.A. Horn and C.R. Johnson. *Matrix Analysis*, 2nd edition. Cambridge University Press, Cambridge, 2013.

[197] L.-Y. Hsieh. On minimum rank matrices having prescribed graph. Thesis (Ph.D.), University of Wisconsin-Madison, 2001.

[198] L.-H. Huang, G.J. Chang, and H.-G. Yeh. On minimum rank and zero forcing sets of a graph. *Linear Algebra Appl.*, 432: 2961–2973, 2010.

[199] IMA-ISU research group on minimum rank (M. Allison, E. Bodine, L.M. DeAlba, J. Debnath, L. DeLoss, C. Garnett, J. Grout, L. Hogben, B. Im, H. Kim, R. Nair, O. Pryporova, K. Savage, B. Shader, A. Wangsness Wehe). Minimum rank of skew-symmetric matrices described by a graph. *Linear Algebra Appl.*, 432: 2457–2472, 2010.

[200] C.R. Johnson and A. Leal Duarte. The maximum multiplicity of an eigenvalue in a matrix whose graph is a tree. *Linear Multilinear Algebra*, 46: 139–144, 1999.

[201] C.R. Johnson and A. Leal Duarte. On the possible multiplicities of the eigenvalues of a Hermitian matrix the graph of whose entries is a tree. *Linear Algebra Appl.*, 348: 7–21, 2002.

[202] C.R. Johnson, A. Leal Duarte, and C. Saiago. Inverse eigenvalue problems and lists of multiplicities of eigenvalues for matrices whose graph is a tree: the case of generalized stars and double generalized stars. *Linear Algebra Appl.*, 373: 311–330, 2003.

[203] C.R. Johnson, R. Loewy, and P.A. Smith. The graphs for which the maximum multiplicity of an eigenvalue is two. *Linear Multilinear Algebra*, 57: 713–736, 2009.

[204] C.R. Johnson and C.M. Saiago. Diameter minimal trees. *Linear Multilinear Algebra*, 64: 557–571, 2015.

[205] C.R. Johnson and C.M. Saiago. *Eigenvalues, Multiplicities and Graphs.* Cambridge University Press, Cambridge, 2018.

[206] C.R. Johnson and T. Wakhare. The inverse eigenvalue problem for linear trees. *Discrete Math.*, 345: 112737 (17 pp.), 2022.

[207] C.R. Johnson, C. Waters. Sign patterns occurring in orthogonal matrices. *Linear and Multilinear Algebra*, 44: 287–299, 1998.

[208] M. Kac. Can One Hear the Shape of a Drum? *American Math. Monthly*, 73: 1–23, 1966.

[209] T. Kalinowski, N. Kamčev, and B. Sudakov. The zero forcing number of graphs. *SIAM J. Discrete Math.*, 33: 95–115, 2019.

[210] C.X. Kang and E. Yi. Probabilistic zero forcing in graphs. *Bull. Inst. Combin. Appl.*, 67: 9–16, 2013.

[211] I. Kim and B. Shader. Smith Normal Form and Acyclic Matrices. *J. Alg. Combin.*, 29: 63–80, 2009.

[212] N.F. Kingsley. Skew propagation time. Dissertation (Ph.D.), Iowa State University, 2015.

[213] W.B. Kinnersley. Bounds on the length of a game of Cops and Robbers. *Discrete Math.*, 341: 2508–2518, 2018.

[214] A. Kotlov, L. Lovász, and S. Vempala. The Colin de Verdère number and sphere representations of a graph. *Combinatorica* 17: 483–521, 1997.

[215] S.G. Krantz and H.R. Parks. *The Implicit Function Theorem: History, Theory and Applications*, Springer, 2013.

[216] T.J. Laffey. Extreme nonnegative matrices. *Linear Algebra Appl.*, 275–276: 349–357, 1998.

[217] J.M. Lee. *Introduction to Smooth Manifolds*, 2nd edition. Graduate Texts in Mathematics, Springer, New York, 2013.

[218] R.H. Levene, P. Oblak, and H. Šmigoc. A Nordhaus-Gaddum conjecture for the minimum number of distinct eigenvalues of a graph. *Linear Algebra Appl.*, 564: 236–263, 2019.

[219] C-S. Liao. Power domination with bounded time constraints. *J. Combin. Optim.*, 31: 725–742, 2016.

[220] J.C.-H. Lin. Odd cycle zero forcing parameters and the minimum rank of graph blowups. *Electron. J. Linear Algebra*, 31: 42–59, 2016.

[221] J.C.-H. Lin. Using a new zero forcing process to guarantee the strong Arnold property. *Linear Algebra Appl.*, 507:229–250, 2016.

[222] J.C.-H. Lin. Zero forcing number, Grundy domination number, and their variants. *Linear Algebra Appl.*, 563: 240–254, 2019.

[223] J.C.-H. Lin. *Sage* code for verifying the strong properties. `https://github.com/jephianlin/publish/blob/master/has_SXP.sage`.

[224] J.C.-H. Lin, P. Oblak, and H. Šmigoc. The strong spectral property for graphs. *Linear Algebra Appl.*, 598: 68–91, 2020.

[225] J.C.-H. Lin, P. Oblak, and H. Šmigoc. On the inverse eigenvalue problem for block graphs. *Linear Algebra Appl.*, 631: 379–397, 2021.

[226] L. Lovász, M. Saks, and A. Schrijver. Orthogonal representations and connectivity of graphs. *Linear Algebra Appl.* 114/115: 439–454, 1989. Corrected in A correction: "Orthogonal representations and connectivity of graphs." *Linear Algebra Appl.* 313: 101–105, 2000.

[227] D. Mayhew, G. Whittle and S. H. M. van Zwam. The structure of graphs with a vital linkage of order 2. *J. Graph Theory*, 71:176–181, 2012.

[228] C. McDiarmid. On the chromatic number of random graphs. *Random Structures Algorithms*, 1: 435–442, 1991.

[229] J. Milnor. Eigenvalues of the Laplace operator on certain manifolds. *Proc. Nat. Acad. Sc.*, 51: 542, 1964.

[230] L.H. Mitchell. On the graph complement conjecture for minimum semidefinite rank. *Linear Algebra Appl.*, 435: 1311–1314, 2011.

[231] L.H. Mitchell, S. Narayan, and A. Zimmer. Lower bounds in minimum rank problems. *Linear Algebra Appl.*, 432: 430–440, 2010.

[232] L.H. Mitchell, S. Narayan, and A. Zimmer. Lower bounds for minimum semidefinite rank from orthogonal removal and chordal supergraphs. *Linear Algebra Appl.*, 436: 525–536, 2012.

[233] K.H. Monfared and B.L. Shader. Construction of matrices with a given graph and prescribed interlaced spectral data. *Linear Algebra Appl.*, 438: 4348–4358, 2013.

[234] K.H. Monfared and B.L. Shader. The nowhere-zero eigenbasis problem for a graph. *Linear Algebra Appl.*, 505: 296–312, 2016.

[235] S. Nikoletseas and P. Spirakis. Near optimal dominating sets in dense random graphs with polynomial expected time. In *Graph Theoretic Concepts in Computer Science* (J. van Leeuwen, editor), Lecture Notes in Computer Science 790, Springer Verlag, Berlin, 1994, pp. 1–10.

[236] M.W. Newman. The Laplacian Spectrum of Graphs. Thesis (M.S.), University of Manitoba, 2000.

[237] R.J. Nowakowski and P. Winkler. Vertex-to-vertex pursuit in a graph. *Discrete Math.*, 43: 235–239, 1983.

[238] P.M. Nylen, Minimum-rank matrices with prescribed graph. *Linear Algebra Appl.*, 248: 303–316, 1996.

[239] K. Owens. Properties of the zero forcing number. Thesis (M.S.), Brigham Young University, 2009.

[240] K.-J. Pai and W.-J. Chiu. A Note on "On The Power Dominating Sets of Hypercubes". *Proceedings of the 29th Workshop on Combinatorial Mathematics and Computation Theory*, Taipei, Taiwan, April 27-28, 2012.

[241] S. Parter. On the eigenvalues and eigenvectors of a class of matrices. *J. Soc. Indust. Appl. Math.*, 8: 376-388, 1960.

[242] G. Perarnau and O. Serra. On the tree-depth of random graphs. *Discrete Appl. Math.* 168: 119–126, 2014.

[243] T. Peters. Positive semidefinite maximum nullity and zero forcing number. Dissertation (Ph.D.), Iowa State University, 2012.

[244] T. Peters. Positive semidefinite maximum nullity and zero forcing number. *Electron. J. Linear Algebra*, 23: 815–830, 2012.

[245] Å. Pleijel. Remarks on Courant's nodal line theorem. *Comm. Pure Appl. Math.*, 9: 543–550, 1956.

[246] P. Prałat. When does a random graph have constant cop number? *Australas. J. Combin.*, 46: 285–296, 2010.

[247] N. Robertson, P. Seymour. Graph Minors. XX. Wagner's conjecture. *J. Comb. Theory Ser. B*, 92: 325–357, 2004.

[248] A. Quilliot. Jeux et pointes fixes sur les graphes. *Thèse de 3ème cycle*, Université de Paris VI [in French], 131–145, 1978.

[249] D. Rosenblatt. On linear models and the graphs of Minkowski-Leontief matrices. *Econometrica*, 2: 325–338, 1957.

[250] M.S. Ross. Throttling processes equivalent to full throttling on trees. *Theoret. Comput. Sci.*, 865: 1–11, 2021.

[251] D.D. Row. A technique for computing the zero forcing number of a graph with a cut-vertex. *Linear Algebra Appl.*, 436: 4423–4432, 2012.

[252] A. Schrijver and B. Sevenster. The Strong Arnold Property for 4-connected flat graphs. *Linear Algebra Appl.*, 522: 153-160, 2017.

[253] S. Severini. Nondiscriminatory propagation on trees. *J. Physics A*, 41: 482002 (10 pp.), 2008.

[254] B. Simon. Sturm Oscillation and Comparison Theorems. In *Sturm-Liouville Theory: Past and Present* (W.O. Amrein, A.M. Hinz and D.B. Pearson, editors), BirkhÃ€user Verlag, Basel, 2005, pp. 29–43.

[255] J. Sinkovic and H. van der Holst. The minimum semidefinite rank of the complement of partial k-trees. *Linear Algebra Appl.*, 434: 1468–1474, 2011.

[256] B. Small. Counterexample to the conjecture that the PSD forcing interval is full. Personal communication. 2020.

[257] T.J. Stieltjes. Recherches sur les fractions continues. *Ann. Fat. Sci. Toulouse*, 8: J1–J122,1894.

[258] K. Symanzik. Dispersion relations and vertex properties in perturbation theory. *Progr. Theoret. Phys.* 20: 690–70, 1958.

[259] A. Takahashi, S. Ueno, and Y. Kajitani. Minimal acyclic forbidden minors for the family of graphs with bounded path-width. *Discrete Math.*, 127: 293–304, 1994.

[260] L. Wang and B. Yang. Positive semidefinite zero forcing numbers of two classes of graphs. *Theoret. Comput. Sci.*, 786: 44–5, 2019.

[261] N. Warnberg. Positive semidefinite propagation time. *Discrete Appl. Math.*, 198: 274–290, 2016.

[262] C. Waters. Sign pattern matrices that allow orthogonality. *Linear Algebra Appl.*, 235: 1–13, 1996.

[263] D. West. *Introduction to Graph Theory*, 2nd edition. Prentice Hall, Upper Saddle River, 2001.

[264] H. Weyl. Über die asymptotische Verteilung der Eigenwerte. *Nachrichten der Königlichen Gesellschaft der Wissenschaften zu Göttingen, Mathematisch-Physikalische Klasse 1911*, 110–117, 1911.

[265] B. Wieland and A.P. Godbole. On the domination number of a random graph. *Electron. J. Combin.*, 8: #P37, 2001.

[266] G. Wiener. Spectral multiplicity and splitting results for a class of qualitative matrices. *Linear Algebra Appl.*, 61: 15–29, 1984.

[267] B. Yang. Fast-mixed searching and related problems on graphs. *Theoret. Comput. Sci.*, 507: 100 113, 2013.

[268] B. Yang. Positive semidefinite zero forcing: complexity and lower bounds. In *Algorithms and Data Structures* (edited by F. Dehne, J.-R. Sack, and U. Stege), Lecture Notes in Comput. Sci. 9214, Springer, Switzerland, 2015, pp. 629–639.

[269] M. Zhao, L. Kang, and G.J. Chang. Power domination in graphs. *Discrete Math.*, 306: 1812–1816, 2006.

[270] S. Zelditch. Inverse spectral problem for analytic domains, II: $\mathbb{Z}_2$-symmetric domains. *Ann. of Math.*, 170: 205–269, 2009.

[271] F. Zhang, *Matrix Theory*, 2nd edition. Springer-Verlag, New York, 2011.

Index

Selected Published Titles in This Series

270 **Leslie Hogben, Jephian C.-H. Lin, and Bryan L. Shader,** Inverse Problems and Zero Forcing for Graphs, 2022

267 **Dragana S. Cvetković Ilić,** Completion Problems on Operator Matrices, 2022

266 **Kate Juschenko,** Amenability of Discrete Groups by Examples, 2022

265 **Nabil H. Mustafa,** Sampling in Combinatorial and Geometric Set Systems, 2022

264 **J. Scott Carter and Seiichi Kamada,** Diagrammatic Algebra, 2021

263 **Vugar E. Ismailov,** Ridge Functions and Applications in Neural Networks, 2021

262 **Ragnar-Olaf Buchweitz,** Maximal Cohen–Macaulay Modules and Tate Cohomology, 2021

261 **Shiri Artstein-Avidan, Apostolos Giannopoulos, and Vitali D. Milman,** Asymptotic Geometric Analysis, Part II, 2021

260 **Lindsay N. Childs, Cornelius Greither, Kevin P. Keating, Alan Koch, Timothy Kohl, Paul J. Truman, and Robert G. Underwood,** Hopf Algebras and Galois Module Theory, 2021

259 **William Heinzer, Christel Rotthaus, and Sylvia Wiegand,** Integral Domains Inside Noetherian Power Series Rings, 2021

258 **Pramod N. Achar,** Perverse Sheaves and Applications to Representation Theory, 2021

257 **Juha Kinnunen, Juha Lehrbäck, and Antti Vähäkangas,** Maximal Function Methods for Sobolev Spaces, 2021

256 **Michio Jimbo, Tetsuji Miwa, and Fedor Smirnov,** Local Operators in Integrable Models I, 2021

255 **Alexandre Boritchev and Sergei Kuksin,** One-Dimensional Turbulence and the Stochastic Burgers Equation, 2021

254 **Karim Belabas and Henri Cohen,** Numerical Algorithms for Number Theory, 2021

253 **Robert R. Bruner and John Rognes,** The Adams Spectral Sequence for Topological Modular Forms, 2021

252 **Julie Déserti,** The Cremona Group and Its Subgroups, 2021

251 **David Hoff,** Linear and Quasilinear Parabolic Systems, 2020

250 **Bachir Bekka and Pierre de la Harpe,** Unitary Representations of Groups, Duals, and Characters, 2020

249 **Nikolai M. Adrianov, Fedor Pakovich, and Alexander K. Zvonkin,** Davenport–Zannier Polynomials and Dessins d'Enfants, 2020

248 **Paul B. Larson and Jindrich Zapletal,** Geometric Set Theory, 2020

247 **István Heckenberger and Hans-Jürgen Schneider,** Hopf Algebras and Root Systems, 2020

246 **Matheus C. Bortolan, Alexandre N. Carvalho, and José A. Langa,** Attractors Under Autonomous and Non-autonomous Perturbations, 2020

245 **Aiping Wang and Anton Zettl,** Ordinary Differential Operators, 2019

244 **Nabile Boussaïd and Andrew Comech,** Nonlinear Dirac Equation, 2019

243 **José M. Isidro,** Jordan Triple Systems in Complex and Functional Analysis, 2019

242 **Bhargav Bhatt, Ana Caraiani, Kiran S. Kedlaya, Peter Scholze, and Jared Weinstein,** Perfectoid Spaces, 2019

241 **Dana P. Williams,** A Tool Kit for Groupoid C^*-Algebras, 2019

240 **Antonio Fernández López,** Jordan Structures in Lie Algebras, 2019

239 **Nicola Arcozzi, Richard Rochberg, Eric T. Sawyer, and Brett D. Wick,** The Dirichlet Space and Related Function Spaces, 2019

For a complete list of titles in this series, visit the
AMS Bookstore at **www.ams.org/bookstore/survseries/**.